SAMPLING AND STATISTICAL METHODS FOR BEHAVIORAL ECOLOGISTS

This book describes the sampling and statistical methods used most often by behavioral ecologists and field biologists. Written by a biologist and two statisticians, it provides a rigorous discussion, together with worked examples, of statistical concepts and methods that are generally not covered in introductory courses, and which are consequently poorly understood and applied by field biologists. The first section reviews important issues such as defining the statistical population when using nonrandom methods for sample selection, bias, interpretation of statistical tests, confidence intervals and multiple comparisons. After a detailed discussion of sampling methods and multiple regression, subsequent chapters discuss specialized problems such as pseudoreplication, and their solutions. It will quickly become *the* statistical handbook for all field biologists.

JONATHAN BART is a Research Wildlife Biologist, in the Biological Resources Division of the U.S. Geological Survey. He is currently based at the Snake River Field Station of the Forest and Rangeland Ecosystem Science Center in Boise, Idaho.

MICHAEL A. FLIGNER is Professor and Vice Chair of the Department of Statistics at the Ohio State University.

WILLIAM I. NOTZ is also a Professor in the Department of Statistics at the Ohio State University, and has served as Associate Editor for the *Journal of the American Statistical Association, Technometrics* and the *Journal of Statistical Education.*

To Susan
Alysha and Karen
Claudia

SAMPLING AND STATISTICAL METHODS FOR BEHAVIORAL ECOLOGISTS

JONATHAN BART, MICHAEL A. FLIGNER,
AND WILLIAM I. NOTZ

CAMBRIDGE UNIVERSITY PRESS
Cambridge, New York, Melbourne, Madrid, Cape Town, Singapore,
São Paulo, Delhi, Dubai, Tokyo

Cambridge University Press
The Edinburgh Building, Cambridge CB2 8RU, UK

Published in the United States of America by Cambridge University Press, New York

www.cambridge.org
Information on this title: www.cambridge.org/9780521457057

First published 1998
Reprinted 2000

A catalogue record for this publication is available from the British Library

Library of Congress Cataloguing in Publication data

Bart, Jonathan.
Sampling and Statistical Methods for Behavioral Ecologists / Jonathan Bart, Michael A.
Fligner, William I. Notz.
p. cm.
Includes index.
ISBN 0 521 45095 0 hardback.
ISBN 0 521 45705 X paperback.
1. Animal behavior – Statistical methods. 2. Sampling (Statistics)
I. Fligner, Michael A. II. Notz, William I. III. Title.
QL751.65.S73B37 1998
590′.1′5195 dc21 97–35086 CIP

ISBN 978-0-521-45095-9 Hardback
ISBN 978-0-521-45705-7 Paperback

Transferred to digital printing 2009

SAMPLING AND STATISTICAL METHODS FOR BEHAVIORAL ECOLOGISTS

JONATHAN BART, MICHAEL A. FLIGNER, AND WILLIAM I. NOTZ

CAMBRIDGE UNIVERSITY PRESS
Cambridge, New York, Melbourne, Madrid, Cape Town, Singapore,
São Paulo, Delhi, Dubai, Tokyo

Cambridge University Press
The Edinburgh Building, Cambridge CB2 8RU, UK

Published in the United States of America by Cambridge University Press, New York

www.cambridge.org
Information on this title: www.cambridge.org/9780521457057

First published 1998
Reprinted 2000

A catalogue record for this publication is available from the British Library

Library of Congress Cataloguing in Publication data

Bart, Jonathan.
Sampling and Statistical Methods for Behavioral Ecologists / Jonathan Bart, Michael A. Fligner, William I. Notz.
p. cm.
Includes index.
ISBN 0 521 45095 0 hardback.
ISBN 0 521 45705 X paperback.
1. Animal behavior – Statistical methods. 2. Sampling (Statistics)
I. Fligner, Michael A. II. Notz, William I. III. Title.
QL751.65.S73B37 1998
590′.1′5195 dc21 97–35086 CIP

ISBN 978-0-521-45095-9 Hardback
ISBN 978-0-521-45705-7 Paperback

Transferred to digital printing 2009

Contents

	Preface		*page* ix
1	**Statistical analysis in behavioral ecology**		**1**
	1.1	Introduction	1
	1.2	Specifying the population	1
	1.3	Inferences about the population	5
	1.4	Extrapolation to other populations	11
	1.5	Summary	12
2	**Estimation**		**14**
	2.1	Introduction	14
	2.2	Notation and definitions	15
	2.3	Distributions of discrete random variables	17
	2.4	Expected value	21
	2.5	Variance and covariance	24
	2.6	Standard deviation and standard error	26
	2.7	Estimated standard errors	26
	2.8	Estimating variability in a population	30
	2.9	More on expected value	32
	2.10	Linear transformations	34
	2.11	The Taylor series approximation	36
	2.12	Maximum likelihood estimation	42
	2.13	Summary	45
3	**Tests and confidence intervals**		**47**
	3.1	Introduction	47
	3.2	Statistical tests	47
	3.3	Confidence intervals	58
	3.4	Sample size requirements and power	65
	3.5	Parametric tests for one and two samples	68

3.6 Nonparametric tests for one or two samples 78
3.7 Tests for more than two samples 81
3.8 Summary 84

4 Survey sampling methods 85
4.1 Introduction 85
4.2 Overview 86
4.3 The finite population correction 97
4.4 Sample selection methods 99
4.5 Multistage sampling 109
4.6 Stratified sampling 124
4.7 Comparison of the methods 131
4.8 Additional methods 132
4.9 Notation for complex designs 137
4.10 Nonrandom sampling in complex designs 139
4.11 Summary 146

5 Regression 148
5.1 Introduction 148
5.2 Scatterplots and correlation 148
5.3 Simple linear regression 154
5.4 Multiple regression 159
5.5 Regression with multistage sampling 174
5.6 Summary 176

6 Pseudoreplication 177
6.1 Introduction 177
6.2 Power *versus* generality 178
6.3 Fish, fish tanks, and fish trials 182
6.4 The great playback debate 185
6.5 Causal inferences with unreplicated treatments 187
6.6 Summary 187

7 Sampling behavior 190
7.1 Introduction 190
7.2 Defining behaviors and bouts 190
7.3 Allocation of effort 192
7.4 Obtaining the data 196
7.5 Analysis 197
7.6 Summary 199

8 Monitoring abundance **200**
8.1 Introduction 200
8.2 Defining 'the trend' 201
8.3 Estimating standard errors 209
8.4 Outliers and missing data 210
8.5 Index methods 211
8.6 Pseudoreplication 216
8.7 Summary 217

9 Capture–recapture methods **219**
9.1 Introduction 219
9.2 Rationale 219
9.3 Capture histories and models 220
9.4 Model selection 222
9.5 Closed population models 222
9.6 Open population models 223
9.7 Summary 226

10 Estimating survivorship **228**
10.1 Introduction 228
10.2 Telemetry studies 228
10.3 Nesting success 231
10.4 Summary 236

11 Resource selection **238**
11.1 Introduction 238
11.2 Population units and parameters 239
11.3 Several animals 243
11.4 Multivariate definition of resources 244
11.5 Summary 246

12 Other statistical methods **248**
12.1 Introduction 248
12.2 Adaptive sampling 248
12.3 Line transect sampling 249
12.4 Path analysis 250
12.5 Sequential analysis 251
12.6 Community analysis 253
12.7 Summary 254

APPENDIX ONE Frequently used statistical methods 257
APPENDIX TWO Statistical tables 279
APPENDIX THREE Notes for Appendix One 311

References 320
Index 328

Preface

This book describes the sampling and statistical methods used most often by behavioral ecologists. We define behavioral ecology broadly to include behavior, ecology and such related disciplines as fisheries, wildlife, and environmental physiology. Most researchers in these areas have studied basic statistical methods, but frequently have trouble solving their design or analysis problems despite having taken these courses. The general reason for these problems is probably that introductory statistics courses are intended for workers in many fields, and each field presents a special, and to some extent unique, set of problems. A course tailored for behavioral ecologists would necessarily contain much material of little interest to students in other fields.

The statistical problems that seem to cause behavioral ecologists the most difficulty can be divided into several categories.

1. Some of the most difficult problems faced by behavioral ecologists attempting to design a study or analyze the resulting data fall between statistics – as it is usually taught – and biology. Examples include how to define the sampled and target populations, the nature and purpose of statistical analysis when samples are collected nonrandomly, and how to avoid pseudoreplication.
2. Some methods used frequently by behavioral ecologists are not covered in most introductory texts. Examples include survey sampling, capture–recapture, and distance sampling.
3. Certain concepts in statistics seem to need reinforcement even though they are well covered in many texts. Examples include the rationale of statistical tests, the meaning of confidence intervals, and the interpretation of regression coefficients.
4. Behavioral ecologists encounter special statistical problems in certain areas including index methods, detecting habitat 'preferences', and sampling behavior.

5. A few mathematical methods of use to behavioral ecologists are generally not covered in introductory methods courses. Examples include the statistical properties of ratios and other nonlinear combinations of random variables, rules of expectation, the principle of maximum likelihood estimation, and the Taylor series approximation.

This book is an attempt to address problems such as those above adopting the special perspective of behavioral ecology. Throughout the book, our general goals have been that behavioral ecologists would find the material relevant and that statisticians would find the treatment rigorous. We assume that readers will have taken one or more introductory statistics courses, and we view our book as a supplement, rather than a substitute, for these courses.

The book is based in part on our own research and consulting during the past 20 years. Before writing the text, however, we undertook a survey of the methods used by behavioral ecologists. We did this by examining every article published during 1990 in the journals *Behavioral Ecology and Sociobiology*, *Animal Behavior*, *Ecology*, and *The Journal of Wildlife Management* and all the articles on behavior or ecology published in *Science* and *Nature*. We tabulated the methods in these articles and used the results frequently in deciding what to include in the book and how to present the examples.

Chapter One describes statistical objectives of behavioral ecologists emphasizing how the statistical and nonstatistical aspects of data analysis reinforce each other. Chapter Two describes estimation techniques, introducing several statistical methods that are useful to behavioral ecologists. It is more mathematical than the rest of the book and can be skimmed by readers less interested in such methods. Chapter Three discusses tests and confidence intervals concentrating on the rationale of each method. Methods for ratios are discussed as are sample size and power calculations. The validity of *t*-tests when underlying data are non-normal is discussed in detail, as are the strengths and weaknesses of nonparametric tests. Chapter Four discusses survey sampling methods in considerable detail. Different sampling approaches are described graphically. Sample selection methods are then discussed followed by a description of multistage sampling and stratification. Problems caused by nonrandom sample selection are examined in detail. Chapter Five discusses regression methods emphasizing conceptual issues and how to use computer software to carry out general linear models' analysis.

The first five Chapters cover material included in the first few courses in statistical methods. In these Chapters, we concentrate on topics that behavioral ecologists often have difficulty with, assuming that the reader has already been exposed to the basic methods and ideas. The subsequent Chapters discuss topics that are generally not covered in introductory statistics courses. We introduce each topic and provide suggestions for additional reading. Chapter Six discusses the difficult problem of pseudo-replication, introducing an approach which we believe might help to resolve the controversies in this area and focus the discussions on biological, rather than statistical, issues. Chapter Seven discusses special statistical problems that arise in sampling behavior. Chapter Eight discusses estimating and monitoring abundance, particularly by index methods. Chapter Nine discusses capture–recapture methods, while Chapter Ten emphasizes the estimation of survival. Chapter Eleven discusses resource selection and Chapter Twelve briefly mentions some other topics of interest to behavioral ecologists with suggestions for additional reading.

Appendix One gives a detailed explanation of frequently used statistical methods, whilst Appendix Two contains a set of tables for reference. They are included primarily so that readers can examine the formulas in more detail to understand how analyses are conducted. We have relegated this material to an appendix because most analyses are carried out using statistical packages and many readers will not be interested in the details of the analysis. Nonetheless, we encourage readers to study the material in the appendices as doing so will greatly increase one's understanding of the analyses. In addition, some methods (e.g., analysis of stratified samples) are not available in many statistical packages but can easily be carried out by readers able to write simple computer programs. Appendix Three contains detailed notes on derivation of the material in Appendix One.

This book is intended primarily for researchers who wish to use sampling techniques and statistical analysis as a tool but who do not have a deep interest in the underlying mathematical principles. We suspect, however, that many biologists will be interested in learning more about the statistical principles and techniques used to develop the methods we present. Knowledge of this material is of great practical use because problems arise frequently which can be solved readily by use of these methods, but which are intractable without them. Basic principles of expectation (by which many variance formulas may be derived) and use of the Taylor series approximation (by which nearly all the remaining variance formulas needed by behavioral ecologists may be derived) are examples of these methods. Maximum likelihood estimation is another statistical method that can be presented without recourse to complex math and is frequently of value to biologists. We introduce these methods in Chapter Two and illustrate their use periodically in the rest of the book. These sections, however, can be skipped without compromising the reader's ability to understand later sections of the book.

Another approach of great utility in developing a deep understanding of the statistical methods we present is to prepare computer programs that carry out calculations and simulations. We encourage readers to learn some programming in an elementary language such as Basic or the languages included in many data bases or statistical packages and then to write short programs to investigate the material we present. Several opportunities for such projects are identified in the text, and all of the examples we mention are listed in the *Index* under the heading 'Computer programming, examples'. We have found that preparing programs in this manner not only ensures that one understands the fine structure of the analysis, but in addition frequently leads one to think much more deeply about how the statistical analysis helps us understand natural systems. Such efforts also increase one's intuition about whether studies can be carried out successfully given the resources available and about how to allocate resources among different segments of the study. Furthermore, data management, while not discussed in this book, frequently consumes far more time during analysis than carrying out the actual statistical tests, and in many studies is nearly impossible without recourse to computer programs. For all of these reasons, we encourage readers strongly to learn a programming language.

The authors thank the staff of Cambridge University Press for their assistance with manuscript preparation, especially our copy editor, Sarah Price. Much of the book was written while the senior author was a member of the Zoology Department at Ohio State University. He acknowledges the many stimulating discussions of biological statistics with colleagues there, especially Susan Earnst, Tom Grubb, and John Harder and their graduate students. JB also acknowledges his intellectual debt to Douglas S. Robson of Cornell University who introduced him to sampling techniques and other branches of statistics and from whom he first learned the value of integrating statistics and biology in the process of biological research.

1
Statistical analysis in behavioral ecology

1.1 Introduction

This Chapter provides an overview of how statistical problems are formulated in behavioral ecology. We begin by identifying some of the difficulties that behavioral ecologists face in deciding what population to study. This decision is usually made largely on nonstatistical grounds but a few statistical considerations are worth discussing. We then introduce the subject of making inferences about the population, describing objectives in statistical terms and discussing accuracy and the general ways used to measure it. Finally, we note that statistical inferences do not necessarily apply beyond the population sampled and emphasize the value of drawing a sharp distinction between the sampled population and larger populations of interest.

1.2 Specifying the population

Several conflicting goals influence decisions about how large and variable the study population should be. The issues are largely nonstatistical and thus outside the scope of this book, but a brief summary, emphasizing statistical issues insofar as they do occur, may be helpful.

One issue of fundamental importance is whether the population of interest is well defined. Populations are often well defined in wildlife monitoring studies. The agencies carrying out such studies are usually concerned with a specific area such as a State and clearly wish to survey as much of the area as possible. In observational studies, we would often like to collect the data throughout the daylight hours – or some portion of them – and throughout the season we are studying.

Sampling throughout the population of interest, however, may be difficult for practical reasons. For example, restricting surveys to roads and observations to one period of the day may permit the collection of a larger sample size. A choice then arises between 'internal and external validity'. If surveys are restricted to roadsides, then smaller standard errors may be obtained, thereby increasing 'internal validity', but we will worry that trends along the roads may differ from trends for the entire area, thus reducing 'external validity'. A similar problem may occur if observations are restricted to certain times of day or portions of the season. When the population of interest is well defined, as in these cases, then the trade-off between internal and external validity is conceptually straightforward, though deciding how to resolve it in specific cases may be difficult.

When there is no single well-defined population of interest, then the situation is a little more complex conceptually. Consider the following example. Suppose we are investigating the relationship between dominance and time spent watching for predators in groups of foraging animals. Dominant individuals might spend more time foraging because they assume positions of relative security from predators. Alternatively, they might spend less time foraging because they obtain better foraging positions and satisfy their nutritional requirements more quickly. Suppose that we can study six foraging groups in one woodlot, or two groups in each of three woodlots. Sampling three woodlots might seem preferable because the sampled population would then be larger and presumably more representative of the population in the general area. But suppose that dominant individuals spend more time foraging in some habitats and less time foraging in others. With three woodlots – and perhaps three habitats – we might not obtain statistically significant differences between the foraging time of dominants and subdominants due to the variation among woodlots. We might also not have enough data within woodlots to obtain statistically significant effects. Thus, we would either reach no conclusion or, by averaging over woodlots, incorrectly conclude that dominance does not affect vigilance time. This unfortunate outcome might be much less likely if we confined sampling to a single woodlot. Future study might then show that the initial result was habitat dependent.

In this example, there is no well-defined target population about which we would like to make inferences. The goal is to understand an interesting process. Deciding how general the process is can be viewed as a different goal, to be undertaken in different studies. Thus, while the same trade-off between internal and external validity occurs, there is much less of a premium on high external validity. If the process occurs in the same way across a large population, and if effort can be distributed across this population without too much reduction in sample sizes, due to logistic costs, then having a relatively large sampled population may be worthwhile. But if such a plan increases logistic costs, or if the process varies across the population, then restricting the population in space, time or other ways may be preferable.

Studies conducted in one location or within 1 year are sometimes criticized on the grounds that the sample size is 1. In some sense, however, nearly all studies have a sample size of 1 because they are carried out in one county, state, or continent. Frequently, those arguing for a distribution of the study across two or more areas or years are really arguing that two or more complete studies should have been conducted. They want enough data to determine whether the results hold in each area or year. This is desirable of course. Two studies are nearly always better than one; but, if the sample size is only sufficient to obtain one good estimate, then little may be gained, and much lost, by spreading the effort over a large area or long period of time.

Superpopulations

Sometimes a data set is collected without any formal random selection – this occurs in many fields. In behavioral ecology, it is most likely when the study is conducted within a well-defined area and all individuals (typically plants or animals) within the boundaries of the area are measured. It might be argued that in such cases we have taken a census (i.e., measured all members) of the population so that calculation of standard errors and statistical tests is neither needed nor appropriate. This view is correct if our interest really is restricted to individuals in the study area at the time of the study. In the great majority of applications, however, we are really interested in an underlying process, or at least a much larger population than the individuals we studied.

In sampling theory, a possible point of view is that many factors not under our control operate in essentially a random manner to determine what individuals will be present when we do our study, and that the individuals present can thus be regarded as a random sample of the individuals that might have been present. Such factors might include weather conditions, predation levels, which migrants happened to land in the area, and so on. In sampling theory, such hypothetical populations are often called 'superpopulations' (e.g., Cochran 1977 p. 158; Kotz and Johnson 1988). We assume that our sample is representative of the superpopulation and thus that statistical inferences apply to this larger group of individuals. If the average measurement from males, for example, is significantly larger than the average from females, then we may legitimately conclude that the average for all males that might have been present in our study area probably exceeds the average for all females. If the difference is not significant, then the data do not support any firm conclusion about which sex has the larger average value. Note that asserting the existence of a superpopulation, describing the individuals it contains, and modeling its relation to our sample require biological or ecological arguments as much as or more than statistical arguments.

The superpopulation concept can also be explained by reference to an 'assignment process'. The word assignment refers to the underlying biological process, not to randomization carried out by the investigator. To illustrate the concept, imagine that we are comparing survival rates of males and females. We might view the individuals of each sex as being 'assigned' to one of two groups at the end of the study, alive and dead, and the process may be viewed as having random elements such as whether a predator happens to encounter a given individual. The question is whether members of one sex are more likely than the other to be assigned to the 'alive' group. The superpopulation is then the set of possible outcomes and inferences apply to the underlying probabilities of survival for males and females. This rationale is appealing because it emphasizes our interest in the underlying process, rather than in the individuals who happened to be present when we conducted the study.

Justifying statistical analysis by invoking the superpopulation concept might be criticized on the basis that there is little point in making inferences about a population if we cannot clearly describe what individuals comprise the population. There are two responses to this criticism. First, there is an important difference between deciding whether sample results might have arisen by chance and deciding how widely conclusions from a study apply. In the example above, if the sample results are not significantly different then we have not shown that survival rates are sex specific for *any* population (other than the sample we measured). The analysis thus prevents our making unwarranted claims. Second, describing the sampled population, in a particular study, is often not of great value even if it is possible. The main value of describing the sampled population is that we can then generalize the results from our sample to this population. But in biological research, we usually want to extend our findings to other areas, times, and species, and clearly the applicability of our results to these populations can only be determined by repeating the study elsewhere. Thus, the generality of research findings is established mainly by repeating the study, not by precisely demarcating the sampled population in the initial study.

Statisticians tend to view superpopulations as an abstraction, as opposed to a well-defined population about which inferences are to be made.

Behavioral ecologists thus must use care when invoking this concept to ensure that the rationale is reasonable. For example, one would probably not measure population size in a series of years and then declare that the years could be viewed as a random sample from a superpopulation of years. Population size at one time often depends strongly on population size in recent years so consecutive years could not legitimately be viewed as an independent sample. Nonetheless, in many studies in the field of behavioral ecology we can imagine much larger populations which we suspect our samples are representative of and to which we would like to make inferences. In such cases statistical analysis is appropriate because it helps guard against unwarranted conclusions.

1.3 Inferences about the population

Objectives

Although biologists study a vast array of species, areas, behaviors, and so on, most of the parameters estimated may be assigned to a small number of categories. Most quantities of interest in behavioral ecology are of two types: (1) means, proportions, or quantities derived from them, such as differences; and (2) measures of association such as correlation and regression coefficients and the quantities based on them such as regression equations. Estimates of these quantities are often called 'point estimates'. In addition, we usually want an estimate of accuracy such as a standard error. A point estimate coupled with an estimate of accuracy can often be used to construct a confidence interval or 'interval estimate', an interval within which we are relatively confident the true parameter value lies. Frequent use is made later in the book of the phrase 'point and interval estimates'.

Definitions

One of the first steps in obtaining point or interval estimates is to clearly understand the statistical terms. In behavioral ecology, the connection between the terms and the real problem is sometimes surprisingly difficult to specify, as will become clear later in the book. Here we introduce a few terms and provide several examples of how they would be defined in different studies.

The quantity we are trying to estimate is referred to as a parameter. Formally, a parameter is any numerical characteristic of a population. In estimating density, the parameter is actual density in the sampled population. In estimating change in density, the parameter is change in the actual densities. The term random variable refers to any quantity whose numerical value depends on which sample we happen to obtain by random selection. The sample mean is thus a random variable as is any quantity calculated from the sample such as a standard deviation or standard error.

A numerical constant is typically a *known* quantity that is not of direct interest and whose value does not depend on the particular sample selected. For example, if we estimate density per m^2 but then multiply the estimate by 10,000 to obtain density per hectare, then the 10,000 is a numerical constant. On the other hand, a parameter is an *unknown* constant whose value does not depend on the particular sample selected but is of direct interest.

In any analysis, one must identify the units in the sample and the measurements taken on each unit. Thus, we may define the sample mean, with respect to some variable as $\bar{y} = \Sigma y_i / n$ where n is the sample size and y_i, $i = 1,...,n$ are the measurements. In this book, we generally follow the tradition of survey sampling in which a distinction is made between the population units and the variables measured on each unit in the sample. Population units are the things we select during random sampling; variables are the measurements we record.

If we capture animals and record their sex, age, and mass, then the population unit is an animal and the variables are sex, age, and mass. If we record behavioral measurements on each of several animals during several 1-h intervals, then the population unit is an animal watched for 1 h, an 'animal-hour', and the variables are the behavioral data recorded during each hour of observation. In time-activity sampling, we often record behavior periodically during an observation interval. The population unit is then an 'animal-time', and the variables are the behaviors recorded. In some studies, plants or animals are the variables rather than the population units. For example, if we record the number of plants or the number of species in each of several plots, then the population unit is a plot, and the variable is 'number of plants' or 'number of species'. In most studies carried out by behavioral ecologists, the population unit is: (1) an animal, plant, or other object; (2) a location in space such as a plot, transect, or dimensionless point; (3) a period or instant of time; or (4) a combination involving time such as an animal watched for 1 h or a location sampled at each of several times.

Nearly all sampling plans assume that the population units are nonoverlapping. Usually this can be accomplished easily in behavioral ecology. For example, if the population units are plots, then the method of selecting the plots should ensure that no two plots in the sample will overlap each other. In some sampling plans, the investigator begins by dividing the population units into groups in such a way that each population unit is in one and only one group. Subdivision in this manner is called a partition of the population. Sample selection is also usually assumed to be without replacement unless stated otherwise. Sampling without replacement implies that a unit cannot be selected twice for the sample, while units could be included two or more times when sampling is with replacement. The names are derived from the practice of physically removing objects from the population, as in drawing balls from an urn and then replacing them or not replacing them.

Application of the 'population unit/variable' approach may seem difficult at first in estimating proportions. If we select 'n' plants and record the proportion that have flowers, what is 'the variable'? Statisticians usually approach such problems by defining the population unit as an individual and the variable as 0 if the individual does not have the trait or condition of interest and 1 if it does. The proportion is thus the mean of the variables in the sample. For example, let y_i refer to the ith plant ($i=1,...,n$) and equal 0 if the plant does not have flowers and 1 if it does have flowers. Then the proportion may be written as $\sum y_i/n$. This principle – that proportions may be thought of as means (of 0s and 1s) – is useful in several contexts. For example, it shows that all results applicable to means in general also apply to proportions (though proportions do have certain properties – described in later Chapters – not shared by all means). Notice that it matters whether we use 0 to mean 'a plant without flowers' or 'a plant with flowers'. The term 'success' is commonly used to indicate which category is identified by a 1. The other category is often called 'failure'. In our example, a 'success' would mean a plant with flowers.

In most studies we wish to estimate many different quantities, and the definitions of population units and variables may change as we calculate new estimates. For example, suppose we wish to estimate the average number of plants/m^2 and seeds/plant. We use plots to collect plants and then count the number of seeds on each plant. In estimating the average number of plants per plot, the population unit is a plot, and the variable is the number of plants (i.e., y_i = the number of plants in the ith plot). In estimating the number of seeds per plant, the population unit is a plant, and the variable is the number of seeds (i.e., y_i = the number of seeds on the ith plant).

The population is the set of all population units that might be selected for inclusion in the sample. The population has the same 'dimensions' as the population units. If a population unit is an animal watched for an observation interval, then, by implication, the population has two dimensions, one for the animals that might be selected, the other for the times that might be selected. The population in this case might be envisaged as an array, with animals that might be selected listed down the side and times that might be selected listed across the top. Cells in the array thus represent population units and the entries in them are the variables. This approach of visualizing the population as a two-dimensional array will be used extensively in our discussions of 'Survey sampling methods' (Chapter Four) and 'Pseudoreplication' (Chapter Six).

Biologists often think of the species as 'the population' they are studying. The statistical population, however, is the set of population units that might enter the sample. If the population units are plots (in which we count animals for instance), then the statistical population is a set of plots. If the population unit is a trap left open for a day, then the statistical population is the set of trap-days that might enter the sample, not the animals that we might catch in them. This is just a matter of semantics, but confusion is sometimes avoided by distinguishing between statistical and biological populations.

Measures of error

The term error, in statistics, has approximately the same meaning as it does in other contexts: an estimate likely to be far from the true value has large error and one likely to be close to the true value has small error. Two kinds of error, sampling error and bias, are usually distinguished. The familiar 'bull's eye' analogy is helpful to explain the difference between them. Imagine having a quantity of interest (the bull's eye) and a series of estimates (individual bullets lodged on the target). The size of the shot pattern indicates sampling error and the difference, if any, between the center of the shot pattern and the bull's eye indicates bias. Thus, sampling error refers to the variation from one sample to another; bias refers to the difference (possibly zero) between the mean of all possible estimates and the parameter.

Notice that the terms sampling error and bias refer to the pattern that would be observed in repeated sampling, not to a single estimate. We use the term estimator for the method of selecting a sample and analyzing the resulting data. Sampling error and bias are said to be properties of the estimator (e.g., we may say the estimator is biased or unbiased). Technically, it is not correct to refer to the bias or sampling error of a single estimate. More important than the semantics, however, is the principle that measures of error reveal properties of the set of all possible estimates. They do not automatically inform us about how close the single estimate we obtain in a real study is to the true value. Such inferences can be made but the reasoning is quite subtle. This point, which must be grasped to understand the rationale of statistical inference, is discussed more in Chapter Three, 'Tests and confidence intervals'.

The quantity most widely used to describe the magnitude of sampling error is called the standard error of the estimate. One of the remarkable properties of modern statistical methods is that standard errors – a measure of the variation that would occur in repeated sampling – can usually be estimated from a single sample. The effects of sampling error can also be described by the coefficient of variation (CV) which expresses the standard error as a percentage of the estimate [i.e., CV=(standard error/estimate)$\times$100%]. Calculation of CV values facilitates comparison of estimates, especially of quantities measured on very different scales. For example, an investigator might report that all the CV values were less than 20%. Sampling error is also sometimes measured by the variance of the estimate, which is the square of the standard error.

Three sources of bias may be distinguished: selection bias, measurement bias, and statistical bias. Selection bias may occur when some units in the population are more likely to be selected than others or are selected but not measured (but the investigator is using a procedure which assumes equally likely selection probabilities). Measurement bias is the result of systematic recording errors. For example, if we are attempting to count all individuals in plots but usually miss some of those present, then our counts are subject to measurement bias. Note that measurement errors do not automatically cause bias. If positive and negative errors tend to balance, then the average value of the error in repeated sampling might be zero, in which case no measurement bias is present. Statistical bias arises as a result of the procedures used to analyze the data and the statistical assumptions that are made.

Most statistical textbooks do not discuss selection and measurement bias in much detail. In behavioral ecology, however, it is often unwise to ignore these kinds of error. Selection of animals for study must often be done using nonrandom sampling, so selection bias may be present. In estimating abundance, we often must use methods which we know do not detect every animal. Many behavioral or morphological measurements are difficult to record accurately, especially under field conditions.

The statistical bias of most commonly used statistical procedures is either zero or negligible, a condition we refer to as 'essentially unbiased', meaning that the bias, while not exactly equal to zero, is not of practical importance. When using newer statistical procedures, especially ones developed by the investigator, careful study should be given to whether statistical bias exists. When estimates are biased, then upper bounds must be placed on the size of the bias or the estimates are of little value. This is often possible using analytical methods for statistical bias. Bias caused by nonrandom selection or measurement errors, however, usually cannot be estimated with statistical methods, a point which has important implications for understanding tests and confidence intervals (see Chapter Three).

A few examples will help clarify the distinctions between sampling error and the various types of bias. Leuschner *et al.* (1989) selected a simple random sample of hunters in the southeastern United States of America and asked them whether more tax dollars should be spent on wildlife. The purpose was to estimate what proportion of all hunters in the study area would answer yes to this question. Sampling error was present in the study because different random samples of hunters would contain different proportions who felt that tax dollars should be spent on wildlife. Selection bias could have been present because 42% of the people selected for the sample were unreachable, gave unusable answers, or did not answer at all. These people might have felt differently, as a group, than those who did answer the question. There is no reason to believe that measurement bias was present. The authors used standard, widely accepted methods to analyze their results, so it is unlikely that their estimation procedure contained any serious statistical bias. Note that the types of error are distinct from one another. Stating, as in the example above, that no measurement or statistical bias was present in the estimates does not reveal anything about the magnitude of sampling error or selection bias.

Otis *et al.* (1978) developed statistical procedures for estimating population size when animals are captured, marked, and released, and then some of them are recaptured one or more times. The quantity of interest was the total number of animals in the population (assumed in these particular models to remain constant during the study). Sampling error would occur because the estimates depend on which animals are captured and this in turn depends on numerous factors not under the biologists' control. Selection bias could occur if certain types of animals were more likely to be captured than others (though the models allowed for certain kinds of variation in capture probabilities). In the extreme case that some animals are so 'trap wary' as to be uncapturable, these animals would never appear in any sample. Thus, the estimator would estimate the population size of capturable animals only and thus systematically underestimate total population size. Measurement bias would occur if animals lost their marks (this was assumed not to occur). The statistical procedures were new, so the authors studied statistical bias with computer simulations. They found little statistical bias under some conditions, but under other conditions the estimates were consistently too high or too low even if all required assumptions were met.

Two other terms commonly used to describe the different components of error are precision and accuracy. Precision refers solely to sampling error whereas accuracy refers to the effects of both sampling error and bias. Thus, an estimator may be described as 'precise but not accurate' meaning it has a small standard error but is biased. Accuracy is defined as the square of the standard error plus the square of the bias and is also known as the mean squared error of the estimator.

1.4 Extrapolation to other populations

Statistical analysis allows us to make rigorous inferences about the statistical population but does not automatically allow us to make inferences to any other or larger population. By 'statistical population' we mean the population units that might have entered the sample. When measurements are complex or subjective, then the scope of the statistical inferences may also be limited to the 'conditions of the study', meaning any aspect of the study that might have affected the outcome. These restrictions are often easy to forget or ignore in behavioral ecology so here we provide a few examples.

If we record measurements from a series of animals in a study area, then the sampled population consists of the animals in the study area at the time of the study and the statistical inferences apply to this set of animals. If we carry out a manipulation involving treatments and controls, then 'the population' is the set of individuals that might have been selected and the inferences apply only to this population and experiment. Inferences about results that would have been obtained with other populations or using other procedures may be reasonable but they are not justified by the statistical analysis. With methods that detect an unknown fraction of the individuals present (i.e., index methods), inferences apply to the set of outcomes that might have been obtained, not necessarily to the biological populations, because detection rates may vary. Attempts to identify causes in observational studies must nearly always recognize that the statistical analysis identifies differences but not the cause of the differences.

One sometimes hears that extrapolation beyond the sampled population is 'invalid'. We believe that this statement is too strong, and prefer saying that extrapolation of conclusions beyond the sampled population must be based on additional evidence, and that this evidence is often largely or entirely nonstatistical. This does not mean that conclusions about a target population are wrong: it only means that the protection against errors afforded by the initial statistical methods is not available and everyone should realize that. For example, if we measure clutch size in one study area and period of time, then the statistical analysis only justifies making inferences about the birds in the study area during the study period. Yet everyone would agree that the results tell us a good deal about likely clutch size in nearby areas and in future or past years. The extent to which conclusions from the study can be extrapolated to larger target populations would be evaluated using biological information such as how clutch size varies in space and time in the study species and other closely related species. This distinction is often reflected in the organization of journal articles. The Results section contains the statistical analysis, whereas analyses of how widely the results apply elsewhere are presented in the Discussion section. Thus, in our view, the reason for careful identification of the sampled population and conditions of the study is not to castigate those who extrapolate conclusions of the study beyond this population but only to emphasize that additional, and usually nonstatistical, rationales must be developed for this stage of the analysis.

1.5 Summary

Decisions about what population to study are usually based primarily on practical, rather than statistical, grounds but it may be helpful to recognize the trade-off between internal and external validity and to recognize that studying a small population well is often preferable to studying the largest population possible. The superpopulation concept helps explain the role of statistical analysis when all individuals in the study area have been measured. Point estimates of interest in behavioral ecology usually are means or measures of association, or quantities based on them such as differences and regression equations. The first step in calculating point estimates is defining the population unit and variable. A two-dimensional array representing the population is often helpful in defining the population. Two measures of error are normally distinguished: sampling error and bias. Both terms are defined with respect to the set of all possible samples that might be obtained from the population. Sampling error is a measure of how different the sample outcomes would be from each other. Bias is the difference between the average of all possible outcomes and the quantity of interest, referred to as the parameter. Three types of bias may be distinguished: selection bias, measurement bias, and statistical bias. Most statistical methods assume the first two types are absent but this is often not a safe assumption in behavioral ecology. Statistical inferences provide a rigorous method for drawing conclusions about the sampled population, but inferences to larger populations must be based on additional evidence. It is therefore useful to distinguish clearly between the sampled population and larger target populations of interest.

2
Estimation

2.1 Introduction

This Chapter describes some of the statistical methods for developing point and interval estimators. Most statistical problems encountered by behavioral ecologists can be solved without the use of these methods so readers who prefer to avoid mathematical discussions may skip this Chapter without compromising their ability to understand the rest of the book. On the other hand, the material may be useful in several ways. First, we believe that study of the methods in this Chapter will increase the reader's understanding of the rationale of statistical analysis. Second, behavioral ecologists do encounter problems frequently that cannot be solved with 'off the shelf' methods. The material in this Chapter, once understood, will permit behavioral ecologists to solve many of these problems. Third, in other cases, consultation with a statistician is recommended but readers who have studied this Chapter will be able to ask more relevant questions and may be able to carry out a first attempt on the analysis which the statistician can then review. Finally, many behavioral ecologists are interested in how estimators are derived even if they just use the results. This Chapter will help satisfy the curiosity of these readers.

The first few sections describe notation and some common probability distributions widely used in behavioral ecology. Next we explain 'expected value' and describe some of the most useful rules regarding expectation. The next few Sections discuss variance, covariance, and standard errors, defining each term, and discussing a few miscellaneous topics such as why we sometimes use 'n' and sometimes '$n-1$' in the formulas. Section 2.10 discusses linear transformations, providing a summary of the rules developed earlier regarding the expected value of functions of random variables. The Taylor series for obtaining estimators for nonlinear transformations is developed in Section 2.11, and the

Chapter closes with an explanation of both the principle and mathematics of maximum likelihood estimation. Examples of how these methods are used in behavioral ecology are provided throughout the Chapter. The mathematics is elementary, mainly involving simple algebra. Derivatives and the solution of simultaneous equations are briefly mentioned in the last few sections.

2.2 Notation and definitions

Throughout this book, we use lower-case letters for quantities associated with the sample and the corresponding upper-case letters for quantities associated with the population. Thus, sample size and population size are typically denoted by n and N respectively, and the sample mean and population mean are typically denoted by $\bar{y}$ and $\bar{Y}$ respectively. The same convention is used for other quantities. Thus, *se* and *cv* are used for the estimated standard error and coefficient of variation, derived from the sample, while *SE* and *CV* are used for the actual values calculated from all units in the population.

Many estimates have the same form as the parameter they estimate, although they involve only sample values and are represented by lower-case letters. The sample mean, $\bar{y}$, is generally used to estimate the population mean, $\bar{Y}$; the proportion of 'successes' in a sample, p, is generally used to estimate the proportion of successes in the population, P. Estimates calculated from samples, which have the same form as the parameter, are referred to as sample analogs. For example, if we are interested in the ratio of two population means, $\bar{X}$ and $\bar{Y}$, we might define the parameter as

$$\frac{\bar{Y}}{\bar{X}}. \tag{2.1}$$

The sample analog of this parameter is

$$\frac{\bar{y}}{\bar{x}}. \tag{2.2}$$

In nearly all cases, we denote means by bars over the symbol as in Eqs. 2.1 and 2.2. In a few of the tables in Appendix One, however, this causes notational problems and a slightly different approach is used (explained in Box 3). A final convention (Cochran 1977 p. 20) is that measurements from single population units are generally symbolized using lower-case letters (e.g., y_i) regardless of whether they are components of an estimate or a parameter. Thus, the sample mean and population mean both use y_i

$$\bar{y} = \frac{1}{n}\sum_{i=1}^{n} y_i \quad \text{and} \quad \bar{Y} = \frac{1}{N}\sum_{i=1}^{N} y_i, \tag{2.3}$$

although with the following distinction. In the sample mean $y_1,...,y_n$ represent the n sample observations (random variables) while for the population mean $y_1,...,y_N$ represent a list of the values (fixed numbers) corresponding to the population units. Thus y_i has a different meaning in the two formulas. It will be clear from the context which is appropriate.

In some cases, particularly with nonrandom sampling and in experimental situations, 'the population' is not well defined. Thus, if we select the first n animals encountered for our sample, or if we carry out ten tests on each of five animals, then 'the population' may be difficult to define which, in turn, causes difficulty in thinking about what the parameters represent. In such cases, we often find it easier to think of the population as a 'much larger sample' (i.e., infinitely large), selected using the same sampling plan, or the parameter as 'the average in repeated sampling'. In most studies we can imagine having obtained many different samples. The notion of a much larger sample or the average of repeated samples may be easier to visualize than a well-defined population from which our data set was randomly selected.

Variables can be classified in several ways depending on the number and kinds of values they take. The broadest classification is discrete or continuous. Discrete variables most commonly take on a finite number of values. In some instances it is convenient to treat discrete variables as though any integer value 0, 1, 2, 3, ... is possible. Even though the number of possible values is infinite in such cases, these variables are still considered to be discrete. The simplest discrete variable that can be measured on a unit is a dichotomous variable which has only two distinct values. Common examples include male/female, young/adult, alive/dead, or present/not present in a given habitat. As already noted, dichotomous variables are usually recorded using '0' for one value and '1' for the other value. When the values of a variable are numerical the variable is said to be quantitative, while if the values are labels indicating different states or categories the variable is called categorical. For a categorical variable, it is often useful to distinguish between those cases in which the categories have a natural ordering and those cases that do not. Examples of categorical variables include sex, which is also a dichotomous variable, and types of behavior or habitat. Position in a dominance hierarchy does imply an order, or rank, and could be considered an ordered, categorical variable. Discrete quantitative variables are often counts, such as number of offspring or number of species in a plot.

Technically, continuous variables are quantitative variables that can take on any value in a given range to any number of decimal places. In practice, of course, such variables are only recorded to a fixed number of decimal places. Examples include age, mass, and duration of a behavior. In each of these examples, the number of possible values for the variable is limited only by the accuracy of our measuring device.

The goals of a study may dictate whether a given variable is treated as categorical or quantitative. For example, year might be used purely as a category in one analysis but as a continuous variable in another analysis (e.g., studying how abundance changes through time).

2.3 Distributions of discrete random variables

The distribution of a discrete variable can be described by a list of the possible values of the variable and the relative frequency with which each value occurs in the population with which the variable is associated. For example, the distribution of a dichotomous variable such as sex just refers to the proportion of males, and of females, in the population. The age distribution in a population may be visualized as a list of ages and the proportion of the population that are each age. Continuous distributions are described by a curve, such as the familiar normal curve. The area under the curve for a given interval is the probability that a randomly selected observation falls in this interval.

In describing distributions, statisticians typically do not distinguish between the population units and the value of the variable measured on each unit. Thus, they may say that a population is normal, or skewed, or symmetrical. Viewed in this manner, the population is the collection of numbers that we might measure rather than the population units (i.e., animals) we might select.

Here are three examples of discrete distributions, each of them developed around the notion of flipping a thumbtack. This tack may land on its side, denoted by 1, or point up denoted by 0. The distribution gives us the probability of obtaining each of the possible results. We will use the letter 'x' to denote a random variable representing the outcome of a flip and the letter K for a specific outcome. Thus, $\mathrm{P}(x=K)$ means 'the probability that we get the result K'. In our example, if $K=0$, then $\mathrm{P}(x=K)$ means the probability that the tack lands point up. We will denote the probability that the tack lands on its side as P, an unknown parameter. Thus $\mathrm{P}(x=1)=P$ and $\mathrm{P}(x=0)=1-P$. Note that the italic letter P is a parameter (the probability that the tack lands on its side) and that

P is an abbreviation meaning 'the probability of ...' Consider the expression

$$P(x=K)=P^{K}(1-P)^{1-K}. \tag{2.4}$$

This expression gives the distribution for the case we are considering. Two outcomes are possible, $K=0$ and $K=1$. If $K=0$, then the expression reduces to $\mathrm{P}(x=0)=1-P$ (because $P^0=1$), which is the probability that the tack lands point up. If $K=1$, the expression reduces to $\mathrm{P}(x=1)=P$ which is the probability that the tack lands on its side. A random variable that can take on only two possible outcomes is often called a Bernoulli random variable, and is said to have a Bernoulli distribution.

Second, suppose we flip the tack n times and count the number of times it lands on its side. The possible outcomes are now $K=0,\ldots,n$. This distribution is called the binomial; its characteristics are that a series of n independent 'trials' occur. On each trial, only two outcomes are possible, often referred to as 'success' and 'failure,' and the probability of a success is the same on each trial. The distribution for a binomial random variable such as x in our example is

$$P(x=K)=\binom{n}{K}P^{K}(1-\mathrm{P})^{n-K}, \tag{2.5}$$

where the term in large parentheses means 'n choose K', the number of distinct ways to select K items from n items. The formula for n choose K is

$$\binom{n}{K}=\frac{n!}{K!\,(n-K)!}, \tag{2.6}$$

where ! means factorial, $n!=n(n-1)(n-2)\ \ldots\ (2)(1)$ and $0!$ is defined to be 1. Notice that Eq. 2.5 reduces to Eq. 2.4 if $n=1$. Thus, statements later about the binomial distribution also apply to the Bernoulli distribution already described with $n=1$. See Moore and McCabe (1995, Chapter 5, Section 1) or Rice (1995, Chapter 1) for a derivation of Eq. 2.6.

Many problems in behavioral ecology can be phrased in terms of the binomial distribution. Notice that the outcome of a single trial is a dichotomous (two-valued) variable. As noted in Section 2.2, behavioral ecologists frequently study dichotomous random variables such as female/male, young/adult, alive/dead, infected/not infected, and so on. Random sampling, in these cases, can be viewed as a series of 'trials'. The probability that the measurement on each unit in the sample is 'success' is the same and equals the proportion of the population that has the attribute of interest. Also, the sample result may be phrased as 'K successes in the sample of size n'. Thus, the outcome of sampling is a binomial random variable. For example, if we select n individuals from a very large population (so that selections are essentially independent) and record how many are female, then the probability that our result equals any specific value K is given by Eq. 2.5 with P = proportion of the population that are female. As noted above, once the distribution of a sample is known, then one can calculate numerous quantities of interest. In this case, knowing that the distribution is binomial, we can easily obtain such quantities as the estimator for P and its standard error (Appendix One, Box 5).

Some of the most interesting applications of the binomial distribution in behavioral ecology involve cases in which it may not be immediately obvious that the data should be treated as binomial. For example, suppose we are studying habitat preferences by recording the habitat type for a series of randomly selected animals. Assume that a single survey is made, and that we want to estimate the proportion of the animals in a given habitat. A sighting may be viewed as a 'trial' and the outcomes as 'success = in the habitat of interest' or 'failure = in some other habitat'. The proportion of animals in the habitat is estimated as the number of successes divided by the total number of trials (i.e., animals observed). Thus, these data can be analyzed using methods based on the binomial distribution.

In other cases, complex measurements may be made on animals, plants, or at sites, but interest centers on the proportion of sites in which a particular pattern was observed or some threshold was exceeded. For example, in a study of northern spotted owls (Thomas *et al.* 1990), one of the questions was whether the owls showed a statistically significant preference for old growth forests as compared to other habitats. The data were collected by radio telemetry and analysis involved a complex effort to delineate home ranges, calculation of the proportion of the home range covered by old growth and determination of whether the owls occurred more often in old growth than would be expected if they distributed themselves randomly across the landscape. This analysis, while complex, yielded a single answer for each owl, 'yes' or 'no'. Once the answer was obtained for each bird, the rest of the analysis could thus be based on the binomial distribution with n = number of animals and K = number of 'yes' answers.

The binomial distribution is sometimes useful when more complex and efficient methods exist but entail questionable assumptions. For example, suppose we select pairs of animals and record some feature such as size for each member of the pair to determine whether the average value for females in the population is larger than the average value for males. Such data can be analyzed using the actual measurements and a t-test or nonparametric test (Chapter Three), but in some cases the required assumptions may not be met. A simpler, albeit less powerful, method is simply to classify the outcome from each pair as 'success' (value for the female was larger) or 'failure'. The data are thus binomial and a very simple procedure (the sign test), based on the binomial distribution and involving fewer assumptions, may be used for the analysis. Note, however, that this analysis addresses a slightly different question. The question addressed by the sign test is whether more than half the paired females are larger than their mates, not whether the mean size of paired females exceeds the mean for their mates. It is possible (though perhaps not likely) that $>50\%$ of the females might be larger than their mates but that the average size of females might be less than the average size of their mates. The biologist must decide whether this distinction is important in the investigation. If it is not, then the sign test may provide a useful alternative to tests requiring more assumptions.

The third distribution we wish to discuss is a generalization of the binomial to more than two outcomes. To explain it, imagine that a 'trial', in our tack flipping example, involves flipping two tacks and counting the number landing on their side. Three outcomes are now possible 0, 1, and 2. Let us use the symbols P_0, P_1, and P_2 for the probabilities of these outcomes. If we select n pairs and determine the number of successes in each, then we will obtain some number of 0s, 1s, and 2s. We will use K_0, K_1, and K_2 for specific numbers of each outcome. This distribution is multinomial; its density is

$$\frac{n!}{K_0!K_1!K_2!}P_0^{K_0}P_1^{K_1}P_2^{K_2}. \tag{2.7}$$

This expression gives the probability that we would obtain K_0 pairs with no successes, K_1 pairs with 1 success and K_2 pairs with 2 successes. If more than 3 outcomes are possible, then the formula has additional K and P values following the same pattern above.

The multinomial distribution is particularly useful, in behavioral ecology, in capture–recapture studies. This application will be explained in more detail in Chapter Nine but, in brief, animals are marked and recaptured on several occasions. In the analysis, marked animals are divided into groups assumed to have similar survival and recapture probabilities. These probabilities may vary through time, but they are the same, at any given time, for all members of the group. The data consist of the numbers of each group recaptured on each occasion or never recaptured. These numbers are the K_i where i refers to a specific group at a specific time. Under the assumption that members of a group have the same survival and recapture probabilities, one may write down the probability of recapturing each individual in each group on each sampling occasion. These are the P_i. The numbers recaptured thus have a multinomial distribution. Methods based on the multinomial distribution can therefore be used to obtain the point and interval estimates of survival or other quantities of interest.

The multinomial distribution may also arise in many other situations. For example, birds are often assigned to the categories 'hatching year', 'second year', and 'after second year'. Animal sightings may be categorized according to the habitat in which the individual was spotted. The number of fertilizations may be restricted to a narrow range such as 0 to 4. The sample outcome, in all of these cases, may follow a multinomial distribution and methods based on this distribution may be useful. Many of these cases, however, are also handled easily by successive application of binomial methods, as indicated by the example of habitat use already discussed. Furthermore, in quantitative cases such as 'number of fertilizations', we are often interested in the mean outcome, rather than the proportion of the outcomes in each category. Thus, use of the multinomial distribution seems to be less common in behavioral ecology with a few conspicuous exceptions such as capture–recapture methods. Other cases in which the multinomial distribution may apply are noted in later Chapters.

Note that the multinomial includes the binomial as a special case in which just two outcomes are possible. If that is true, then $P_1 = 1 - P_0$ and the expression with factorials may be written 'n choose K_0'. The expression thus reduces to Eq. 2.5 for the binomial distribution. The three distributions, Bernoulli, binomial, and multinomial, are thus an increasingly general series, all involving the same notion of independent trials on which the possible outcomes and the probability of each outcome are the same from trial to trial.

2.4 Expected value

Many concepts discussed in this book involve the concept of 'expected value'. We explain the meaning of expected value here briefly, and only for discrete values, and provide a first few useful properties. This material may prove difficult for readers who have not previously encountered the concept. It can be skipped during a first reading of the Chapter. Study of expected value at some point, however, will increase one's understanding of statistical procedures and make the answers to many problems that behavioral ecologists encounter in real work easier to derive and understand.

The idea of an expected value is most easily understood in the context of sampling from a finite population consisting of the values $y_1,\ldots,y_N$. The expected value of a statistic, y, calculated from a sample, $y_1,\ldots,y_n$, is simply its average value in repeated sampling. More precisely, suppose that the number of possible distinct samples that could be drawn from a specified population with a specified sampling plan is N^*, and that the N^* different samples are all equally likely. The expected value of y, denoted $E(y)$, is

$$E(y)=\frac{1}{N^*}\sum_{i=1}^{N^*} y_i, \tag{2.8}$$

where y_i = the value of the statistic calculated from the ith sample.

The term 'distinct sample' refers to the set of population units that is included in the sample (not the values of the response variable). As a simple example, if the population size was 4 and we selected a sample of size 2 without replacement, then the number of distinct possible samples is 6 (units 1,2; 1,3; 1,4; 2,3; 2,4; and 3,4). More generally, if we sample randomly without replacement, ensuring each time we select a unit that every unit still in the population has the same probability of being selected (i.e., we select a simple random sample), then the number of samples is

$$N^* = \binom{N}{n} = \frac{N!}{n!\,(N-n)!}. \tag{2.9}$$

If the distinct samples are not all equally likely, as occurs with some sampling plans (see Chapter Four), then we define the expected value of y as the weighted average of the possible sample results with the weight, f_i, for a given sample result, y_i, equal to the probability of obtaining y_i

$$E(y)=\sum_{i=1}^{N^*} f_i y_i. \tag{2.10}$$

Notice that Eq. 2.10 is a more general version that includes Eq. 2.8. In the case of equal-probability samples all $f_i = 1/N^*$, and Eq. 2.10 reduces to Eq. 2.8. The expected value of y is thus a particular type of 'average', the special features being that all possible samples are included in the average and that weighting is equal if the samples are equally likely and equal to the selection probabilities if the samples are not all equally likely.

Here is a simple example of calculating expected value. Suppose we flip a coin once and record '0' if we obtain a head and '1' if we obtain a tail. What is the expected value of the outcome? The notion of 'all possible samples' is not readily applicable to this example, but if the coin is 'fair' then the probability of getting a 0 and the probability of getting a 1 are both 0.5. The expected value of the outcome of the coin flip is therefore $(0.5\times 0+0.5\times 1)=0.5$, the sum of the possible outcomes (0 and 1) weighted by the probabilities with which they occur.

Rules

A few useful rules regarding expected values are now given.

1. The expected value of a constant, a, times a random variable, y, is the constant times the expected value of the random variable

$$E(ay) = aE(y).$$

2. The expected value of a sum is the sum of the expected values. For example, with two random variables, y and x

$$E(y + x) = E(y) + E(x).$$

3. The expected value of the product of random variables, the ratio of random variables, or of a random variable raised to a power other than 0 or 1 is, in general, not equal to the same function of the expected values. For example

$$E(yx) \neq E(y)E(x)$$

$$E(y/x) \neq E(y)/E(x)$$

and

$$E(y^a) \neq [E(y)]^a,$$

if $a \neq 0$ or 1. The term 'in general' means that we cannot assume that equality always holds. It might hold in a specific case, depending on the values or other attributes of y and x, but often it does not hold. One special case of rule 3 is worth noting. If two random variable, x and y, are independent then $E(xy) = E(x)E(y)$.

The above rules help identify conditions under which estimators are unbiased. For example, suppose we have measured numbers per 1–m^2 plot and we wish to express the results on another scale, for example numbers per hectare. We will use the subscript 'm' to indicate number per meter and 'h' for number per hectare. Also, assume that our sample mean / 1-m^2 plot, $\bar{y}_m$, is an unbiased estimate of the population mean per 1-m^2 plot, $\bar{Y}_m$, that is $E(\bar{y}_m) = \bar{Y}_m$. A hectare equals 10,000 square meters, so the true mean per hectare is

$$\bar{Y}_h = 10{,}000\ \bar{Y}_m.$$

According to rule 1, if we multiply our estimate, $\bar{y}_m$ by 10,000, then we may write

$$E(10{,}000\bar{y}_m) = 10{,}000\ E(\bar{y}_m) = 10{,}000\ \bar{Y}_m = \bar{Y}_h.$$

This demonstrates that changing the scale at which an unbiased estimate is reported (i.e., multiplying the estimate by a constant) produces an unbiased estimate of the original parameter multiplied by the same constant. The same reasoning shows that if we have an unbiased estimate of density per plot in a study area, then we can obtain an unbiased estimate of population size by multiplying our sample mean times the number of plots in the study area. In this example, the term 'a' becomes the number of plots in the study area.

Rule 2, i.e., that the expected value of the sum is the sum of the expected values, shows that if we have unbiased estimates of two quantities, we can simply add the estimates to obtain an unbiased estimate of the sum of the parameter values. For example, if a study area is divided into two habitats and we have unbiased estimates of the number of animals in each, then we can add them to obtain an unbiased estimate of total population size. The principle is also useful in evaluating expressions such as $E(\bar{y})$. Thus, according to this principle, $E(\bar{y}) = 1/n \sum E(y_i)$ which is often easier to evaluate than $E(\bar{y})$.

The third principle indicates the situatons in which we may not be able to use the kind of reasoning already discussed. One of the most common examples in which this is important for behavioral ecologists is in estimating ratios. For example, suppose we want to estimate proportional change in population size between two years, $\bar{Y}_2/\bar{Y}_1$, where $\bar{Y}_1$ and $\bar{Y}_2$ are the true population sizes in years 1 and 2. Assume that we have unbiased estimates, $\bar{y}_1$ and $\bar{y}_2$, of $\bar{Y}_1$ and $\bar{Y}_2$. It would be natural to assume that $\bar{y}_2/\bar{y}_1$ would be an unbiased estimate of actual change, $\bar{Y}_2/\bar{Y}_1$. In this case, however, rule 3 cautions us that the expected value of this quantity may not be equal to the same expression with parameters in place of estimates. That is, $\bar{y}_2/\bar{y}_1$ may be a biased estimator of $\bar{Y}_2/\bar{Y}_1$, and we must be careful if we use this estimator to ensure that the bias is acceptably small. Later in the Chapter we describe a method (the Taylor series approximation) for estimating the magnitude of the bias in specific cases such as this one.

2.5 Variance and covariance

Consider a population consisting of N numbers, y_i, where $i = 1,\ldots,N$. The variance of the y_i, referred to as the 'population variance', is usually defined as

$$V(y_i) = \frac{\sum_{i=1}^{N}(y_i - \bar{Y})^2}{N}, \tag{2.11}$$

where $\bar{Y}$ is the mean of $y_1,\ldots,y_N$ (Cochran 1977 p. 23). Variance is thus the average of the 'squared deviations', $(y_i - \bar{Y})^2$.

The variance of random variables is usually defined using expectation. The variance of a random variable, y, with expected value Y, is $E[(y-Y)^2]$ where expectation is calculated over all possible samples. If N^* distinct samples are possible, all of them equally likely, then

$$V(y) = \frac{\sum_{i=1}^{N^*}(y_i - Y)^2}{N^*}, \tag{2.12}$$

where y_i, $i=1,\ldots, N^x$, is the value of the random variable from sample i.

If the samples are not equally likely, as occurs with some sampling plans (Chapter Four), then $E[(y-Y)^2]$ is calculated by weighting each distinct sample by its selection probability as indicated in Eq. 2.10

$$V(y) = \sum_{i=1}^{N^*} f_i(y_i - Y)^2,$$

where f_i = the probability of drawing sample i.

The random variable, y, may be a single population unit, the mean of a sample of units, or a derived quantity such as a standard deviation or standard error. For example, the variance of the sample mean, $\bar{y}$ (assuming simple random sampling from a 'large' population) is

$$V(\bar{y}) = \frac{\sum_{i=1}^{N^*}(\bar{y}_i - \bar{Y})^2}{N^*}. \tag{2.13}$$

where $\bar{y}_i$ is the mean from sample i, $\bar{Y}$ is the population mean and is known, in this case, to be the expected value of $\bar{y}$.

Now suppose our population consists of N pairs of numbers, (x_1,y_1), (x_2,y_2), $(x_3,y_3),\ldots,(x_N,y_N)$. The covariance of the pairs (x_i,y_i) is usually defined as

$$\text{Cov}(x_i,y_i) = \frac{\sum_{i=1}^{N}[(x_i - \bar{X})(y_i - \bar{Y})]}{N}, \tag{2.14}$$

where $\bar{X}$ is the mean of $x_1,\ldots,x_N$ and $\bar{Y}$ is the mean of $y_1,\ldots,y_N$. Covariance is thus the average of the 'cross-products', $(x_i - \bar{X})(y_i - \bar{Y})$.

The covariance of random variables, like the variance of random variables, is defined using expectation. $Cov\ (x,y) = [E(x-\bar{X})(y-\bar{Y})]$ where $\bar{X}$ and $\bar{Y}$ are the expected values of x and y $[E(x)=\bar{X}$ and $E(y)=\bar{Y}]$, and

Cov (x,y) is calculated as the simple average of the cross-product terms $(x_i - \bar{X})(y_i - \bar{Y})$ in repeated sampling if all samples are equally likely, and as a weighted average if the samples are not all equally likely.

Covariance formulas are often complex because they involve two random variables. We present one result that is useful in many contexts. Suppose a simple random sample of n pairs is selected from the population and the sample means of the x values and y values are denoted $\bar{x}$ and $\bar{y}$. Then

$$\mathrm{Cov}(\bar{x},\bar{y}) = \frac{1}{N^*}\sum_{i=1}^{N^*} (\bar{x}_i - \bar{X})(\bar{y}_i - \bar{Y}). \tag{2.15}$$

2.6 Standard deviation and standard error

The standard deviation of any random variable, y, is the square root of the variance of y

$$SD(y) = \sqrt{V(y)}. \tag{2.16}$$

If the random variable is a sample mean, $\bar{y}$, then we have

$$SD(\bar{y}) = \sqrt{V(\bar{y})}. \tag{2.17}$$

The same relationship applies to quantities derived from samples such as correlation and regression coefficients.

The standard deviation of an estimate is frequently referred to as the standard error. Thus, the standard error of an estimate is its standard deviation (in repeated sampling) which is the square root of its variance. For example, with sample means

$$SE(\bar{y}) = SD(\bar{y}) = \sqrt{V(\bar{y})}, \tag{2.18}$$

and if b is the usual least-squares estimate of the slope in simple linear regression

$$SE(b) = SD(b) = \sqrt{V(b)} \tag{2.19}$$

2.7 Estimated standard errors

Formulas in the preceding sections define parameters which are important in sampling theory. We now turn to the estimation of these quantities. The general approach is to rewrite the formula for the true standard error, $SE(\bar{y})$, in a simpler form, and then to derive an estimator that is unbiased. We omit proofs but include enough details so that the meaning of the various quantities can be explained. This also lets us explain, at the end of this section, why 'variances' are defined as 'average squared deviations' but are then often written with $N-1$ or $n-1$, rather than N and n, in the denominator.

It can be shown (e.g., Cochran 1977 p. 23) that the variance of the sample mean, with simple random sampling from a finite population, can be written

$$V(\bar{y}) = E(\bar{y} - \bar{Y})^2 = \frac{(N-n)}{Nn} \frac{\sum_{i=1}^{N}(y_i - \bar{Y})^2}{N-1}, \tag{2.20}$$

where N and n are the population and sample size, respectively. Thus, we do not have to obtain all N^* different samples to calculate the variance of these means, we can use the simpler formula in Eq. 2.20. Notice that $V(\bar{y})$ is a simple function of the quantity $\sum(y_i - \bar{Y})^2/(N-1)$. It is customary to use the term S^2 for this quantity

$$S^2 = \frac{\sum_{i=1}^{N}(y_i - \bar{Y})^2}{N-1}. \tag{2.21}$$

It also can be shown (e.g., Cochran 1977 p. 26) that the sample analogue of S^2, $s^2 = \sum(y_i - \bar{y})^2/(n-1)$, is an unbiased estimate of S^2. That is

$$E\left(\frac{\sum_{i=1}^{n}(y_i - \bar{y})^2}{n-1}\right) = \frac{\sum_{i=1}^{N}(y_i - \bar{Y})^2}{N-1}, \tag{2.22}$$

or more compactly $E(s^2) = S^2$. The quantity s^2 is often referred to as the sample variance. Note, however, that $E(s^2) = S^2$ is not the population variance (Eq. 2.11) which has N, not $N-1$, in the denominator.

From Eqs. 2.20 and 2.22, and since n and N are known constants, we may write

$$E\left(\frac{N-n}{Nn}s^2\right) = \frac{N-n}{Nn}E(s^2) = \frac{N-n}{Nn}S^2 = V(\bar{y}). \tag{2.23}$$

Thus, the term on the left is an unbiased estimator of $V(\bar{y})$ and we may use it to estimate the standard error of $\bar{y}$

$$se(\bar{y}) = \sqrt{v(\bar{y})} = \sqrt{\frac{N-n}{Nn}s^2} \tag{2.24}$$

In most cases, population size, N, is so much larger than sample size, n, that $(N-n)/N = 1-n/N$ is very close to 1.0 and may be omitted. This leads to a simple formula for the estimated standard error

$$se(\bar{y}) = \frac{s}{\sqrt{n}}, \tag{2.25}$$

where s is the sample standard deviation

$$s = \sqrt{\frac{\sum_{i=1}^{n}(y_i - \bar{y})^2}{n-1}}. \tag{2.26}$$

We emphasize that these equations do not necessarily apply to sampling plans other than simple random sampling. This issue is discussed at greater length in Chapter Four.

In writing computer programs to calculate variances and covariances, two algebraic identities are useful

$$\sum_{i=1}^{n}(y_i - \bar{y})^2 = \sum_{i=1}^{n} y_i^2 - n\bar{y}^2, \tag{2.27}$$

and

$$\sum_{i=1}^{n}(y_i - \bar{y})(x_i - \bar{x}) = \sum_{i=1}^{n} x_i y_i - n\bar{x}\bar{y}. \tag{2.28}$$

Thus, for example

$$s^2(y_i) = \frac{\sum_{i=1}^{n}(y_i - \bar{y})^2}{n-1} = \frac{\left(\sum_{i=1}^{n} y_i^2\right) - n\bar{y}^2}{n-1},$$

and

$$cov(x_i, y_i) = \frac{\sum_{i=1}^{n}(x_i - \bar{x})(y_i - \bar{y})}{(n-1)} = \frac{\sum_{i=1}^{n} x_i y_i - n\bar{x}\bar{y}}{n-1}.$$

Readers may be interested to note that while s^2 is an unbiased estimate of S^2 [and thus $v(\bar{y})$ is an unbiased estimate of $V(\bar{y})$], the same cannot be said of s and S (or $se(\bar{y})$ and $SE(\bar{y})$]. The reason for this can be seen by recalling the discussion of expected values. We noted there that, in general, the expected value of a random variable raised to a power other than 0 or 1 is not equal to the parameter raised to the same power. Thus, for any random variable g, with expected value G, $E(g^{0.5}) \neq G^{0.5}$. In this case, $g = s^2$ and thus we have $E[(s^2)^{0.5}] \neq (S^2)^{0.5}$ or $E(s) \neq S$. Thus, the usual estimators of the standard deviation and standard error are slightly biased. This does not affect the accuracy of conclusions from tests or construction of confidence intervals, however, because the effect of the bias has been accounted for in the development of these procedures. Readers able to write computer programs may find it instructive to verify that sample estimates of the variance of a sample mean are unbiased but that estimates of the standard error of the mean are slightly biased (Box 2.1).

Box 2.1 A computer program to show that the estimated variance of the sample mean is unbiased but that the estimated standard error is not unbiased.

The steps in this program are: (1) create a hypothetical population, (2) determine the true variance and standard error of the sample mean, and (3) draw numerous samples to determine the average estimates of these quantities. We assume that the reader knows a programming language and therefore do not describe the program in detail. A note about Pop1() and Pop2() may be helpful however. Sample selection is without replacement, so we must keep track, in drawing each sample, to ensure that we do not use the same population unit twice. This is accomplished by the use of Pop2. The program below is written in TruBasic and will run in that language. It can be modified easily to run under other similar languages and could be shortened by calling a statistical function to obtain the mean and SD.

```
!Program.1 - Creates a population of size N1 and takes nreps
!samples, each of size n2, to evaluate bias in the estimated
!variance and standard error of the sample mean.

Let N1=1000                                !Declare pop'n and sample sizes
Let n2=10
Let nreps=1000                             !Number of samples
Dim Pop1(0), Pop2(0), y(0)                 !Declare arrays
Mat Redim Pop1(N1), Pop2(N1), y(n2)        !Dimension them
Randomize                                  !New random seed
For i=1 to N1                              !Create the pop'n
  Let Pop1(i)=rnd                          ! rnd=a random number (0-1)
Next i
For i=1 to N1                              !Calculate the pop'n S^2
  Let sum=sum+Pop(i)
  Let ssq=ssq+Pop(i)^2
Next i
Let PopMn=sum/N1
Let PopS2=(ssq - N1*PopMn^2)/(N1-1)
Let TruVarMn=[(N1-n2)/(N1*n2)] * PopS2     !True v(M¯)
Let TruSEMn=TruVarMn^.5                    ! and SE(M¯)
For k=1 to nreps                           !Begin drawing samples
  Mat Pop2=Pop1                            !Complete pop'n
  Let ct, sum, ssq=0                       !Set counters to 0
  DO                                       !Draw a sample w/o repl
    Let v1=int(rnd*N2)+1                   !A random integer, 1-N2
    If Pop2(v1) <> 0 then                  !Use this unit only if it
      Let ct=ct+1                           has not been selected.
      Let y(ct)=Pop2(v1)                    Track sample size and
      Let Pop2(v1)=0                        exit when it=n2
```

Box 2.1 (*cont.*)

```
      If ct=n2 then exit DO
    End if
  Loop
  For i=1 to n2                                  !sample s²
    Let sum=sum+y(i)
    let ssq=ssq+y(i)^2
  Next i
  Let SamMn=sum/n2
  Let SamS2=(ssq -n2*samMn^2)/(n2-1)
  Let EstVarMn=[N1-n2)/(N1-n2)]*SamS2            !Est'd v(ȳ)
  Let EstSEMn=EstVarMn^.5                        !  and se(ȳ)
  Let SumEstVarMn=SumEstVarMn+EstVarMn           !Keep totals
  Let SumEstSEMn=SumEstSEMn+EstSEMn              !  of the ests.
Next k
Let AveEstVarMn=round(SumEstVarMn/nreps,4)       !Get ave. ests
Let AveEstSEMn=round(SumEstSEMn/nreps,4)         !round
Print " Actual Ave. est."                        !Print results
Print " Variance "; TruVarMn, AveEstVarMn
Print "Standard error '; TruSEMn, AveEstSEMn
END
```

Two traits of s^2 and $se(\bar{y})$ with simple random sampling are worth noting. First, s^2 is an unbiased estimate of S^2 regardless of sample size. Thus, we do not expect s^2 to change in any consistent manner with increasing sample size. On the other hand, $se(\bar{y}) = s/\sqrt{n}$, does change with sample size; it declines, with the decline being proportional to $\sqrt{n}$. Many people confuse s and $se(\bar{y})$ or at least do not understand that s is: (1) a measure of variability in the population; (2) an ingredient in $se(\bar{y})$; and (3) a quantity whose expected value does not change with sample size, whereas $se(\bar{y})$ is a measure of precision (how much $\bar{y}$ would vary in repeated sampling), and thus its value does change (decreases) with increasing sample size.

We once encountered a biologist who spent an entire summer estimating the proportion of twigs on a large study area browsed by moose. He consistently made the error described above of calculating the standard deviation instead of the standard error, so that he was estimating s, rather than $se(\bar{y})$. With each few additional weeks of data he expected his estimated 'standard error' to decrease, since his sample size was getting larger. But it never did, and he finally abandoned the project, a remarkable example of why it is important to understand the difference between the standard deviation and the standard error.

2.8 Estimating variability in a population

Sometimes we are interested in how variable the observations in a population are. Suppose, for example, that we are studying the time required to immobilize animals using different chemicals. We want this time to be as short as possible because animals may injure themselves or escape if the chemical requires a long time to take effect. In evaluating different chemicals, one must estimate the mean immobilization time, but also must know how consistent the time is from animal to animal.

Various approaches exist for describing the spread or variability in a population. The simplest approach is to use percentiles such as the 0th, 25th, 50th, 75th, and 100th. The pth percentile is the quantity such that p % of the values are less than or equal to the quantity. Notice that the 0th percentile is the smallest value in the population, the 50th percentile is the median, and the 100th percentile is the largest. Reporting the 0th and 100th percentiles identifies the interval encompassing all of the values. The difference between them is referred to as the range. Reporting the 25th (called the first quartile) and 75th (called the third quartile) percentiles identifies the interval containing the middle 50% of the values. When a sample is used to estimate these population parameters, the population percentiles are typically estimated by the corresponding sample percentiles. With small samples, specifying a given percentile may require some care in order to avoid ambiguity. For example, in a sample of size 10, the 25th percentile is often found by interpolation.

If the distribution of values in the population is approximately normal, then the population percentiles can be calculated using only the standard deviation and the mean. One uses a table showing percentiles from the standard normal distribution. For example, 68% of the area under a normal curve lies within 1 standard deviation of the mean, 80% lies within 1.28 standard deviations of the mean and 95% of the observations lie within 1.96 standard deviations of the mean. Thus, given estimates of the mean and standard deviation (sd), and assuming that the population is normal, one can conclude that approximately 68% of the population values are in the interval (mean $\pm$ 1 sd), 80% are in the interval (mean $\pm$ 1.28 sd), and 95% are in the interval (mean $\pm$ 1.96 sd). Methods of determining whether observations follow a normal distribution and of calculating intervals estimated to contain any given proportion of the observations in the population are provided by many computer packages and statistics texts (e.g., Moore and McCabe 1995 pp. 67–78). We emphasize that this approach depends strongly on assuming that the population values have a normal distribution.

For normal populations the coefficient of variation can also be used to express variability in the observations. For this purpose, its formula is standard deviation/mean rather than standard error/mean. A report that the *cv* of the observations was 15% would tell us that approximately 68% of the observations were within 15% of the mean and approximately 80% of them were within 19% (1.28 × 15%) of the mean. Thus the same multipliers are used as in the previous paragraph, with the coefficient of variance replacing the standard deviation.

2.9 More on expected value

In Section 2.4 we presented three general rules regarding the expected values of random variables. Since $V(y_i) = E(y_i - \overline{Y})^2$ we can apply these rules to obtain a parallel set of principles about variances:

1. The variance of a constant (a) times a random variable (y) is the square of the constant times the variance of the random variable

$$V(ay) = a^2\, V(y), \tag{2.29}$$

and therefore

$$SD(ay) = a\; SD(y).$$

Similarly, if the random variable is a mean, then $SE(a\overline{y}) = a\; SE(\overline{y})$. A similar rule holds for covariances

$$Cov\,(ax,by) = ab\,Cov\,(x,y), \tag{2.30}$$

where a and b are constants and x and y are random variables.

2. The variance of the sum of random variables is the sum of the variances plus twice the covariances. For example, with two variables, y and x

$$V(x+y) = V(x) + V(y) + 2Cov(x,y). \tag{2.31}$$

The covariance terms include all pairwise covariances. Thus, with three random variables, x, y, and z, we have

$$\begin{aligned} V(x+y+z) &= V(x) + V(y) + V(z) + \\ &2[Cov\,(x,y) + Cov\,(x,z) + Cov\,(y,z)]. \end{aligned} \tag{2.32}$$

In general, with k terms in a sum, the number of pairs is 'k choose 2' $= k(k-1)/2$ so the number of covariance terms increases rapidly with k. This point will become important when we consider sampling plans in which selection of some units is not independent. When estimates are independent, their covariance is 0.0 so the covariance terms drop out. Thus, for independent estimates, the variance of the sum is the sum of the variances. Note, however, that the analogous statement about standard deviations or standard errors is not true. Instead, the standard deviation or standard error of a sum must be defined using the square root of the corresponding variance

$$SD(x + y) = \sqrt{V(x + y)}, \qquad (2.33)$$
$$= \sqrt{V(x) + V(y) + 2Cov(y,x)}$$

and

$$SE(\bar{x} + \bar{y}) = \sqrt{V(\bar{x} + \bar{y})}, \qquad (2.34)$$
$$= \sqrt{V(\bar{x}) + V(\bar{y}) + 2Cov(\bar{y},\bar{x})}$$

3. The variance of the product of random variables (or more generally of a nonlinear function of random variables) is not easily expressed as a function of the variances and covariances of the random variables. For example, in general

$$V(\bar{y}\,\bar{x}) \neq V(\bar{y})\,V(\bar{x}),$$
$$V(\bar{y}\,/\,\bar{x}) \neq V(\bar{y})\,/\,V(\bar{x}),$$

and

$$V(y^a) \neq [V(\bar{y})]^a.$$

if $a \neq 0$ or 1. One special case is worth noting. If two random variables, x and y, are independent and have expected values X and Y respectively, then

$$V(xy) = X^2 V(y) + Y^2 V(x) + V(x)V(y). \qquad (2.35)$$

The principles above are useful in calculating variances (and thus standard errors) of several simple quantities of interest to behavioral ecologists. For example, we discussed changing the scale at which results are reported in Section 2.4, noting that if the initial estimate of density is unbiased then the usual estimate of density at the new scale is also unbiased. The first principle shows that multiplying an estimate by a constant produces a corresponding change in the standard error of the estimate. Working through this result for a particular case may be helpful. Recall that we used the subscripts m and h to denote density on a per m^2 and per hectare basis. The conversion of the density/m^2 to density/ha was

$$\bar{y}_h = 10{,}000\,\bar{y}_m,$$

so we may write

$$V(\bar{y}_h) = V(10{,}000\,\bar{y}_m) = 10{,}000^2\ V(\bar{y}_m),$$

and therefore

$$SE(\bar{y}_h) = \sqrt{V(\bar{y}_h)} = 10{,}000\ SE(\bar{y}_m).$$

The second principle is widely used because sums of random variables are widely used (e.g., in calculating means). Furthermore, the standard error of a difference, $\bar{y} - \bar{x}$, for example, can be derived using this result as shown below:

$$SE(\bar{y} - \bar{x}) = \sqrt{V(\bar{y} - \bar{x})}, \tag{2.36}$$

and (using rule 1)

$$\begin{aligned} V(\bar{y} - \bar{x}) &= V[\bar{y} + (-1)\bar{x}] \qquad (2.37) \\ &= V(\bar{y}) + V[(-1)\bar{x}] + 2Cov[\bar{y}, (-1)\bar{x}] \\ &= V(\bar{y}) + (-1)^2 V(\bar{x}) + (-1)2Cov(\bar{y},\bar{x}) \\ &= V(\bar{y}) + V(\bar{x}) - 2Cov(\bar{y},\bar{x}). \end{aligned}$$

Therefore

$$SE(\bar{y} - \bar{x}) = \sqrt{V(\bar{y}) + V(\bar{x}) - 2\mathrm{Cov}(\bar{y},\bar{x})}. \tag{2.38}$$

If $\bar{y}$ and $\bar{x}$ are independent then their covariance is zero, so we obtain

$$SE(\bar{y} - \bar{x}) = \sqrt{V(\bar{y}) + V(\bar{x})}. \tag{2.39}$$

Computer packages sometimes give standard errors of means but not their variances. In such cases, a slightly different version of Eq. 2.39 may be helpful

$$SE(\bar{y} - \bar{x}) = \sqrt{SE(\bar{y})^2 + SE(\bar{x})^2}. \tag{2.40}$$

The third principle tells us when not to use the sample analogs in calculating standard errors. Thus, the variance of $\bar{y}/\bar{x}$ is not $\mathrm{V}(\bar{y})/\mathrm{V}(\bar{x})$ and so we cannot use the corresponding expression with sample estimates $\mathrm{v}(\bar{y})$ and $\mathrm{v}(\bar{x})$ even if these estimates are themselves unbiased estimates of $\mathrm{V}(\bar{y})$ and $\mathrm{V}(\bar{x})$. More complex methods (Section 2.11) must be used.

2.10 Linear transformations

The principles listed in Section 2.9 can be used to identify two broad classes of functions of random variables that have fairly simple statistical properties. The classes are referred to as 'linear transformations (or linear combinations) of random variables' and 'affine transformations of random variables'. Here, we will define the terms linear transformation and affine transformation, contrast them with nonlinear transformations (which usually have more complex statistical properties), and begin to explore the statistical properties of each group. The subject will be taken up again in later sections.

The formula for an affine transformation, X (say), involves a set of random variables, $y_1, y_2, y_3,\ldots,y_k$ and two sets of constants, $a_1, a_2, a_3,\ldots,a_k$, and $b_1, b_2, b_3,\ldots,b_k$. The general equation is

$$X = \sum_{i=1}^{k}(a_i + b_i y_i). \tag{2.41}$$

For a linear transformation the general equation is

$$X = \sum_{i=1}^{k} b_i y_i. \tag{2.42}$$

These equations are designed to include cases that may be quite complex. In most cases of interest to behavioral ecologists k, a_i , and/or b_i take on simple values such as n (= sample size), 1 or 0. For example, the sample mean is a linear transformation of the units in the sample (y_i) with $k=n$, and all $b_i = 1/n$. The estimated population total, $N\bar{y}$, is a linear transformation of the sample mean with $k=1$, and $b_1 = N$. In contrast, the sample variance

$$v(y_i) = \frac{\sum_{i=1}^{n}(y_i - \bar{y})^2}{n-1}$$

is not a linear or affine transformation of the units in the sample (y_i) because of the squared terms. There is no way to rearrange $v(y_i)$ so that it has the form of Eq. 2.41.

The rules of expectation previously used to discuss bias and standard errors of transformed estimates can now be applied to provide very general results pertaining to linear or affine transformations. First, however, we need two other results. If a is a constant, then in repeated sampling $E(a)=a$ and $V(a)=0$. Both statements just say that a does not vary from sample to sample. The value of these results will be made clearer below. Three rules regarding linear or affine transformations are:

1. $E(a+bY)=a+b\ E(y)$. Thus, if y is an unbiased estimate of Y, then $a+by$ is an unbiased estimate of the corresponding transformation of the parameter, $a+bY$.

2. $V(a+by)=0+b^2V(y)$, and thus $SE(a+by)=b\text{SE}(y)$. Thus, the standard error of a linear or affine transformation of an estimate equals the multiplicative constant (if any) times the standard error of the estimate.
3. In contrast, no simple, general principles apply to nonlinear transformations. As noted previously, in general, the expected value of the function is not equal to the function of the expected values and the variance of the function is not equal to the function of the variances. In other words, sample analogs of the parameters are likely to produce biased point and interval estimates; other quantities must be used.

These results are simply a restatement, in briefer form, of the three rules about expected values in general (Section 2.4) and expected values of variances (Section 2.9). They consolidate this information into a compact form and will be referred to many times in later Chapters.

2.11 The Taylor series approximation

The preceding Sections raise the question of how one approximates the bias and estimates the variance when using nonlinear combinations of random variables as estimators. Several approaches are possible, but one, called the Taylor series expansion or the Delta Method, is particularly useful and is therefore described below. It is somewhat complex both algebraically and conceptually and requires knowledge of elementary calculus. Some readers may wish to skip this section and later sections in which examples of using the approach are presented.

The steps in using this approach are

1. Express the estimator of interest as a Taylor series.
2. Use lower-order terms to obtain expressions for approximate bias and variance.
3. Replace parameters in the expressions with sample estimates to obtain the desired estimate of the bias or variance.

Consider a collection of random variables x_i each with expected value X_i. In most cases of interest to behavioral ecologists, the x_i values are estimates and we have only one or two in the collection. We present the method in general notation however. Assume that the random variables are combined in some nonlinear manner. The formula might be a single random variable raised to a power, x^a, a ratio, x_1/x_2, a product of three variables, $x_1x_2x_3$ or any other nonlinear expression. If n random variables, $x_1,\ldots,x_n$, are involved, then $f(x_1,\ldots,x_n)$ will denote the nonlinear function of interest and $f(X_1,\ldots,X_n)$ will denote this function evaluated at the expected values. Thus, if $f(x)=x^a$, then $f(X)=X^a$, or if $f(x_1,x_2)=x_1x_2$ then $f(X_1,X_2)=X_1 X_2$.

In the following, all partial derivatives are evaluated at $(X_i,\ldots, X_n)$. The Taylor series expansion of the random variables is

$$f(x_1,\ldots,x_n)=f(X_1,\ldots,X_N)+\sum\left(\frac{\partial f}{\partial X_i}\right)(x_i-X_i) \tag{2.43}$$

$$+\frac{1}{2}\sum_i\sum_j\left(\frac{\partial^2 f}{\partial X_i \partial X_j}\right)(x_i-X_i)(x_j-X_j)+\ldots.$$

The ellipsis on the right indicate the presence of additional terms. The next term involves three deviations multiplied together and third-order derivatives, and so on. If x_i is an estimate of, and hence fairly close to, X_i then the higher order terms should be sufficiently small that we can ignore them in approximating the bias and variance. In most cases of interest to behavioral ecologists, this assumption is reasonable. Nonetheless, scrutiny of the higher order terms is a part of using this method (which is why we present the derivation).

With a single variable [e.g., $f(x)=x^a$] all of the sums drop out and the expression becomes

$$f(x)=f(X)+\frac{df}{dX}(x-X)+\frac{d^2f}{dX^2}\frac{(x-X)^2}{2}+\ldots.. \tag{2.44}$$

With more than one variable, all sums extend over all variables. In the double sum, when $i=j$ we take the derivative twice with respect to the same variable, and the deviations become $(x_i-X_i)^2$. We can therefore re-write the expression in Eq. 2.43 as

$$f(x_1,\ldots,x_n)=f(X_1,\ldots,X_n)+\sum\left(\frac{\partial f}{\partial X_i}\right)(x_i-X_i)+\frac{1}{2}\sum_i\left(\frac{\partial^2 f}{\partial X_i^2}\right)(x_i-X_i)^2 \tag{2.45}$$

$$+\sum_i\sum_{j>i}\left(\frac{\partial^2 f}{\partial X_i \partial X_j}\right)(x_i-X_i)(x_j-X_j)+\ldots.$$

Step 2 involves eliminating the higher-order terms. The terms eliminated vary depending on whether approximations for the bias or variance are to be developed. For the bias approximation, we customarily ignore terms of third or higher order. To develop the approximation, first move $f(X_1\ldots,X_n)$ to the left side of Eq. 2.45 and then take expected values of both sides.

This gives us $E[f(x_1,\ldots,x_n)-f(X_1,\ldots,X_n)]$, which is bias, on the left. On the right side, $E(x_i - X_i) = 0$ (since $E(x_i) = X_i$), $E(x_i - X_i)^2 = V(x_i)$ and $E(x_i - X_i)(x_j - X_j) = Cov(x_i x_j)$. The expression for approximate bias is thus

$$\text{Bias} = E[f(x_1,\ldots,x_n)-f(x_1,\ldots,x_n)]. \tag{2.46}$$

$$\simeq \frac{1}{2}\sum_i \frac{\partial^2 f}{\partial X_i^2} V(x_i) + \sum_i \sum_{i>j} \frac{\partial^2 f}{\partial X_i \partial X_j} Cov\,(x_i,x_j).$$

This estimate of bias is often referred to as the 'leading term' in the bias because only the first (nonzero) term in the Taylor series expansion is used in its derivation. When the function has a single parameter X, then Eq. 2.44 simplifies to

$$Bias = 0.5\left(\frac{d^2 f}{dX^2}\right)V(x). \tag{2.47}$$

Incidentally, note that if $f(x_1,\ldots,x_n)$ is a linear or affine combination of the random variables $x_1,\ldots,x_n$, then the second derivative of $f(X_1,\ldots,X_n)$ is zero. All higher derivatives are also zero. This provides another demonstration that linear or affine combinations of random variables provide unbiased estimates of the corresponding functions of the parameters.

To obtain the approximate variance we consider only the first term (not any of the higher-order terms), and re-write the expansion, collecting constants, as

$$f(x_1,\ldots,x_n) \simeq f(X_1,\ldots,X_n) + \sum\left(\frac{\partial f(X)}{\partial X_i}\right)x_i - \sum\left(\frac{\partial f}{\partial X_i}\right)X_i, \tag{2.48}$$

$$\simeq \sum\left(\frac{\partial f}{\partial X_i}\right)x_i + \text{constants} + \ldots.$$

We thus may write

$$V[f(x)] \simeq V\left[\sum\left(\frac{\partial f}{\partial X_i}\right)x_i\right], \tag{2.49}$$

because the constants drop out (variance of constant is 0). Using expression 2.31 for the variance of a sum, the approximate variance thus becomes

$$V[f(X_1,\ldots,x_n)] \simeq \sum\left(\frac{\partial f}{\partial X_i}\right)^2 V(x_i) + 2\sum_i\sum_{j>i}\left(\frac{\partial f}{\partial X_i}\right)\left(\frac{\partial f}{\partial X_j}\right)Cov(x_i,x_j). \tag{2.50}$$

Note that the expression contains squared deviations (variances) and cross-products (covariances). If we had included higher-order terms then the expression for variance would include deviations to the fourth order (or higher). We noted above that third-order and higher terms are generally ignored in the approximation. When the function has a single parameter X, then Eq. 2.49 simplifies to

$$V[f] = \left(\frac{df}{dX}\right)^2 V(x). \qquad (2.51)$$

While the right-hand side of Eq. 2.49 gives an approximate expression for the desired variance, it depends on the unknown parameters $V(x_i)$ and $Cov(x_i, x_j)$. Thus the third step is to replace these quantities with sample estimates. Since in many cases appropriate sample estimates are complicated, it is probably best to consult a statistician concerning what sample estimates should be used. As usual, standard errors of the estimates are simply the square roots of the variances of the estimates.

Examples

Here are two examples of applying this method to obtain expressions for approximate bias and variance. Suppose we are studying nesting success using the 'Mayfield method' (see Chapter Ten) in which survivorship is assumed to be constant through time and an estimate is made of the daily survival rate, p say. Survival to fledging, or throughout some shorter period of interest such as incubation, is thus p^k where k equals the average length of a successful attempt and is known or estimated from other information and is thus treated as a constant. The same problem might arise in studying contests or efforts to avoid predation. Our interest might center on the probability of succeeding in all k trials. In these cases, assuming a constant probability of success might be more questionable but we might initially model success as constant. Sometimes interest centers on producing 'at least one success'. In such cases, we can define p as a failure and estimate the probability of $1-p^k$, that is 1 minus the probability of failing all k times which is the probability of succeeding 1 or more times. This is essentially the same problem since the bias and variance of $1-p^k$ can easily be calculated from estimates of the bias and standard error of p^k. We now derive expressions for the approximate bias in p^k used to estimate P^k and for $V(p^k)$.

We assume that p is an unbiased estimate of the true probability, P [i.e., $E(p)=P$], we have an estimate of $V(p)$, and our objective is to obtain point and interval estimates of P^k. To obtain expressions for the estimated bias and standard error we need the first and second derivatives with respect to the parameter, P

$$\frac{\mathrm{d}(P^k)}{\mathrm{d}P} = kP^{k-1}, \tag{2.52}$$

and

$$\frac{\mathrm{d}^2(P^k)}{\mathrm{d}P^2} = k(k-1)P^{k-2}. \tag{2.53}$$

The estimated bias from Eq. 2.47 is thus

$$\text{bias} \simeq 0.5k(k-1)p^{k-2}v(p), \tag{2.54}$$

and the estimated variance from Eq. 2.51 is

$$v(p^k) \simeq (kp^{k-1})^2 v(p), \tag{2.55}$$

so that

$$\text{se}(p^k) \simeq kp^{k-1}\text{se}(p). \tag{2.56}$$

Note that bias decreases at a rate proportional to n [because $v(p)$ decreases at this rate]. This is frequently – though not always – the case with bias, a relationship described as 'the leading term in the bias is of order 1 over n'.

The effects of bias are generally negligible if bias/SE is less than 10% and of relatively little importance if bias/SE is less than 20% (Cochran 1977 pp. 12–15). We may therefore be interested in how large a sample size is needed for bias/SE to be less than 10% or 20%. $V(p)$ is needed for both bias and SE and depends on the sampling plan used to collect the data. If p is estimated by observing a series of n trials and recording the number of successes then (Appendix One) $V(p) = P(1-P)/n$ so

$$\frac{\text{Bias}(p^k)}{SE(p^k)} = \frac{0.5k(k-1)P^{k-2}V(p)}{kP^{k-1}SE(p)} = \frac{0.5(k-1)SE(p)}{P} \tag{2.57}$$

$$= \frac{0.5(k-1)\sqrt{(1-P)/P}}{\sqrt{n}}$$

If bias/SE is going to less than a given value, R say, then n must satisfy

$$= \frac{0.5(k-1)\sqrt{(1-P)/P}}{\sqrt{n}} < R, \tag{2.58}$$

or

$$n > \left(\frac{0.5}{R}\right)^2 (k-1)^2 \left(\frac{1-P}{P}\right). \tag{2.59}$$

In nesting studies the daily survival rate (P) is often about 5% so $(1-P)/P=1/19$. If the period length is 20 then $k-1$ is 19 so n must satisfy

$$n>\frac{19}{4R^2} \tag{2.60}$$

Obtaining $R<0.2$ thus requires $n>118$. Virtually all studies of nest success using the Mayfield approach achieve n (= number of 'nest days', not nests) >118. Thus, bias is seldom a problem in these studies, even though the estimators are slightly biased.

For our second example, consider the ratio of sample means, $\bar{y}/\bar{x}$. This case arises in so many situations that we give the formulas for bias and standard errors in Appendix One. Nonetheless, we derive the equations below as an example.

The first and second derivatives with respect to the parameters $\bar{X}$ and $\bar{Y}$ are

$$\frac{\partial(\bar{Y}/\bar{X})}{\partial\bar{Y}}=\frac{1}{\bar{X}} \quad \frac{\partial(\bar{Y}/\bar{X})}{\partial\bar{X}}=-\frac{\bar{Y}}{\bar{X}^2}, \tag{2.61}$$

and

$$\frac{\partial^2(\bar{Y}/\bar{X})}{\partial\bar{Y}^2}=0 \quad \frac{\partial^2(\bar{Y}/\bar{X})}{\partial\bar{Y}\partial\bar{X}}=-\frac{1}{\bar{X}^2} \quad \frac{\partial^2(\bar{Y}/\bar{X})}{\partial\bar{X}^2}=\frac{2\bar{Y}}{\bar{X}^3}, \tag{2.62}$$

and thus the estimated bias, from Eq. 2.46, is

$$\begin{aligned} bias &\approx \frac{1}{2}\left[0+\frac{2\bar{y}}{\bar{x}^3}v(\bar{x})\right]+\left(-\frac{1}{\bar{x}^2}\right)\mathrm{cov}(\bar{x},\bar{y}), \\ &\approx \frac{1}{\bar{x}^2}\left[\frac{\bar{y}}{\bar{x}}v(\bar{x})-\mathrm{cov}(\bar{x}\bar{y})\right], \end{aligned} \tag{2.63}$$

and, from Eq. 2.50

$$\begin{aligned} v\left(\frac{\bar{y}}{\bar{x}}\right) &= \left(\frac{1}{\bar{x}}\right)^2 v(\bar{y})+\left(-\frac{\bar{y}}{\bar{x}^2}\right)^2 v(\bar{y})+2\left(\frac{1}{\bar{x}}\right)\left(-\frac{\bar{y}}{\bar{x}^2}\right)cov(\bar{x},\bar{y}) \\ &= \frac{v(\bar{y})}{\bar{x}^2}+\frac{\bar{y}^2 v(\bar{x})}{\bar{x}^4}-\frac{2\bar{y}}{\bar{x}^3}cov(\bar{x},\bar{y}) \\ &= \left(\frac{\bar{y}}{\bar{x}}\right)^2\left[\frac{v(\bar{y})}{\bar{y}^2}+\frac{v(\bar{x})}{\bar{x}^2}-\frac{2cov(\bar{x},\bar{y})}{\bar{x}\bar{y}}\right]. \end{aligned} \tag{2.64}$$

This expression is rather complex but if advance estimates of the needed terms are available they may be substituted to obtain the minimum sample size needed for bias to be safely ignored.

2.12 Maximum likelihood estimation

In some applications we can write down the probability distribution for the set of possible sample outcomes. Sample outcomes that have Bernoulli, binomial, and multinomial distributions provide examples (Section 2.3). Whenever the distribution can be written out, then a method known as maximum likelihood estimation may provide a useful way to obtain point and interval estimates. Although maximum likelihood estimators are sometimes quite difficult to derive and may require special numerical methods to calculate, the principle of the method is simple and elegant. In this section we explain the principle and provide a few examples, restricting our attention to discrete distributions. Our goal is that readers understand the rationale for the use of maximum likelihood estimates. Consultation with a statistician is advised for deriving maximum likelihood estimators.

The conceptual basis of the method is explained with a simple example. Suppose we were told that a computer program had been prepared to generate 0s and 1s with a constant probability, P, of producing a 1. In addition, imagine that a sample of 10 0s and 1s contained 6 1s, and we were asked to make our best guess about the value of P. Given only this information, we would probably select 0.6 as our guess, perhaps noting our uncertainty given the rather small sample. Now consider the question, 'why is 0.6 a better estimate, in this case, than some other value, such as 0.3?' The answer to this question may not be obvious, but consider the following rationale. If the value of P is actually 0.3, then it seems unlikely that we would get as many as 6 successes in a sample of size 10. In contrast, the probability of getting 6 1s if $P=0.6$ is considerably higher. Furthermore, if we calculated the probability of getting 6 1s for all possible values of P, we would find that the maximum probability occurs with $P=0.6$. Thus, the estimate 0.6 is the value that maximizes the probability of obtaining the observed outcome.

The rationale above summarizes the maximum likelihood approach: write down the probability distribution for the possible sample outcomes. The expression, called the 'likelihood', will include one or more unknown parameters (P in our example). Find the values of the unknown parameters that maximize the likelihood and take those as the parameter estimates. Since the likelihood is a distribution, and thus gives the probability of obtaining various sample outcomes, we may replace the symbols for specific values with the more general symbol for any outcome. In this example, we replace X with x. The likelihood is thus the probability of obtaining x

successes in n trials with constant probability of success P, which, from Section 2.3, is

$$\binom{n}{x} P^x(1-P)^{n-x}. \tag{2.65}$$

Various approaches may be used to find the maximum of the likelihood. One approach is to take derivatives, set them equal to zero, and solve the resulting equations for the unknown parameters. Recall from calculus that the solutions are either maxima, minima, or saddle points. To determine which, additional methods (such as checking second derivatives) must be used. In practice, statisticians usually take the natural logarithm (ln) of the expression before calculating the derivatives. This gives the same maximum and simplifies the mathematics.

In our example, the ln of the likelihood is

$$\ln\binom{n}{x} + x\ln(P) + (n-x)\ln(1-P), \tag{2.66}$$

and taking derivatives with respect to P yields

$$\frac{x}{P} + \frac{n-x}{1-P}(-1). \tag{2.67}$$

Setting this expression equal to zero and solving for P, we obtain $P = x/n$. It is easily verified that this is a maximum by checking second derivatives. Thus, if p is our symbol for the estimate, then $p = x/n$ is the maximum likelihood estimate of the parameter P.

This procedure, while fairly general, does not always work. For example, in some cases the maximum is at one of the extreme values of the parameter and the derivative is not equal to 0. However, anytime we can write down an expression for the probability of obtaining each possible sample result, expressed in terms of unknown parameters and perhaps some constants, then we can use maximum likelihood methods to obtain expressions for the estimates. The mathematics are complex in some cases, and consultation with a statistician is definitely recommended. The important point for behavioral ecologists is to understand how maximum likelihood can be used to develop estimates.

Standard errors

One other feature of maximum likelihood estimation should be described. Formulas for the asymptotic variances (i.e., variances if the sample size is 'large') can usually be obtained from the second derivatives of the likelihood. The calculations can easily become complex, and are usually carried out by computers, so we do not describe them in detail. Behavioral ecologists sometimes consider cases with only one or two parameters, however, and may be interested in writing short computer programs to obtain the estimates so we outline the procedure briefly. With a single parameter, R say, and maximum likelihood estimator r, the formula for V(r), is

$$V(r)=\left[-E\left(\frac{d^2 lnL}{dR^2}\right)\right]^{-1} \tag{2.68}$$

where L is the likelihood. In words, this expression says, take the second derivative of the log of the likelihood with respect to R, calculate the expected value of this expression, and then invert it. The result is V(r). In our first example, with $R=P$ and $r=p$, the second derivative is

$$\frac{\mathrm{d}^2 lnL}{\mathrm{d}P^2}=-xP^{-2}-(n-x)(1-P)^{-2}. \tag{2.69}$$

The only random variable is x, which has expectation nP (this is easy to prove using the principles of expectation in Section 2.4). Thus

$$E\left(\frac{\mathrm{d}^2 lnL}{\mathrm{d}P^2}\right)=-nP^{-1}-n(1-P)^{-1}. \tag{2.70}$$

Taking the negative of this expected value and inverting the result yields the large-sample variance of p

$$V(p)=\frac{P(1-P)}{n}. \tag{2.71}$$

Notice that this is also the formula for the actual variance although this need not be the case in general. The formula for an unbiased estimate of the variance is not necessarily the sample analogue. In this case, the sample analog $[p(1–p)/n]$ would be a biased estimate. An unbiased estimate may be obtained by replacing n with n–1. This should not be surprising given that proportions may be viewed as means (Section 2.2) and that the unbiased estimate of the variance of a sample mean has n–1, not n, in the denominator (Section 2.7).

When the likelihood has more than one parameter, the method for finding maximum likelihood estimates and their asymptotic (i.e., large sample) variances is more complex. If calculus is used partial derivatives with respect to each parameter must be taken, set to 0, and the resulting system of equations solved. For finding variances the second derivatives of the ln of the likelihood must be taken with respect to all possible pairs of parameters. This yields a matrix of second derivatives. We calculate (−1) times the expected value of each element in the matrix and invert the resulting matrix. The result is the asymptotic 'variance–covariance matrix', an array in which the diagonal elements are the variances and the rest of the elements are the covariances. For more information on maximum likelihood methods see textbooks on statistical theory such as Mood *et al.* (1974, Chapter 7) or Rice (1995, Chapter 8).

As already noted, the critical requirement for maximum likelihood estimation is specifying the probability distribution for the sample outcomes. To do this, assumptions must be made about the distribution of the observations. In the example given this meant assuming that the probability of 'success' on each trial was the same. This allowed us to write down the expression (Eq. 2.65) for the probability of any number of successes, x, $x=0,\ldots,n$, in the n trials.

Capture–recapture methods generally use maximum likelihood methods. In these studies, animals are marked and resighted on several occasions. For the analysis, the marked individuals are divided into groups that are assumed to have similar survival and recapture probabilities. These probabilities may vary through time, but they are the same, at any given time, for all members of the group. Under this assumption, one may write down a general, multinomial expression for the probability of re-sighting each individual in each group on each sampling occasion. Examples are provided in Chapter Nine. Another example is provided by the Mayfield method for estimating nesting success mentioned earlier. In this case the observations are 'success' or 'failure' of nests during observation intervals which vary in length. The 'trials' thus have different probabilities of success, depending on their length, yet, under the assumption of constant daily survival, we can write down an expression for the probability of observing any set of successes given a set of n trials of lengths l_i, $i=1,\ldots,n$. Maximum likelihood methods can then be used to obtain the estimates (though this is not the way that Harold Mayfield originally approached the problem).

2.13 Summary

Four approaches for calculating point and interval estimates are discussed in this Chapter. First, and in the great majority of cases encountered by behavioral ecologists, reasonable estimators have already been worked out and are available in books or other publications. Many of the most useful ones are found in this book in Appendix One. Second, for linear combinations of random variables, Section 2.4 provides simple rules that anyone can learn with a little practice. Third, the Taylor series approximation can be used to derive estimators for nonlinear combinations of random variables. This approach is more complex, and we recommend consultation with a statistician to be sure that there are no errors and for assistance in evaluating higher-order terms (see Section 2.11). Fourth, when a reasonable parametric model expressing the probability of obtaining each possible sample outcome can be written down, then maximum likelihood methods can be used to obtain the estimators. As with the Taylor series, we recommend seeking assistance from a statistician in using this method, but anyone who studies the material in this Chapter can learn to recognize cases when the method may be applicable and to make a good first attempt at the derivations. Frequent references are made later in the book to cases in which each of these approaches might be useful.

3

Tests and confidence intervals

3.1 Introduction

We begin this Chapter by reviewing the meaning of tests and statistical significance and providing guidance on when to use one- and two-tailed tests in behavioral ecology. Confidence intervals are then discussed including their interpretation, why they provide more complete information than tests, and how to decide when to use them. We also discuss confidence intervals for the ratio of two random variables such as estimates of population size in two consecutive years. Sample size and power calculations are then reviewed emphasizing when and how they can be most useful in behavioral ecology. The rest of the Chapter discusses procedures for carrying out tests. We first discuss parametric methods for one and two samples including paired and partially paired data. A discussion is included of how large the sample size must be to use *t*-tests with data that are not distributed normally. The Chapter ends with some simple guidelines for carrying out multiple comparisons.

3.2 Statistical tests

Carrying out hypothesis tests to determine whether a parameter is different from a hypothesized value, or to determine whether two parameters are different from each other, is undoubtedly the most common statistical technique employed by behavioral ecologists. Understanding what these tests reveal is thus important, but is also surprisingly difficult. In this section we review some general principles briefly and identify a few subtleties that are easily overlooked in behavioral ecology. These principles are introduced in the context of testing hypotheses about a proportion. Procedures for carrying out these tests are reviewed briefly in this section and described in more detail in Sections 3.6 and 3.7

Suppose we are studying costs of parental care in swans. As part of the study we wish to determine whether pairs with young (i.e., families) have an advantage (or disadvantage) in fights with pairs that do not have young. We are studying a flock of several hundred swans including many families and pairs. To collect the data, we use scan sampling, watching each family encountered until it fights once with a pair (without young). We repeat the scans until we have watched 100 fights, and we view the data as a simple random sample. If the probability of winning is the same for pairs with and without young, then families should win about 50% of the fights we observe. In fact, in our sample families win 66% of their fights with pairs. The following question now arises: Do these results show that the probability of a family winning a fight with pairs is different from 50%? Using the language of hypothesis testing, we can phrase our question as a test of H_0: $P=0.5$ *versus* H_1: $P\neq 0.5$, where P is the probability of a family winning a fight with pairs, H_0 is the null hypothesis and H_1 is the alternative hypothesis.

A significance test provides a way of answering this question by determining the probability of observing a difference as large as the one we observed if, in fact, overall, families win 50% of their fights with pairs. In other words, how unusual or how difficult is it to explain our observed data under the null hypothesis? If data such as ours would occur only rarely under the null hypothesis, then we regard this as strong evidence in favor of the alternative. The quantity we are going to compute is commonly called the *p*-value, or occasionally the observed significance level of the test.

The *p*-value is the probability of getting a result that deviates from the value specified by the null hypothesis as much as, or more than, the value we actually obtained. Deviations can be either positive or negative in our example. The set of positive deviations, in this example, includes 66 or more wins. The set of negative results that deviate as much as, or more than, the sample outcome, includes 34 or fewer wins. The *p*-value in this case is thus the probability of 0 to 34 wins, or 66 to 100 wins, if the true probability of winning is 0.5. The probability of 66 to 100 wins, if the probability in each trial is 0.5, turns out to be 0.0009. The probability of 0 to 34 wins is the same (since $P=0.5$ and so the distribution is symmetrical about 50) so the overall value is $2(0.0009)=0.0018$. Thus, if the null hypothesis is true, then the probability of obtaining a proportion that deviates from 0.5 as much as, or more than, our sample did is only 0.18%, less than one-fifth of 1%. Thus we have two possible explanations for our data. Either we have observed a rare event, or the null hypothesis is not true. The choice of explanation is ours, and should be based on our scientific knowledge regarding the problem, as well as these computations. Using the computations alone,

0.18% would typically be considered quite small and we would therefore be inclined to believe that the data are better explained by assuming that families in the population we watched had a higher probability of winning fights than pairs. Typical thresholds for concluding that the p-value is implausibly small, and thus the null hypothesis should be rejected, are 5% and 1%. The rationale for these choices will be given shortly.

Another possible approach for the analysis, given the relatively large sample size of 100, is to use the normal approximation to the binomial. Most statistical packages include this option. The steps involved in deciding whether the sample size is large enough and, if so, in carrying out the test are explained in the Boxes in Appendix One. We are testing a proportion so the key in Box 1 directs us to Box 5. Part A of Box 5 defines terms. From Part B we find that the sample size is large enough that the normal approximation may be used. The test statistic as

$$\frac{|p-P|}{\sqrt{P(1-P)/n}} - \frac{c}{\sqrt{nP(1-P)}}. \tag{3.1}$$

In our example, $p=0.66$, $P=0.5$, $n=100$ and c is defined in Box 5, Part C1b as depending on the fractional part of $n|p-P|$ which in our case is 16.0. The 'fractional part' of $n|p-P|$ is zero so c also equals zero. Thus, the computed value of the test statistic is 3.2, and the area to the right of 3.2 on the normal curve is 0.0007. The large sample approximation to the p-value is then $2\times(0.0007)=0.0014$, which is in fairly good agreement with the exact value of 0.0018 based on calculations using the binomial distribution. The reason for this is that the sample size is large enough for the normal distribution to provide a good approximation to binomial probabilities.

The computation of a p-value is probably the most common approach for summarizing the results of a statistical test. It may be regarded as a measure of the strength of the evidence against the null hypothesis.

A second approach is typically referred to as fixed α-level testing. The two approaches are not completely unrelated and their relationship will be explained shortly. In fixed α-level testing the experimenter chooses an α, or significance level, for the test before collecting the data. The significance level corresponds to the probability of incorrectly rejecting the null hypothesis if it is true, and is also referred to as the probability of a type I error. The rejection region, corresponding to this significance level, is then defined and if the test statistic falls in the rejection region the investigator concludes that the result is statistically significant at this α-level.

In the swan example, if we selected a 5% level of significance then the null hypothesis would be rejected if the computed value of the test statistic exceeded 1.96. Since the test statistic equals 3.2, we would reject H_0. The reader may wonder why we chose 0.05 rather than 0.01 or some other value for α. Choosing a particular value of α corresponds to setting the maximum type I error rate we are willing to tolerate, and is usually more appropriate in a decision-making context, than in the type of problems considered by a behavioral ecologist. In addition, the choice of levels such as 0.05 and 0.01 is arbitrary and corresponds to what we might think of as 'unlikely' to occur by chance under H_0: namely less than one in 20 times or less than one in 100 times.

The use of fixed α-levels is probably most appropriate in behavioral ecology as a first step in the analysis. Thus, the Methods section of a paper may assure readers that null hypotheses were not rejected unless the observed p-value was smaller than 0.05 or 0.01. This approach is particularly useful when a substantial number of test results are being reported but reporting each one separately is not warranted. A simple statement such as 'all differences were significant' then provides a convenient capsule summary of the results.

For more important results, however, noting only that H_0 was rejected at $\alpha = 0.05$ is unsatisfactory to most behavioral ecologists since the evidence against H_0 may have been much stronger than $\alpha = 0.05$. In the example above, we would reject H_0 at $\alpha = 0.01$ and even at $\alpha = 0.005$.

An interpretation of the p-value related to the α-level is that it is the smallest α-level at which we would reject H_0. Since the p-value is 0.0018, we would reject at any α-level larger than 0.0018 and fail to reject at any smaller α-level. So, just from knowing the p-value it follows immediately that we would not reject at $\alpha = 0.001$ since this is less than 0.0018, but we would reject at $\alpha = 0.005$. Since the p-value both measures the strength of the evidence against the null hypothesis, and tells you what your decision would be at any α-level, it is the number given by most computer packages when conducting a statistical test. Note that the p-value is a sample result. It varies depending on which sample we happen to select. In contrast, the α-level is not dependent on the sample; it is chosen prior to sample selection.

Interpretation of significance

Having described the basis for deciding whether or not to reject the null hypothesis, we now discuss what conclusion is appropriate if the null hypothesis is rejected in favor of a two-sided alternative. Two interpretations are possible. The first is simply that the true value is different from the value specified by the null hypothesis. In our example, we would conclude that the probability that families win fights is different from 0.50 which corresponds to the alternative hypothesis. In response, however, biologists are likely to say that this is not what they want to know. What they want to know is whether families are more likely, or less likely, to win fights than pairs without young. The second possible conclusion, after rejecting the null hypothesis, is that the sample results provide a reliable indication of whether the parameter's actual value is larger or smaller than the value under the null hypothesis (i.e., the sample results allow us to specify a direction in rejecting H_0). In our example, we would conclude that families are more likely to win fights than pairs without young. This sort of conclusion is more useful because it tells us something we really did not know – or at least could not demonstrate convincingly – before doing the test.

There is a possible objection, however. If families actually win more fights than pairs without young (i.e., $P>0.5$), we might obtain a sample which suggests statistically that families won fewer fights (i.e., $P<0.5$). This introduces a new type of error, deciding $P<0.5$ when in fact $P>0.5$. If we interpret the sample results as indicating that the parameter is smaller than 50% then we would be correct in rejecting the null hypothesis but incorrect in our conclusion that $P<0.5$. The chance of making this error depends on how different the parameter is from the value under the null hypothesis. For example, if families really win 60% of their fights, then the probability of obtaining a sample result in which the number of fights won is significantly less than 50% is quite small. The chance of making this type of error increases as the true difference between the parameter and its value under the null hypothesis decreases. The limiting case occurs when the two values are essentially equal. Suppose the true value is larger than 50% but so close to it that we can evaluate the case by assuming it equals 50%. Also, let us assume that p-values of less than 5% will be declared statistically significant. Since the true value and the value under the null hypothesis are essentially equal, the probability of obtaining a significant result is only 0.05. Furthermore, we have a 50:50 chance of obtaining a significantly positive result – in which case we reach the correct conclusion. Thus, the chance of making the type of error we are discussing is only 0.025. Keep in mind that this is the worst possible scenario. With true differences large enough to be interesting, the chance of making this type of error about whether the parameter is larger or smaller than its value under the null hypothesis is extremely small. It is always less than $\alpha/2$ and typically is close to 0.

The issue of whether one can legitimately conclude that the parameter is larger than, or smaller than, its value under the null hypothesis, in the two-sided case when a statistically significant result has been obtained, has always been a source of concern among practitioners. Those involved in practical uses of statistics tend to encourage analysts to assume that statistically significant sample results do indicate the direction of the true value (e.g., Snedecor and Cochran 1980). In behavioral ecology, the null hypothesis is usually known to be wrong and thus uninteresting (e.g., in our example it is unlikely that families and pairs without young would have exactly the same probability of winning fights), so the test is of little use if we can only conclude what we already knew – that the null hypothesis is false. This fact, in combination with the rarity of incorrect conclusions about whether the parameter is larger or smaller than its value under the null hypothesis, leads us to recommend that investigators view statistically significant outcomes as indicating whether the parameter is larger or smaller than its value under the null hypothesis. Thus we would interpret the example as showing that families win more fights than pairs without young in this population.

Two-sample tests

We now turn to the rationale of two-sample tests, again using an example. Suppose we are trying to determine whether juvenile males differ from juvenile females in how much time they spend in close proximity to a novel stimulus. We record the time for ten juveniles of each sex, and find that the average for females is 35 s more than the average for males. Individuals varied in how much time they spent close to the stimulus, and this raises the possibility that we might have obtained very different results had we been able to test a different sample. Perhaps the mean for females, in a different sample, would be equal to or even smaller than the mean for males.

To investigate this possibility, we calculate the probability of obtaining our observed difference (or a larger one) if the true means (the population means or the means from an indefinitely large sample) were equal. If this probability is sufficiently small, then we reject the null hypothesis that the true means are equal and conclude that the true mean for females is almost certainly larger than the true mean for males, as suggested by the sample data. As with one-sample tests, the significance level may be set beforehand at 0.05 or 0.01 or we may report the p-value and thereby indicate the smallest α-level at which the null hypothesis would be rejected.

The rationale in the two-sample test is similar to that in a one sample test. If the true means are equal, then obtaining a difference as large as the one we observed would be very unlikely, whereas if the true mean for females really is larger than the true mean for males, then a difference similar to the one we observed could be quite likely to occur. Therefore, we choose to assume that the mean for females really is larger. The statistical analysis does not prove that the mean for females is larger. Perhaps the mean for females is actually less than the mean for males and we just happened to obtain a rare sample in which the reverse was true. There might even be additional information suggesting that such a rare event really did happen, in which case it might be reasonable to doubt the statistical analysis. But if no other information is available, then the rationale above seems reasonable as a way of deciding whether the observed results provide a reliable indication of which parameter is larger.

These calculations, done in practice by software packages, may also be illustrated using Appendix One. We start again with Box 1. Let us assume that we will use a t-test. We thus are directed to Box 6. Part A explains the notation. The test statistic is

$$\frac{|\bar{y}_1 - \bar{y}_2| - (|\bar{Y}_1 - \bar{Y}_2|)}{\mathrm{se}(\bar{y}_1 - \bar{y}_2)} \tag{3.2}$$

where $\bar{y}_1$ and $\bar{y}_2$ are the estimates of $\bar{Y}_1$ and $\bar{Y}_2$, and $\bar{Y}_1 - \bar{Y}_2$ is 0.0 under the null hypothesis that $\bar{Y}_1 = \bar{Y}_2$. The standard errors and degrees of freedom are calculated with the formulas in Box 2 since we used simple random sampling to obtain the data. As explained in Chapter Four, if a different sampling plan had been used, then other formulas for the means and standard errors might be needed. Some latitude exists in precisely what formulas to use in the test (see Section 3.5), and it is thus important to ensure that the computer is using the formulas you want it to.

Additional issues

A few other, miscellaneous points about statistical tests and significance are worth mentioning. Investigators sometimes describe the results of tests without making reference to the population values (i.e., the parameters). Thus, they may report that 'the sample means were not different'. In fact, of course, the sample means probably were different. The summary statement is an abbreviated way of saying that the difference between the estimates was not statistically significant, and thus the statistical analysis does not support a strong conclusion about which parameter was larger. It is well to keep in mind, however, that the purpose of the test is to make inferences about the parameters not the samples.

Biologists often decide whether two estimates are significantly different by examining plots that show the means (or other estimates) and error bars. In most cases the error bars either represent one standard error or the 95% confidence interval. It is widely assumed that these plots are a kind of graphical t-test, and that the difference is significant if and only if the bars do not overlap. This belief, however, is mistaken. If the estimates are based on paired data, then virtually nothing can be deduced about whether the difference is significant merely on the basis of the error bars (see Section 3.5). Even if the estimates are independent, one cannot easily determine whether the difference is significant using a *t-test* from whether the bars overlap. For example, if the bars represent 95% confidence intervals and do overlap, it is still possible that the difference is significant. Conversely, if the bars represent one standard error and do not overlap, it is still possible that the difference is not significant. A rule we have occasionally found useful is that if the difference between estimates exceeds three times the length of the larger standard error, then the difference is significant at the 95% level. The rule holds only if the estimates are independent and the total sample size $(n_1 + n_2)$ is at least 16. This rule, however, is hard to remember unless one examines bar plots frequently. The best rule to remember about whether differences are significant is that only the t-test or nonparametric alternative tells for certain.

Two other points about tests should be kept in mind. First, p-values produced by statistical tests do not provide any information about how much larger one parameter is than another. The true difference could be extremely small – so small, for instance, that it is of little biological importance. But we still might obtain a statistically significant result. Furthermore, if one comparison is just significant and another is highly significant, we should not assume that the first difference is smaller than second because other factors, in addition to the size of the true difference, influence p-values.

Second, failing to detect a statistically significant difference does not mean that no difference exists. As already noted, population means are usually different; we may simply have too small a sample to detect the difference. Furthermore, failing to find a significant difference does not mean that the two populations are similar enough to each other that no difference of biological importance exists. It is entirely possible for two populations, for example males and females, to differ in biologically important ways, but for this difference to be undetectable from the data collected. The most direct way to decide how similar two parameters are is by constructing a confidence interval for the difference (see Section 3.3).

These two errors – assuming that statistical significance indicates biological significance and that nonsignificance indicates equality of parameters – are extremely common in behavioral ecology (and other fields). Focusing on the practical use of tests – deciding which mean (or other statistic) is larger – may be helpful in this regard. Under this line of reasoning, failure to reject the null hypothesis leads to the conclusion 'we have not demonstrated which population mean is larger' rather than the (nearly always incorrect) conclusion, 'the population means are equal'. Similarly, when a statistically significant result is obtained, it might be more appropriate to use statements such as 'analysis indicated ($p=0.032$) that the average size of males was larger than the average size of females in this population' rather than statements such as 'males in this population were significantly ($p<0.05$) larger than females'. The second statement is more impressive because the word 'significant' is used, but it runs the risk that statistical significance will be mistaken for biological significance and it may imply that males 'in general' were larger than females whereas inferences extend only to the average values not to how consistently males outweighed females.

Degrees of freedom

In this Section we explain why degrees of freedom arise in tests and confidence intervals, and provide some level of insight into the way they work. For purposes of discusssion we consider only two-sided tests. First, consider an unrealistic example in which we are testing a hypothesis about the mean of a normal population and know the true population standard deviation, and thus the true standard error of the estimate. Let the sample value be $\overline{x}$ and the population value, under the null hypothesis, be $\overline{X}$. The test statistic is

$$\left|\frac{\overline{x}-\overline{X}}{SE(\overline{x})}\right|,$$

where $SE(\overline{x})$ is the actual standard error under the null hypothesis. The critical value for the test statistic is 1.96, meaning that if the null hypothesis were true, then the probability of obtaining a test statistic larger than 1.96 is 5%, assuming all statistical assumptions are met. Now consider the more usual case in which the standard deviation is estimated from the sample data. Because the standard deviation (and hence the standard error) is estimated, it might be too small, which would increase the value of the test statistic, for a given value in the numerator. Of course, it might also be too large, which would have the opposite effect. This uncertainty about the standard deviation increases the uncertainty in the test statistic, or, to put it another way, makes the test statistic more variable. Thus, it will no longer be the case that the probability of obtaining a test statistic larger than 1.96 is 5%. An adjustment to the critical value must be made which accounts for the additional variation in the test statistic.

The required adjustment arises from the need to incorporate the variation introduced by using an estimate of the standard deviation. The estimate is a random variable. The distribution of this random variable (or, more precisely, the square of this estimate, which is just the sample variance) is proportional to the chi-square distribution which in turn depends on the sample size, n, and the number of additional parameters (in this case the sample mean) that must be estimated to compute the variance. In particular, the distribution depends on the sample size minus the number of additional parameters that must be estimated. In this case, this quantity is n–1 and is called the degrees of freedom of the distribution. In other cases, such as regression, the number of parameters estimated is equal to the number of regression parameters (including the intercept) and the degrees of freedom are n–(number of regression parameters). Readers familiar with matrix and linear algebra may wish to consult Seber (1977, Chapter Two) for a more precise development of the chi-square distribution and the notion of degrees of freedom.

Returning to our example, our test statistic contains an estimate of the mean in the numerator and an independent estimate of the variance (in the form of the standard error) in the denominator. The sample mean is normally distributed, and the sample variance has a distribution proportional to the chi-squared distribution with n–1 degrees of freedom. A statistician can show that, as a result, the distribution of the test statistic has the t distribution with n–1 degrees of freedom (and, more generally, the degrees of freedom are the same as those of the chi-squared distribution associated with the term in the denominator). For a given significance level, the t-value increases as the degrees of freedom decreases. For example, with $\alpha = 0.05$ and a large sample size (so our estimate of the standard deviation should be essentially equal to the population standard deviation), the t-value is 1.96, the same as the critical value when the population standard deviation is known in the test statistic (and which is the critical value for the normal distribution). With smaller sample sizes and thus fewer degrees of freedom, the t-value increases. For example, with 50 degrees of freedom the t-value is 2.008, with 20 degrees of freedom the t-value is 2.086, and with 5 degrees of freedom it is 2.571. With fewer degrees of freedom, its value increases rapidly, reaching 12.706 with just one degree of freedom.

Degrees of freedom are also used with other distributions that are associated with random variables based on samples from normal populations and which are functions of the sample variance. An example is the F distribution, which arises from random variables that are ratios of independent sample variances. Also, the t table is sometimes used for test statistics that follow very complex distributions that can be approximated by the t distribution. In such cases, an 'effective' number of degrees of freedom may be calculated which permits use of the t table as an approximation. The formulas for effective degrees of freedom in such cases may be very complex.

One- and two-tailed tests

In most one-sample tests, the alternative hypothesis is two-sided, i.e., that the true value is *either* smaller or larger than the value under the null hypothesis. Similarly, in most two-sample comparisons, the alternative hypothesis is two-sided, i.e., that the difference between population means is either positive or negative. Occasionally, however, investigators feel that only one of the alternatives to the null hypothesis is realistic or interesting, i.e., we have a one-sided alternative. For example, in comparing survival rates of animals with and without a radio transmitter, the investigator might assume that animals with transmitters would not survive at a higher rate than those without. In such cases, the alternative hypothesis may be restricted to the values (either positive or negative) deemed possible. The test is then called 'one-tailed' in contrast to the more typical 'two-tailed test' under which the true value may be either larger or smaller than the value under the null hypothesis. One-tailed tests are more powerful than two-tailed tests, because we have (or are claiming to have) more information about the range of possible values for the parameter. We account for this additional information by adjusting the threshold value of the test statistic at which the null hypothesis will be rejected. The procedures are explained in Box 2 (t-tests) and Box 8 (nonparametric tests) in Appendix One.

One-tailed tests should only be used if one is *certain* that the true difference between the parameter and its value specified under the null hypothesis cannot be positive (or that it cannot be negative). This assumption is rarely justified in behavioral ecology. In the telemetry example already described, for instance, the one-tailed test might at first seem reasonable. We know of a case, however, in which birds with transmitters sharply restricted feeding activities. As a result, they were less subject to predation and, during the course of the study, survived at a higher rate (later on they fared poorly due to malnutrition). The assumption that transmitters would not increase survivorship during the study period was thus invalid.

One-tailed tests are sometimes appropriate when only one result is interesting. In testing a new procedure (e.g., a drug), the purpose may be to decide whether it should be adopted, and this might only be appropriate if it performs better than existing options. Such reasoning might lead to adoption of a one-tailed test procedure. In our experience, however, cases such as this one are in rare in behavioral ecology. In nearly all cases, two-tailed tests are more appropriate. One check to use before employing a one-tailed test is to ask, 'suppose I obtain a result that would be significant in the unexpected direction – would I want to report it as significant?' If the answer is 'yes', then the two-tailed test should generally be used.

3.3. Confidence intervals

Confidence intervals provide a way of describing how precisely we have estimated a parameter – the population quantity of interest. Having constructed a 95% confidence interval for a mean, for example, we would like to say that there is a 95% probability that the mean lies within the computed interval. We must be clear, however, about what is meant here by probability. To define probability in the context of confidence intervals, consider a large series of additional samples selected from the same population using the same sampling plan. A 95% confidence interval could be calculated for each of these samples. If the statistical assumptions are met, then 95% of these confidence intervals would include the true mean and 5% would not include it. Thus, *probability refers to the proportion of times in repeated sampling that the confidence interval would contain the true mean*. Once an interval is calculated it either does or does not contain the mean. Hence statisticians prefer to use the word confidence rather than probability in referring to a particular interval.

Construction of a confidence interval begins with deciding how confident one wants to be that the interval will contain the value of the parameter of interest. The most common level of confidence is 95%, but any other level can be selected. The higher the level of confidence the more likely the resulting interval will contain the true value of the parameter. A higher level of confidence, however, tends to increase the width of the interval. With a given confidence level, increasing the sample size is the primary way to reduce the width of the resulting intervals.

Confidence intervals can be constructed for any population parameter including means, population totals, survival rates, and parameters representing relationships such as the slope from a regression analysis. The approach depends on several factors including whether the data are continuous or discrete, the underlying distribution of the data, the parameter to be estimated, and the sample size. Detailed guidance is provided in Appendix One. For example, suppose the data can be viewed as a series of independent trials in which the outcomes are of only two types, success or failure, and the probability of success is the same on each trial. We wish to construct a confidence interval for the probability of success. In this case, and if the sample size is not too large, exact methods, which guarantee that the coverage probability is at least the value specified, based on the binomial distribution may be used. Alternatively, approximate methods based on the normal distribution may be used (see Box 5, Appendix One). Note that with approximate methods we are only guaranteed that the coverage probability will be close to the correct value for large sample sizes. For any given sample size we do not know whether the coverage probability will be too small or too large.

For moderate to large samples sizes, if the parameter of interest is a mean, proportion, or, more generally, any parameter which is estimated by an average or weighted average of data (this includes regression parameters), a simple formula is usually sufficient. A $(1-\alpha)\times 100\%$ confidence interval for G is

$$g \pm crit_{\alpha} se(g), \tag{3.3}$$

where g is the estimate of the parameter G, $se(g)$ is the estimated standard error of g, α is the level of significance, and crit_{α} is the critical value which depends on the sampling distribution of g. For example, when sampling from normal populations crit_{α} is often $t_{\alpha,df}$. Formulas for the standard error and the appropriate crit_{α}, for the situations encountered most often in behavioral ecology, are given in Chapter Four and in Appendix One. Many statistical packages provide estimates and their standard errors and, in conjunction with Eq. 3.3, can be used to obtain confidence intervals.

For an example of a confidence interval that does not have the form of Eq. 3.3, consider the swan data discussed in Section 3.2. The data in this example are dichotomous and might be treated as binomial with $n=100$. The parameter of interest, P, is the true proportion of times that families win fights with pairs. Suppose we wish to construct a 95% confidence interval for this proportion. The endpoints are

$$\text{lower endpoint} = [1 + F_{(\alpha/2,|\nu_1,\nu_2)}\,(q + 1/n)/p]^{-1},$$

where $\nu_1 = 2(nq+1)$ and $\nu_2 = 2np$, and

$$\text{upper endpoint} = \left[1 + \frac{q}{(1/n + p)F_{(\alpha/2,|\nu_1,\nu_2)}}\right]^{-1} \tag{3.4}$$

where $\nu_1 = 2(np+1)$, $\nu_2 = 2nq$ and $q = 1-p$. In this example, $\alpha = 0.05$, $p = 0.66$, and $q = 1-p = 0.34$. For the lower endpoint, ν_1 and ν_2 are 70 and 132, $F_{(\alpha/2|70,132)} = 1.4909$, and the value of the lower endpoint is 0.56. For the upper endpoint, ν_1 and ν_2 are 134 and 68, $F_{(\alpha/2|134,68)} = 1.5369$, and the value of the upper endpoint is 0.76. The resulting 95% confidence interval is $0.56 < p < 0.76$. Most statistical packages provide values of $F_{(\alpha/2|\nu_1,\nu_2)}$ and are necessary for larger sample sizes. The package MINITAB (Minitab Inc., PA, USA) was used to obtain the values of $F_{(\alpha/2|\nu_1,\nu_2)}$ above.

Confidence intervals may also be calculated for the difference between two population means. The meaning of the interval may be difficult to grasp at first, but it has exactly the same interpretation as the confidence interval for a single estimate. Each of the two populations has a true but unknown mean (or other quantity of interest); we are estimating the difference between these means, and may refer to it as the 'true difference'. We consider a large series of samples selected from the same populations using the same sampling plan. A 95% confidence interval for the true difference could be calculated from each of these samples. If the statistical assumptions are met, then 95% of these confidence intervals would include the true difference and 5% would not include it. As with confidence intervals for single population parameters, 'probability' in the two-sample case refers to the proportion of times in repeated sampling that the confidence interval would contain the true difference. Once an interval is calculated it either does or does not contain the true difference. Hence statisticians prefer to use the word confidence rather than probability in referring to a particular interval.

The explanation above may help show the importance of bias in statistical analysis. If our point estimate is biased, then a nominal 95% confidence interval of the form (3.3) would probably cover the population mean less often than 95% in repeated sampling. Thus, construction of confidence intervals entails the assumption that the point estimate is unbiased.

Utility

Confidence intervals have two practical uses. When the null hypothesis of no difference has been rejected, the confidence interval tells us the largest and smallest-values that are realistic for the true difference. Such a conclusion can be of great-value. For example, suppose masses of males and females were 12 kg and 8 kg, and the difference was statistically significant. Given only this information, we cannot determine whether the study shows that males are much heavier than females, or whether they might be only slightly heavier. But suppose we were also told the confidence interval on the difference was ± 3.5. The difference was 4.0, so the lower endpoint for the confidence interval is 0.5. The average mass for males might be only 0.5 kg more than the average for females. In contrast, a confidence interval of

± 0.7 would show that the average for males was at least 3.3 kg more than the average for females. Thus, providing the confidence interval helps the reader, and perhaps the investigator, to evaluate the biological importance of the observed difference.

Confidence intervals can also be useful when the estimates are not significantly different. Suppose, for example, that the masses in this example were 12 kg for males and 11.8 kg for females and the result was not significant. Do these results show that males and females have about the same average mass? Again, the reader cannot answer this question without additional information. But if the confidence interval on the difference was ± 3.5 kg, then the confidence interval would be -3.3 to 3.7 kg. The study thus only shows that the difference in masses is less than 3.7 kg, which we might have known before starting the study. Conversely, if the confidence interval was ± 0.3, the confidence interval would be -0.1 to 0.5 kg, so we could conclude that the average weights of males and females were within about half a kilogram of each other, a conclusion which might represent a gain in knowledge.

In reporting estimates, researchers commonly provide the standard error rather than a confidence interval. Readers can then construct their own confidence interval (if the statistical assumptions are met) at whatever level of confidence they feel is appropriate provided intervals of the form in Eq. 3.3 are applicable. In many cases knowledge of the standard errors also allows one to construct a test of whether two estimates are significantly different. Alternatively, particularly in describing the precision of several estimates, authors may report the coefficient of variation (cv = standard error/mean). Using the cv readers can quickly determine how large the confidence interval is compared to the estimate. For example, if the cv is 10%, and if we assume the *t-value* is 2, we may infer that the 95% confidence interval is the mean plus and minus 20% of the mean [because CI = mean $\pm$ 2 se = (1 $\pm$ 2 se /mean)mean = (1 $\pm$ 2 cv)mean]. This gives us a very different picture of how accurate the estimate is than, say, a cv of 40% which tells us that the 95% confidence interval is approximately the mean plus and minus 80% of the mean.

Relationship to hypothesis tests

In many cases $(1-\alpha)100\%$ confidence intervals and two-sided hypothesis tests give consistent results if all statistical assumptions are the same. If the test leads to rejection of the null hypothesis, then the confidence interval will not include the value of the parameter specified under the null hypothesis. If the test does not lead to rejection of the null hypothesis, then the confidence interval will include the value specified under the null hypothesis.

However, there are cases where we choose to make slightly different assumptions about the population variances in carrying out tests and constructing confidence intervals. Under the null hypothesis that two population means are identical, we may actually believe that the two populations have the same distribution and hence that the variances are also identical. We thus must only estimate one variance (generally accomplished by the 'pooled' estimate of variance). If we reject the null hypothesis and conclude that the population means are different, then we often will no longer wish to assume that the variances are equal because in behavioral ecology means and variances are generally related (i.e., populations with larger means have larger variances). Under the assumption that the variances are not the same, we have two variances to estimate, and this causes a slight loss in precision. It is thus mathematically possible to reject the null hypothesis (using a pooled estimate of variance) yet obtain a confidence interval (estimating separate variances for each population) that includes the parameter value under the null hypothesis. As already noted, this only happens due to the slightly different statistical assumptions and procedures used. Of course, one could use the separate variance approach for the test as well as in constructing the confidence interval. This approach gives slightly less precision for the test (due to estimating an extra parameter) but always yields consistent results. This approach is also favored by many statisticians because the approach using separate variance estimates is more robust to failures in the assumption that the two populations have the same variance, especially when sample sizes are small and/or unequal.

The relationship between one-sided tests and two-sided confidence intervals is more complex. If we wish to test

$$H_0\text{: parameter} = \text{constant } \textit{versus } H_1\text{: parameter} > \text{constant}$$

at level α and the constant specified in the hypotheses is smaller than the values in the corresponding $(1-\alpha)\times 100\%$ confidence interval, we would reject the null hypothesis at level $\alpha/2$. An analogous result holds for testing

$$H_0\text{: parameter} = \text{constant } \textit{versus } H_1\text{: parameter} < \text{constant.}$$

If a one-sided hypothesis is actually of interest, statisticians recommend that should one construct one-sided confidence intervals. these are, for example, of the form 'we are 95% confident that $P > 0.02$'.

The connection between results using confidence intervals and two-sided tests suggests that confidence intervals are more informative than the corresponding tests. Confidence intervals 'include' the results of tests in the sense that we can determine from the interval what the outcome of the test would have been at a particular significance level. Reporting confidence intervals for differences is generally much more informative than just reporting test results. This is particularly true for paired data (or any other nonindependent estimates) because in such cases one cannot calculate the confidence interval on the difference even if the two standard errors are reported. The covariance between the estimators is also needed (Section 2.9, expression 2.38).

Although confidence intervals have important advantages over tests, the results of tests can be reported more compactly and this is sometimes advantageous. For example, in many studies several comparisons are made, and the main conclusions of the study hinge on the pattern of results, rather than on the size of the difference observed in any single comparison. In such situations, investigators usually rely on hypothesis tests, rather than confidence intervals. The rationale is that we are interested in determining if there is a consistent pattern in the direction of the differences rather than in estimating the magnitude of these differences. Furthermore, if the differences are nearly all significant, then we have good reason to conclude that the pattern did not happen by chance. If there are actually no differences (i.e., if the population parameters being compared are actually all equal), then the chance that all would be declared significant at individual level α is smaller than α (generally much smaller). In contrast, if most of the differences are not significant, the data provide no grounds for such an expectation. A completely different pattern of results – perhaps leading to different biological conclusions – might have been obtained had different samples been selected from the populations. When several comparisons are made, reporting which were significant is relatively simple, especially if they all were significant and if the direction of the differences were all consistent with a single biological explanation. In contrast, reporting and interpreting the sizes of the confidence intervals for each comparison may be cumbersome.

Ratio of two random variables

In some studies, the ratio of two estimates is a more appropriate measure of how different the two values are than their difference. For example, in studying change in abundance through time, it often seems more meaningful to estimate the percentage increase or decrease, rather than the absolute change. Thus, if numbers/km^2 increased from 10 to 12, the investigator may prefer describing the change as a 20% increase rather than an increase of two animals/km^2. Weight gain, difference in reproductive success between dominant and subdominant individuals, and many other quantities seem better described by proportional change than by absolute change.

If the goals of the statistical analysis are restricted to showing whether the change is positive or negative, then the significance test for a difference provides the answer. There is no need to structure the test around the ratio of estimates because testing whether this ratio equals 1.0 is equivalent to testing whether the absolute difference equals 0.0. If confidence intervals are to be constructed, however, then it is worth considering which measure of change – absolute change or proportional change – is more meaningful, because procedures for calculating the confidence intervals on differences and ratios are not the same. Methods for proportional change (i.e., ratios) are explained in Appendix One, Box 9.

Consider this example in which estimating the confidence interval for a ratio of estimates would be appropriate. Suppose we have recorded the number of animals per survey route in each of 2 years. An index method of some sort was used to collect the data. We assume that the same proportion of the animals present in each year was detected (aside from effects of sampling error), but we do not know what this proportion is. Thus we could not estimate the absolute change in density even if we wished to; proportional change is the only option. We find that the number per survey route was 15% higher in year 2. After following the guidelines in Box 9, we summarize the results with a statement such as 'numbers increased 15% between years ($p = 0.027$, 95% CI = 9–21%)'.

Care is sometimes needed in deciding whether a quantity is the ratio of two random variables. The sample mean, $\Sigma y_i/n$, is not a ratio of random variables because n would not vary in repeated sampling. On the other hand, suppose we select several plots and record some measurement on several plants in each one as a way of estimating the population mean. If the number of plants varies from plot to plot, then the total number of plants is a random variable. Methods for this case are explained in Section 4.5.

As noted in Section 2.4 ratios are nonlinear combinations of random variables and, accordingly, have more complex statistical properties. In particular, point and interval estimates for ratios based on the ratio of unbiased estimators are generally biased. Recall that $\bar{y}$ is unbiased for estimating $\bar{Y}$. Suppose we plan to estimate $1/\bar{Y}$ using $1/\bar{y}$. Consider a population of just three values, 1, 5, and 9, and suppose we have drawn a sample of size 1 to estimate the quantity $1/\bar{Y}$. The parameter, $1/\bar{Y}$, equals $1/5 = 0.20$. The possible estimates are 1/1, 1/5, and 1/9 whose average is 0.43. Thus $1/\bar{y}$ is a biased estimator of $1/\bar{Y}$ in this case.

The bias in ratio estimates tends to be small in cases encountered by behavioral ecologists. Here, bias is proportional to the reciprocal of the sample size and thus decreases as sample size increases. In the example above, if the sample size is 2, then the possible estimates are (1,5), (1,9), and (5,9), producing estimates of 0.33, 0.20, and 0.14. The mean of these estimates is 0.22, so the bias is only 0.02, compared to 0.23 when $n = 1$. The dramatic decrease by factor of about 10 is, in this case, largely due to the fact that a sample of size 2 is nearly a sample of the entire population. If the population had been larger, the decrease would not have been as dramatic.

Box 9, Appendix One gives the formula for the estimated standard error of the ratio of two random variables. The formula is based on the Taylor series expansion and provides only an approximate estimate (Section 2.11). When the estimates are 'typical' quantities, such as means, proportions, survival rates, or slopes from regression, and when the estimates are precise enough to be interesting, then the formula nearly always works quite well. Care (and consultation with a statistician) is needed, however, in some cases. For example, if 0 is a possible value in the denominator then the expected value (the mean of all possible values) is not even defined.

3.4 Sample size requirements and power

Investigators planning to carry out statistical tests often wonder how large their sample sizes should be. The answer to this question depends on several quantities including the purpose of the study, attributes of the populations being compared, and the magnitude of the effect one wishes to detect. In practice, investigators often have a 'feel' from past experience or studies carried out by other researchers for how large the sample sizes should be to detect an effect or difference of practical significance, and frequently little gain is achieved by employing formulas to estimate sample size requirements. For the 'standard' methods to be exactly valid, they must be used at the beginning of a study, although they require estimates of unknown population parameters. However, they are often used when a study is partly completed and the investigator is trying to decide how long to continue the study or whether to change sample sizes during the remainder of the study. The data already collected can be viewed as a pilot study and used to estimate unknown population parameters. These estimates can then be used in formulas for power in place of the unknown population parameters. This section explains the procedures in cases most likely to be needed by behavioral ecologists. Box 10, Appendix One, provides the needed formulas and additional guidelines.

Sample size requirements are difficult to discuss without first discussing the concept of power in more detail. We do this in the context of testing whether the parameters for two populations differ. Power is the probability of obtaining a significant result in a statistical test when the null hypothesis is false and a particular alternative is true. Thus power is a function of the actual difference in the parameters plus other quantities such as sample size and population variances. If the difference is increased, power increases. It is thus unwise (or at least incomplete) to say that the power of a test is, for example, 0.8 because power depends on the actual difference.

As noted in Section 3.2, population means (or other parameters being compared) are seldom exactly equal. The question is, what magnitude of the difference do we wish to detect? Suppose that we decide, at the start of a study, to sample two populations whose means actually differ by D, and to test whether the difference between sample means is significant. We might obtain many different pairs of samples some of which would be significantly different and some of which would not. Power equals the proportion of the differences that we could obtain in which the sample means are significantly different when the actual difference is D. To put this another way, we can imagine a population of differences, somewhat like a jar full of green or white marbles. Each green marble represents two samples that produce a significant difference; each white marble represents two samples that produce a nonsignificant difference. Selecting the two samples in our study is like drawing one marble from the jar after the marbles have been mixed thoroughly. Power is the probability of drawing a green marble.

Power is higher when: (1) the true difference between the parameters is larger; (2) the sample size is larger; and (3) the populations being sampled are more uniform (i.e., have smaller variances and hence estimates have smaller standard errors). Power also increases if we increase the significance level. Thus, it is harder to obtain significance at $\alpha=0.01$ than when $\alpha=0.05$. There is an exact mathematical relationship between the true difference, the sample size, variability in the two populations, level of significance, and power. The relationship can be solved either for power or for sample size if all the remaining values are known. Power and sample size are often discussed at the same time because of this close relationship between them.

Estimating sample size requirements is useful when some flexibility exists in how large a sample to collect. For example, suppose you are comparing responses to playbacks using conspecific songs of a neighbor and a non-neighbor. You score the results of each trial as 'did' or 'did not' respond. The objective is to determine whether individuals respond more often to the familiar or unfamiliar song. You have collected 20 trials using each song type, the difference in response rates is biologically – but not statistically – significant, and you wish to estimate how much more data are needed to give a high probability that the difference will be significant if the true difference is as large as the difference in your sample. In examples like this one, the pilot study can be used to estimate the true difference and the other quantities needed for the calculations (see Box 10, Appendix One). You might calculate the required sample size using a few different powers – for example 75% and 90% – before selecting the final sample size.

Power calculations are sometimes useful in deciding whether to continue a study. For example, suppose you have studied a species for 2 years and are trying to decide whether to continue for another year. One of your goals is to compare time spent foraging by males before and after they obtained mates. You could use power calculations to help decide whether collecting data for another year was worthwhile. To carry out the calculations, you would need to estimate how many more individuals you could watch during another year, and you would need to make some assumption about the value of the true difference between the proportion of time spent foraging before and after pair formation. One possibility would be to assume that the true difference equalled the difference observed in the data already collected. To be conservative, you might decrease this difference by one standard error. You would also need to estimate how variable males are with respect to time spent foraging (see Box 10, Appendix One), and decide on the significance level. These data could then be used with the formulas in Box 10 to calculate power. You might find, for example, that currently you have relatively little power, say 40%, for detecting a particular difference deemed to be biologically significant, but that with one more year of additional data, the power would increase appreciably, say to 80%. These calculations are only approximate – for example the true difference might be larger or smaller than your cstimate – and the decision on how much data, or more data, to collect seldom hinges solely or even primarily on power calculations. Nonetheless, power calculations often provide useful information when one is trying to decide how much data to collect. Numerous authors have advocated that power calculations be carried out more often in behavioral ecology (e.g., Cohen 1988; Greenwood 1993; Taylor and Gerrodette 1993; Steidl *et al.* 1997).

Power and sample size calculations can be made on software designed specifically for this purpose. Typically one estimates the desired sample size or power for many different combinations of the quantities of interest. This process is of great-value in understanding which quantities have the greatest impact on power and gives one an excellent basis for making a final decision about total effort and allocation of effort among different segments of the study. Those who know a programming language can easily produce the needed code, and may find this option easier than understanding exactly what the packages are doing.

Calculating power after the data have been collected

The use of power calculations, when the null hypothesis was not rejected, has occasionally been used as a way to determine how large the true difference might be. If only a very large difference could have been detected with high probability, then we might conclude that the study should be regarded as inconclusive. The objective of using the power formula in this way is thus identical to the objective of calculating a confidence interval for the difference (Section 3.3). If the observed difference is positive, then the upper endpoint of the confidence interval gives us the largest plausible value for the true difference. We may infer, with 95% confidence of being correct (if $\alpha = 0.05$), that the true difference is less than this value. Inverting the power calculation, however, does not always yield the same values as obtained by the confidence interval calculations and has little theoretical justification. We therefore recommend that you do not use power analysis in this way. Stick with confidence intervals when you want to know how large the true difference might be.

3.5 Parametric tests for one and two samples

Section 3.2 described the purpose, rationale, and interpretation of statistical tests in general. In this Section we discuss procedures for one category of tests – parametric tests for one and two samples – in more detail. Subsequent Sections discuss nonparametric tests for one and two samples (Section 3.6), and discuss tests for comparing more than two samples (Section 3.7). Most of the discussion in this section pertains to t-tests.

Suppose we have an estimator, g, for a parameter, G, and are testing the null hypothesis H_0: $G = G_0$ where G_0 is a specified value. Let $se(g)$ be the estimated standard error of g assuming the null hypothesis is true. The importance of the phrase, 'assuming the null hypothesis is true', will become clearer later.

Many estimators, including maximum likelihood estimators, are approximately normal with large sample sizes. In such cases, an approximate test of H_0, based on the statistic

$$t = \frac{g - G_0}{\mathrm{se}(g)}, \tag{3.5}$$

is obtained by comparing t to the appropriate critical value found in a standard normal table. The degrees of freedom depends on the sampling plan and the random variables used to calculate g. For example, if g is the mean of a simple random sample, then degrees of freedom are one less than the sample size. In general we refer to the standardized (or Studentized) form in Eq. 3.5 as a t-statistic, even in those cases where the critical value comes from a standard normal table.

Depending on the goals of a study, the quantity estimated and its standard error may vary widely. In many cases the value of the parameter, G, under the null hypothesis is zero, i.e., $G_0 = 0$, and thus drops out of the formula for the t-statistic. The estimate, g, may be a single quantity such as a mean, proportion, or correlation coefficient, or it may be the difference between two random variables such as means or proportions. There may also be a 'correction for continuity' (see later) which slightly modifies the general formula. The critical value in the t-table depends on the significance level (usually 0.05 or 0.01) and the sample size, sampling plan, and quantity being estimated through the degrees of freedom. If we know the true value of the standard error (assuming the null hypothesis is true) then we set degrees of freedom equal to infinity. This is equivalent to using the critical value of the normal distribution and the resulting test is sometimes referred to as the normal-theory test or z-test. This situation arises when we know that g is approximately normally distributed as already discussed.

Single estimate

In behavioral ecology, the t-statistic based on a single estimate is probably used most widely to test whether an observed proportion differs significantly from 0.5. A particularly common example involves observing contests between members of different groups (e.g., male/female, old/young, resident/intruder) to determine whether members of one group tend to win more often. It might seem that we are comparing two groups in these cases, and, from a biological standpoint, we are. But from a statistical standpoint, it is easiest to focus on the members of one group. In each trial, the member of this group either wins or loses the encounter (assuming there are no ties). The population unit is thus an encounter, the variable is 'won' or 'lost', and we view the observed encounters as a simple random sample from a large population. The quantity of interest is the proportion of encounters that would be won in a very large sample (i.e., the 'probability' of winning). Our estimate of this quantity is the proportion of fights in our sample that members of our group won. Under the null hypothesis, the two populations are equally likely to win these encounters. The true value of the proportion under the null hypothesis is thus 0.5.

The formula for the standard error in this case (see Box 5, Appendix One) is $P(1-P)/n$ or $0.25/n$. Note that we calculate the standard error under the null hypothesis, and thus use $P=0.5$ rather than $p=$ the observed proportion in our sample. One unusual feature of the t-statistic in this situation is that under the null hypothesis we know the value of the standard error, and we do not have to estimate it. We therefore set degrees of freedom equal to infinity (equivalently, use the critical value of the normal distribution). This one-sample t-test, used with an estimated proportion, is often called the binomial test, the approximate binomial test, or the binomial test based on the normal approximation.

The t-test requires either that the distribution of the observations is normal (which is rare in practice) or that the sample size is 'large' so that the estimator (a mean or proportion) can be assumed to be approximately normally distributed. Guidelines are presented in Box 5, Appendix One, for deciding how large the sample size must be to use the t-test. The t-test assumes that the distribution of the estimate (in repeated sampling) is continuous, whereas the estimate of a proportion can take only a finite number of values ($n+1$ values where $n=$ the sample size) and hence is discrete. A 'correction for continuity' is therefore recommended by many authors. There is some controversy about whether this correction is needed or useful; we follow the recommendations of Fleiss (1981) and Snedecor and Cochran (1980). Instructions for including the correction for continuity, and for deciding whether the observed value is significantly different from the value under the null hypothesis, are included in Box 5, Appendix One.

We now return to the question of why we should not treat the data from 'encounters' as two separate estimates. For example, if one group wins 30% of the encounters and the other wins 70%, why would it be improper to test whether 0.3 is significantly different from 0.7 with a two-sample t-test? The reason is that the usual two-sample t-test requires the assumption that the two estimates are independent of each other. But if one estimate is 0.3, then we know that the other estimate has to be 0.7; there is no other possibility. The t-test for independent estimates thus could not be used. If we developed the appropriate version of a t-test acknowledging the dependence between estimates, it would turn out to be identical with the t-test for one sample, as already explained. This point may be clearer after we have explained procedures for a t-test with paired data.

Single estimate *t-tests* may be used to test proportions other than 0.5. In genetics experiments, for example, one can often calculate the expected frequency (generally different from 0.5) of a genotype based on the assumption of random mating and perhaps other assumptions. It may be of interest to determine whether the observed frequency is significantly different from the expected frequency. Single estimate t-tests are also often used to test whether the slope in a linear regression is significantly different from zero. In such an analysis, we must estimate the standard error – we do not know its value exactly even if we assume the null hypothesis of zero slope is true. The degrees of freedom are set to account for the additional uncertainty introduced by having to estimate the standard error by first estimating the intercept and the slope. More generally, the one-sample t-test is likely to be appropriate anytime we wish to test whether an observed value of an estimator is significantly different from some hypothesized value established on a priori grounds. Statistical textbooks often present examples in which a new treatment of some kind (e.g., a drug or fertilizer) is being compared to an established one of known effectiveness. Such situations seem to occur rarely in behavioral ecology.

Two independent estimates

The t-test for comparing two independent estimates is probably the most widely used statistical method in behavioral ecology. It may be used to compare means, proportions, survival rates, regression or correlation coefficients, and many other quantities. The only requirements for the test are that the estimates are based on random sampling from a normal population (but not necessarily one-stage or simple random sampling), or that the sample size is large enough. 'Large enough', as explained later in this Chapter, varies according to the distribution of the variable measured, but for estimating a mean or proportion in many cases encountered by behavioral ecologists sample sizes of 10–15 are sufficiently large for the t-test to be valid.

Statistics texts present the formulas for carrying out a t-test in a variety of ways. Some of them apply only if the estimates are sample means and simple random sampling has been used to collect the data. Other formulas may be difficult to understand if the reader has not recently studied a statistics text. Our formulas (Boxes 6 and 7, Appendix One) are general in the sense that they describe how to carry out the test regardless of what quantity is being estimated or what sampling plan was used to collect the data. The formulas are expressed in terms of the standard errors of the two estimates. Of course, the specific formulas for these standard errors and the associated degrees of freedom do depend on what quantity is being estimated and what sampling plan is used. Calculation of these values, with different parameters and sampling plans, is explained in other boxes (especially Boxes 2 and 3, Appendix One).

Special methods are needed for comparing proportions estimated with simple random sampling. With small sample sizes Fisher's Exact *test* is recommended (see Box 7, Appendix One). With larger samples, the *t*-test may be used. Two alternatives to the *t*-test are the chi-square test, which is identical to the *t*-test, and the *G*-test (see Sokal and Rohlf 1981) which, while not identical to the others, usually yields similar results. We prefer the *t*-test because it extends more easily than the chi-square approach to construction of confidence intervals on the difference. One disadvantage of the *t*-test approach is that it does not extend as nicely to comparing more than two proportions, whereas the chi-square approach does. Most statistics texts (e.g., Section 7.11 in Snedecor and Cochran 1980) discuss the chi-square approach as well as the *t*-test approach.

Paired data

The phrase 'paired data' usually means that two observations were collected on every population unit in the sample. The quantity of interest usually is the difference between the means of the two variables being measured. In behavioral ecology, most examples of paired data seem to involve measuring the same animal, plot, or other subject at two times, such as before and after a treatment, or in two periods such as years. Data may also be paired in space, as for instance when we record two measurements from a series of randomly selected plots (e.g., number of males, number of females). In these examples the data are clearly paired in the sense that we have a single random sample of units from the population of interest, two random variables are measured on each unit in the sample, and the goal is to estimate the difference between the population means of these variables.

Two classes of paired data account for most cases one is likely to encounter in practice. The first case is dichotomous data. Methods for this case are explained in Box 7, Appendix One. The second case is where the mean of the difference of the paired observations is of interest. In this setting we proceed as follows: (1) create a new variable for each unit in the sample defined as the difference between the observations for the unit; (2) calculate the mean and standard error of this variable; and (3) use the guidelines in Boxes 2–4 to test the null hypothesis that the true mean is 0.0. If the null hypothesis is rejected (or the confidence interval does not cross 0.0), then the means of the two variables measured are significantly different.

Partially paired data

Occasionally, some of the observations are paired and others are not. This can arise when the initial sampling plan calls for all data to be paired, but in collecting the data some observations (e.g., one member of a pair) are missing. For example, abundance may be estimated in 2 years with the same survey routes, but some routes might not be used each year, or data might be collected on animals in summer and winter, but some of the animals might die before winter. Numerous other examples could be cited in which the observations are largely, but not completely, paired. Several approaches are available for analyzing such data. We provide some general recommendations. Partially paired data often require careful analysis so consulting a statistician for advice may be worthwhile.

The most important consideration in analyzing paired data with missing observations is whether the missing *differences* would have been larger or smaller, on average, than differences that were recorded. To consider an obvious example, imagine that we are trying to determine whether substantial weight loss occurs in a species during the winter. At the start of the winter we weigh and mark a sample of individuals. At the end of the winter we recapture most but not all of them. We also capture and weigh some unmarked individuals at the end of the winter. The data set is thus largely, but not completely, paired. We should be suspicious, in this case, that some of the individuals we failed to recapture died of starvation or some other weight-related cause. If these are the reasons for the missing data then our estimate of weight loss based on the observed data is likely to be biased, perhaps severely. There is no easy answer to what one should do in such cases but it may be best to admit that weight loss cannot be estimated from this data set.

On the other hand, suppose we believe that missing individuals did not differ, as a group, in weight loss. For example, they may have emigrated or died of causes unrelated to weight loss, or they may have been present but were not captured just by chance. If such an assumption is reasonable, then we can proceed with the analysis. We discuss three different ways to analyze the data. They vary in complexity and in how efficiently they use the data.

First, we could ignore the animals captured only once and base the analysis solely on the paired data as explained in Appendix One, Box 7. We would then calculate a new variable from each individual (the difference in weights in our example), calculate the mean of these differences and its standard error, and test the null hypothesis that the true mean was equal to 0.0. This is the simplest and safest approach, but fails to use all the data collected.

The second approach makes use of the general formula for the difference between two random variables (Section 2.9). Calculate the two means, $\bar{y}_1$ and $\bar{y}_2$, using all the data available from each sample and then take the estimated difference as $\bar{y}_1 - \bar{y}_2$. Let n_1 and n_2 be the number of observations used in computing $\bar{y}_1$ and $\bar{y}_2$ respectively. The standard error of the difference (Eq. 2.38) is,

$$se(\bar{y}_1 - \bar{y}_2) = \sqrt{se(\bar{y}_1)^2 + se(\bar{y}_2)^2 - 2cov(\bar{y}_1,\bar{y}_2)} \quad (3.6)$$

where, in this case, $se(\bar{y}_1)$ and $se(\bar{y}_2)$ are calculated using n_1 and n_2 observations respectively and the usual formula for one-stage sampling (Eq. 2, Box 2 in Appendix One). The covariance term in this case is

$$cov(\bar{y}_1,\bar{y}_2) = \frac{n_p}{n_1 n_2} cov(y_{p1i}, y_{p2i}) \quad (3.7)$$

$$= \frac{n_p}{n_1 n_2}\left[\frac{\sum_{i=1}^{n_p} y_{p1i} y_{p2i} - n_p \bar{y}_{p1} \bar{y}_{p2}}{n_p - 1}\right],$$

where the subscript p indicates that only the paired observations are used. Thus, n_p is the number of units measured twice, $\bar{y}_{p1}$ is the mean of the first observations from this group, and $\bar{y}_{p2}$ is the mean of the second observations from this group. If all units are paired, then $n_p = n_1 = n_2$ and this formula reduces to the usual formula for $cov(\bar{y}_1,\bar{y}_2)$ (Section 2.5, Eq. 2.14). Most calculators and statistical software provide the covariance for pairs of data. Check to make sure that yours uses n–1 in the denominator.

The third approach is more complex but uses the data in the most efficient way. Calculate one estimate of the mean difference, d_p, using only the paired data, and another estimate, d_u, using only the unpaired data. In the notation of the previous paragraph, $d_p = \bar{y}_{p1} - \bar{y}_{p2}$ while $d_u = \bar{y}_{1u} - \bar{y}_{2u}$, where $\bar{y}_{1u}$ is the mean from the units measured only on the first occasion and $\bar{y}_{2u}$ is the mean from the units measured only on the second occasion. The final estimate is then taken as a weighted average

$$\text{estimated difference} = wd_p + (1 - w)d_u, \quad (3.8)$$

where the estimated weight is

$$w = \left[1 + \frac{[\text{se}(d_p)]^2}{[\text{se}(\bar{y}_{1u})]^2 + [\text{se}(\bar{y}_{2u})]^2}\right]^{-1} \quad (3.9)$$

The standard errors are calculated with the usual formula for simple random sampling (Eq. A1.2, Box 2, Appendix One). For $se(d_p)$, based on the paired data, the terms in the standard deviation are the differences, and for the other two standard errors they are the individual (unpaired) observations.

Most introductory tests do not present the derivation (Eq. 3.9) so we explain it briefly. Since d_p and d_u are both unbiased estimates of the true difference, assuming observations are missing at random, the weighted average of d_p and d_u is also unbiased with any constant W. To find the value of W which minimizes the variance of the estimated difference we write the formula for the variance of the weighted average

$$V[Wd_p + (1-W)d_u] = W^2V(d_p) + (1-W)^2V(d_u). \tag{3.10}$$

There is no covariance term because d_p and d_u are independent. We wish to find the value of W that minimizes this expression. We therefore differentiate the right-hand side with respect to W, set the derivative equal to zero and solve for W. This yields

$$W = \frac{V(d_u)}{V(d_u) + V(d_p)} = \frac{1}{1 + \dfrac{V(d_p)}{V(d_u)}} \tag{3.11}$$

To obtain an estimate, w, we use sample estimates of the variances. Substituting the more detailed formulas for $V(d_p)$ and $V(d_u)$ yields Eq. 3.9.

Decide beforehand which approach you intend to use. Trying all three and using the one which happens to produce the smallest standard error is not appropriate. The first approach, ignoring the unmatched observations, is the easiest and performs well if the unused data comprise a small proportion of a large sample. The second approach uses all the data, can be carried out easily with a pocket calculator, and will often be nearly as efficient as the third method. The third method, which uses the data in the most efficient way, may appeal most to investigators who are familiar with computer programming and are thus able to calculate Eq. 3.9 with relative ease.

Performance when assumptions are not met

As noted in Section 3.2, the conditions under which t-tests are exactly valid are seldom met in real data sets collected by behavioral ecologists. In this section we discuss validity of the t-test with non-normal data. In the context of hypothesis testing, validity refers to the true significance level. Although we claim to be testing a hypothesis at a value such as $\alpha = 0.05$, when the statistical assumptions underlying the procedure are not met the actual probability of rejecting the null hypothesis (H_0) may be different from 0.05. In the context of confidence intervals, validity refers to the actual proportion of the confidence intervals (in repeated sampling) that would include the parameter. Because of the close relationship between hypothesis tests and confidence intervals, many statements made regarding the validity of hypothesis tests refer to the validity of the associated confidence intervals as well.

Concern over the validity of *t*-tests when the data do not have a normal distribution is probably the most common reason that biologists turn to nonparametric analyses. On the other hand, statistics texts point out that with a sufficiently large sample, *t*-tests about means are valid regardless of how non-normal the underlying population is (e.g., Lehmann 1975 p. 77). We wish, therefore, to gain some impressions of how large the sample must be, with specific distributions, before *t*-tests are valid.

In the Sections that follow, we first consider one-sample and paired-data procedures, and then consider the case of two independent samples. The one-sample and paired-data cases are similar in that both involve a series of observations, $y_1,\ldots,y_n$. In the one-sample problem the y_i values are single measurements made on the units in the sample whereas in the paired data case the y_i are the differences between the observations made on each unit in the sample. In both cases, however, the validity of the *t*-test depends on sample size and the distribution of the y_i in the population. Note that in the paired-data case, validity does not depend directly on the distributions of the two variables measured on each unit.

Validity is difficult to evaluate theoretically, so statisticians often use simulations with hypothetical populations and various samples sizes. We illustrate this approach with a series of populations similar to those we have encountered in consulting and in personal research. Our hypothetical populations are illustrated in Fig. 3.1. In Fig. 3.1a 10% of the response variables are distributed evenly between 0.0 and 0.99; 20% are distributed evenly between 1.00 and 1.99, and so on. In each plot, the sum of areas under the curve equals 1.0. We shifted the curves horizontally in different simulations to create differences between populations. We arbitrarily considered sample sizes acceptable if they produced true significance levels no higher than 6%.

For one-sample tests, the simulation showed that in symmetrical (Fig. 3.1a,b,e,f) or nearly symmetrical (Fig. 3.1c) populations, sample sizes as small as five or ten achieved satisfactory levels of significance (Table 3.1). With highly skewed populations (Fig. 3.1d), a sample size of 20 was required before significance fell to 6%. Populations even more skewed would require larger sample sizes for *t*-tests to be valid.

For independent estimates, the distribution of the difference between sample means (under the null hypothesis that the two populations have the same distribution) is symmetrical regardless of the shape of the underlying distribution. To understand this, imagine drawing two samples from the population. Let the difference in their means be a specific value, *d*. Since the

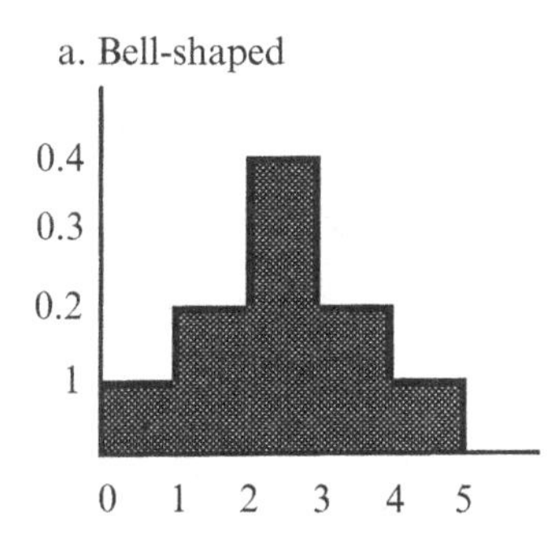

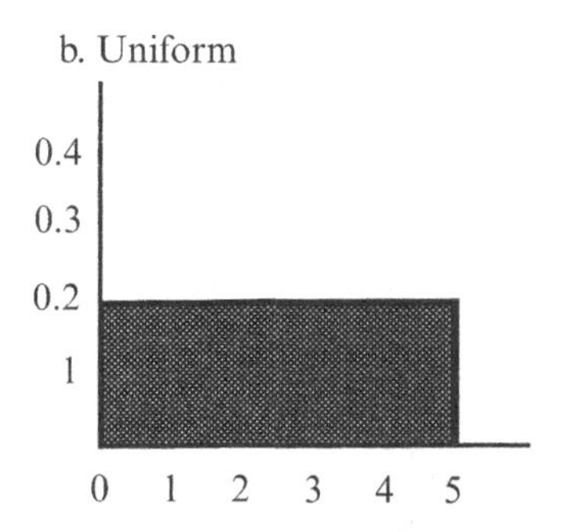

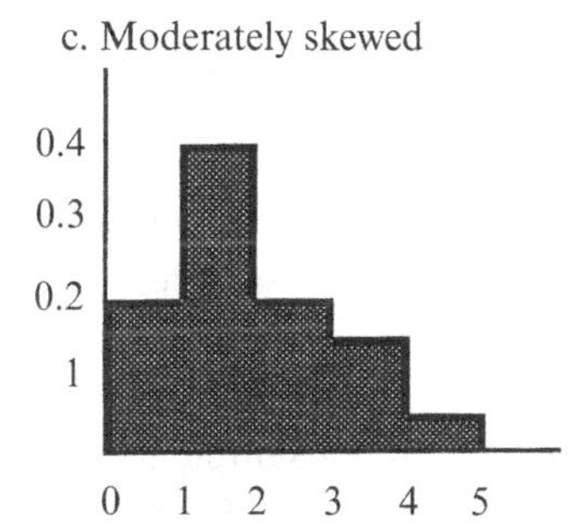

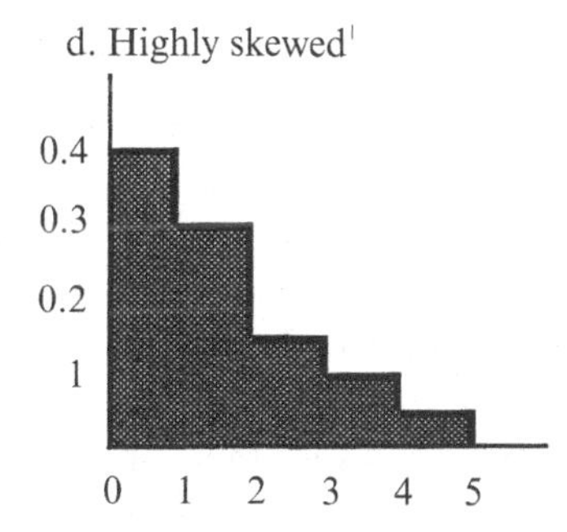

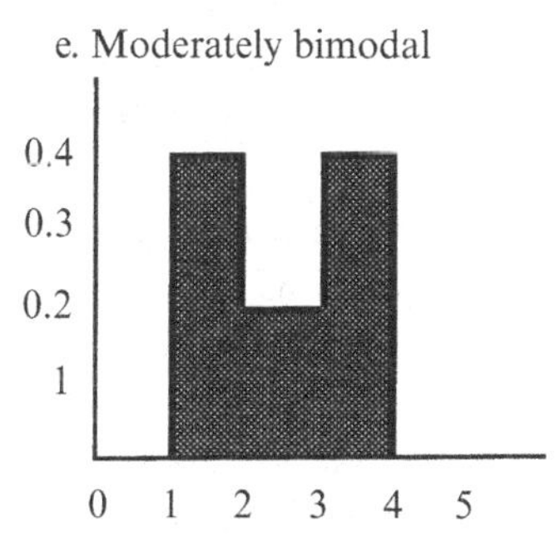

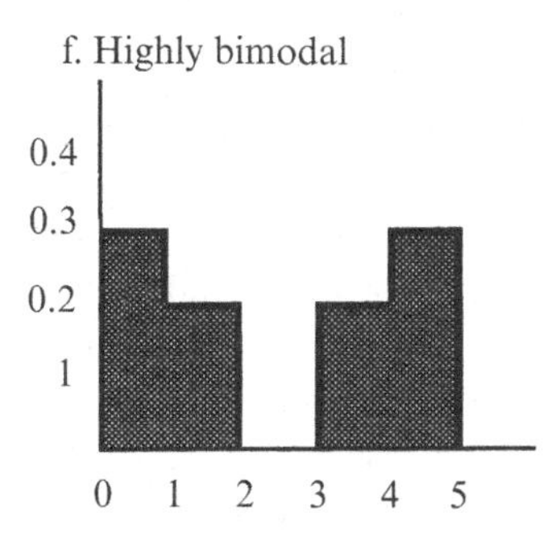

Fig. 3.1. Artificial populations used to investigate the validity of *t*-tests with small sample sizes. The distribution of values is shown for each population (e.g., in the bell-shaped population 10% of the values are uniformly distributed between 0.0 and 0.999, 20% are uniformly distributed between 1.0 and 1.999 and so on.

Table 3.1. *Validity of* t-*tests for single samples or paired data. Entries are the probabilities of rejecting the null hypothesis (with $\alpha=0.05$) when it was true (i.e., the true mean was 0.0). The populations are shown in Fig. 3.1*[a]

Population	True mean	Sample size (n) 5	10	20	40	Minimum n for $\alpha \leq 0.06$
Bell-shaped	0.0	0.05	0.05	0.05	0.05	≤5
Uniform	0.0	0.07	0.06	0.05	0.05	10
Moderately skewed	0.0	0.07	0.06	0.06	0.05	10
Highly skewed	0.0	0.08	0.07	0.06	0.05	20
Moderately bimodal	0.0	0.07	0.05	0.05	0.05	10
Highly bimodal	0.0	0.07	0.06	0.05	0.05	10

Note:
[a] Each estimated probability is based on 10,000 replicates.

samples come from the same population, there is an equal chance that d will be positive or negative. Furthermore, in a large number of samples, one would get equal frequencies of positive and negative values of d. The same can be said for any value of d; positive and negative values are equally likely. This shows that the distribution of the difference between sample means must be symmetrical if the null hypothesis is true. As was pointed out in the discussion of paired data, this symmetry alone ensures that levels of significance will be close to the nominal levels.

To illustrate this point empirically, we drew two samples from each of our populations and tested the null hypothesis of no difference. As before, we recorded the frequency with which the null hypothesis was rejected. In no case was this frequency greater than 5%.

In conclusion, in our populations t-tests were valid with independent samples (if $n>5$) and with paired samples unless populations were drastically non-normal and sample sizes were less than 20. With only minor departures from normality, sample sizes of ten or even five also performed well. On the other hand, t-tests cannot be expected to perform well on populations with extreme outliers. For example, if the distribution was similar to any of the ones we studied except that a small fraction of the population had extreme values (e.g., >10), then we could not expect t-tests to perform well.

3.6 Nonparametric tests for one or two samples

Nonparametric or 'distribution-free' methods provide alternative procedures for the one- and two-sample problems which were described in

Section 3.5, as well as for some more complicated problems. They are particularly useful when populations are distinctly non-normal and sample sizes are so small that t-tests are unlikely to perform well. One- and two-sample nonparametric tests are relatively simple to understand and carry out. This section makes several general points about nonparametric methods; details of how to carry out the tests are provided in Box 8, Appendix One. Nonparametric methods are included in most statistical computer packages.

The most familiar nonparametric methods are for testing hypotheses but confidence intervals can also be constructed. Nonparametric methods do not generalize easily to more complicated experimental designs or sampling plans. For example, exactly valid nonparametric methods do not exist for multiple regression problems or the two-factor analysis of variance with interaction.

In contrast to t-tests, which permit inferences about means, many nonparametric tests permit inferences about quantities related to medians. The median of a group of numbers (e.g., a sample or a population) is the 'middle number', the value such that half the values are smaller and half the values are larger. If the group contains an odd number of distinct numbers, then a single value fulfills this condition; when the group contains an even number of numbers, the median is usually taken as the average of the two middle values. For example, the median of the numbers 2, 3, 4 is 3, and the median of the numbers 2,3,4,5 is 3.5.

Sometimes the median represents our concept of average – in the sense of 'middle' – better than the mean. For example, suppose we expose animals to a stimulus and record how close each of several individuals approaches. If most individuals approach quite closely, but a few stay far away, then the sample of distances might be 1, 3, 2, 4, 25. In this case, the mean, 7.0, is considerably larger than all but one of the measurements, and the median, 3, may seem to provide a more useful measure of the middle or 'typical' value.

Despite the utility of medians in some cases behavioral ecologists contemplating use of a nonparametric method should think carefully about the distinction between means and medians. Means (and proportions) are often much more useful than medians as a foundation for further analysis. For example, from knowing the mean number present per unit area, total population size can be calculated. Knowing the mean number of young produced per adult means that demographic analyses can be carried out. Medians, in general, do not support this type of analysis. Furthermore, one should not assume that statements about the relative size of population medians can automatically be extended to analogous statements about means. For example, consider the following two populations. One population consists of the numbers 2, 3, and 4 in equal frequencies; it has a mean of 3 and a median of 3. The second population consists of the numbers 1, 2, and 9 in equal frequencies; it has a mean of 4 and a median of 2. Thus, in the second population, the mean is larger, but the median is smaller, than in the first population. Clearly in this case one would not wish to assume that deciding which population has the larger median reveals which population has the larger mean.

With paired data, it is also important to recognize that inferences apply to the median of the differences and this is not, in general, equal to the difference of the medians. Thus, if the population of paired observations consisted of the pairs (1,5), (3,2), and (2,0) in equal frequencies, then the differences would be -4, 1, and 2, with a median of 1. The population medians, however, are each 2 so the difference in medians is 0.0. This issue does not arise with t-tests; the mean of the differences is equal to the difference of the means. In the example above, both values are -0.33. Notice the range in values depending on which definition of 'typical' or 'middle' value is used: difference of the means (-0.33), difference of the medians (0.0), median of the differences (1.0). This example also shows that care must be exercised before assuming that conclusions about medians can be extended to conclusions about means.

A common misconception regarding nonparametric methods is that no assumptions are required (except no selection or measurement bias). In the one-sample and paired-data cases, the sign procedures require that the population is continuous (rather than normal as in the parametric case). The signed rank procedures require that the population is symmetrical as well. In the two-sample case, the populations do not have to be normal (as is required for exact validity in the t-test), but they must both be continuous, have the same variance, and their distributions must have the same shape differing only by a 'shift'. As with t-tests, small departures from these assumptions are not serious, and large sample sizes help reduce errors caused by failure of the assumptions.

Two other caveats about the interpretation of results from nonparametric tests should be made, though we do not discuss them in detail. First, when the assumptions (e.g., equal variance) are not met, but a nonparametric test is employed anyway, then the definition of the parameter about which inferences are being made can be surprisingly complex. Second, when assumptions are met and the null hypothesis is rejected, the exact nature of the conclusion one should reach is more difficult to understand. In practice, one usually assumes that the test indicates which population has the larger median (Snedecor and Cochran 1980 p. 145).

Useful additional comments on nonparametric methods in behavioral ecology are provided by Johnson (1995), Smith (1995), and Stewart-Oaten (1995).

3.7 Tests for more than two samples

In most behavioral ecology studies, investigators are interested in making comparisons between several populations and drawing overall conclusions from these multiple comparisons. Weights of males and females may be compared at each of several times, home range size may be compared in several different cohorts (e.g., older *versus* younger, paired *versus* unpaired), density of a species may be compared before and after a treatment on each of several plots, and so on. Several issues warrant consideration in such cases. First, investigators must guard against reaching unwarranted conclusions simply because of the number of tests carried out. Even if no differences actually exist, we would still expect about 1 in 20 comparisons (made at the 5% level of significance) to be statistically significant. That is the meaning of the 5% significance level: even if populations are not different, there is a 5% probability of (incorrectly) rejecting the null hypothesis. Thus, there is a difference between the level of significance for each individual comparison (say 5%) and the overall level of significance or probability of incorrectly deciding that one or more of several comparisons are statistically significant when, in fact, none of them differs. When many comparisons are to be made, the overall level of significance can be substantially higher than the level of significance used for individual comparisons.

To further complicate matters, the notion of the overall level of significance can be defined in several ways. As already described, it might be defined as the probability of incorrectly deciding that one or more of the comparisons made are statistically significant when, in fact, none are. Alternatively, it could be defined as follows. Among only those comparisons which do not actually differ, it is the probability of incorrectly deciding that one or more of these comparisons are statistically significant. In the first definition, we are protecting against incorrectly deciding that two of the populations differ when in fact no populations differ. In the second definition we are only interested in protecting against incorrectly deciding that two populations differ among those subsets of the populations which do not differ. The difference between these (and other possible) definitions is subtle, but many of the methods for multiple comparisons differ in what sort of overall level of significance they guarantee.

The second issue to keep in mind is the ability of a procedure to detect differences that actually exist, that is, the power of a procedure. Several methods (described later) guarantee that the overall level of significance (when in fact no populations differ) will be less than or equal to a prespecified value such as 5%. However, these methods differ in their overall power and generally we would prefer to use the method with the highest power.

Many different procedures have been suggested for maintaining the overall significance level when several comparisons are made. Some of the most widely used methods are Tukey's procedure, the Student–Neuman–Keuls procedure, Duncan's multiple range test, and the Bonferroni procedure. These and other methods differ in a number of ways, including the assumptions they make, the type of overall level of significance they provide, and the power of the procedure (in other words, how much protection they give against various types of errors). Because of these differences, some of which are quite subtle, it is difficult to compare the methods. We suspect that many biologists would welcome a simple and general guideline as to which multiple comparison procedure to use. With this in mind, we describe two approaches.

The simplest method, and the one requiring the fewest assumptions, is the Bonferroni method. If M comparisons (hypothesis tests) are to be made and one wishes to guarantee that the overall level of significance is no more than α, carry out each comparison using a significance level of α/M. This method is very general and works regardless of the nature of the individual comparisons. For example, some of these may be t-tests, some may involve nonparametric procedures, and some may be exact binomial tests. The drawback is that the actual overall level of significance is usually less than α and the power is low if more than a few tests are made. However, the overall level of significance associated with the Bonferroni test is the 'second definition' (bottom of p. 81) which is preferred by many statisticians.

A second approach is to begin with a comprehensive test of whether all populations are equal. This test is often made using a one-way analysis of variance (ANOVA), a chi-square test, or the Kruksal–Wallis test. If the result is nonsignificant, then no further tests are carried out and the investigator concludes that the samples do not permit detection of any differences between the populations. If the comprehensive test statistic is significant, then pairwise tests are made *using the same level of significance as was used in the comprehensive test*.

If the level of significance used for the overall test of equality is α, then we are guaranteed that the overall level of significance of this method is also α. This is because the overall test will be significant with a probability of α if none of the populations differ. Thus, if none of the populations differ, we will only proceed to carry out the pairwise comparisons (and thus perhaps incorrectly reject one or more of the null hypotheses) with a probability of α. The method is referred to as the 'protected' least significant difference (LSD) method.

Occasionally in behavioral ecology, the pairwise tests of interest are independent. When this is true, the comprehensive test may be carried out using a very simple procedure based on the binomial distribution. The procedure may have smaller power than an ANOVA or chi-square test, but may be easier to apply if a complex sampling design was used to obtain the point estimates. The procedure makes use of the fact that if all null hypotheses are true, in the pairwise tests of interest, then the probability of achieving a significant result in each test is α. Suppose that we carry out n tests and that k of them yield significant results. The probability of achieving exactly k significant results, when the probability of a significant result is α on each test, is

$$\binom{n}{k}\alpha^k(1-\alpha)^{n-k}, \tag{3.12}$$

and the probability of achieving k or more significant results is

$$\sum_{k^*=k}^{n}\binom{n}{k^*}\alpha^{k^*}(1-\alpha)^{n-k^*}. \tag{3.13}$$

Thus, if the test statistic in Eq. 3.13 is large, we would be inclined to view the various significant results (pairs declared significantly different) as arising by chance. If the value of Eq. 3.13 is small, say less than α, then we may regard the comprehensive test as having been rejected and the pairwise tests may be interpreted in the same way as if we had carried out a comprehensive test such as an ANOVA and obtained a significant result. We emphasize that this approach may be invalid if the tests are not independent. Consultation with a statistician to decide whether this requirement is satisfied in a particular application may be advisable.

Biologists often combine the two approaches described here, first applying a comprehensive test and then, if results are significant, using the Bonferroni approach to adjust α for the pairwise comparisons. As already note, however, if the Bonferroni approach is used, there is no need to carry out an initial comprehensive test (e.g., Neter and Wasserman 1974 p. 146;

Seber 1977 p. 465, Snedecor and Cochran 1980 p. 116). This point is important because carrying out the comprehensive test, when complex survey designs (Chapter Four) have been used, can be quite difficult. Use of the Bonferroni approach avoids this difficulty.

In specific situations, a particular multiple comparison procedure (such as Tukey's procedure) may be preferred to either of the procedures described here. However, because of the many subtleties associated with multiple comparisons, we refrain from making further recommendations about other procedures here. If you wish to use other procedures, we suggest that you first consult a statistician. Review of Steward-Oaten (1995), who questions the utility of most multiple comparisons in behavioral ecology, may also help provide a deeper understanding of the complexities.

3.8 Summary

Statistical tests provide one way to determine whether a parameter, such as a population mean, is larger or smaller than a hypothesized value, or which of two parameters, such as two population means, is larger. Confidence intervals provide similar information and also establish upper bounds on the difference between the value under the null hypothesis and the true value. Sample size and power calculations provide a method for estimating how large a sample is needed, or how likely we are to obtain a statistically significant result given a specified difference between the null hypothesis and the true parameter values. The *t*-test for one or two samples applies to numerous cases in behavioral ecology, though exact procedures depend on many factors including whether the data are paired, partially paired or independent. In most cases the test may be safely used, even when the data are distinctly non-normal as long as sample sizes exceed five or ten. Nonparametric alternatives to the *t*-test are attractive when the median is a more meaningful measure of the 'middle' or 'typical' measurement than the mean or when the distribution is highly skewed and sample size is extremely small. When several pairwise tests are to be made, a procedure should be followed that protects the investigator from reaching unwarranted conclusions simply because numerous tests have been made. We note that this issue is complex but provide a general recommendation that is simple to follow and will suffice in most cases.

4
Survey sampling methods

4.1 Introduction

Consider the following sampling problem, taken from a real consulting session we encountered. A biologist was studying the density of a large mammal in different habitats in a large study area. He had defined eight habitat types. The species was always close to water bodies, so the biologist viewed water bodies as the sampling unit. He had delineated the habitats along each water body in the study area. During each of three field seasons, aerial surveys of most or all of the study area were flown two to four times per year. Individual animals were plotted on maps and the habitat each was in, at the time it was sighted, was subsequently determined. The biologist wished to determine which habitats were used disproportionately. More specifically, he wished to estimate density in each habitat and then test the null hypothesis that actual density was the same in each habitat. Ideally, this step would be carried out using a comprehensive test. If the null hypothesis was rejected, then pairwise comparisons would be made as explained in Section 3.7.

This example presents us with a host of problems from an analytical standpoint. Habitat patches varied in size and care must be taken in defining the population units if density is the characteristic of interest. It is problematic whether we have a random sample of habitats since the entire study area was covered in at least some of the surveys. The variable, number of individuals, was recorded at specific times so the population unit is an area-time, but we do not have a simple random sample of area-times within each year because there were only two to four surveys per year. Thus, the biologist probably should not assert that the sample of area-times is a simple random sample, and if he does his colleagues are likely to accuse him of pseudoreplication (Chapter 6). But he does not see how else the data could have been collected, and, in any case, he already has the data and wants to analyze it in a way that will produce meaningful statistical results.

The example above is typical of the data sets that behavioral ecologists often need to analyze. No simple analytical approach seems completely justified, and deciding how to analyze the data is difficult. Part of the difficulty in this example is that we clearly do not have a simple random sample. It is also unclear which parts of the sample selection were random, or should be viewed as random sampling.

This Chapter provides the background needed to better understand the problems presented in this example. We explain alternatives to simple random sampling, provide guidelines on when they are most useful, and describe how these techniques can be used to understand problems caused by nonrandom sample selection. In general, we approach problems of this sort in two stages: (1) decide what assumptions will be made in analyzing the data, and (2) identify the appropriate analytical techniques given the assumptions. If the material in this Chapter is understood, then there should usually be little doubt about the appropriate analytical techniques, given the set of assumptions adopted. This means that debate about the analysis is restricted to the assumptions. Such debates frequently hinge almost entirely on biological rather than statistical issues. Reasonable people may disagree, but we have found it helpful to distill out the statistical issues since there is often little if any controversy over them once assumptions have been separated from the rest of the analysis. Thus, biologists can argue about biology, rather than about statistics.

4.2 Overview

In Chapter 2, it was assumed that the sample is a simple random sample (see also Section 4.4) from the population, although the issue of how such a sample would be selected was not discussed. Simple random sampling, however, is only one of several possible ways to obtain a sample, and in behavioral ecology this approach is often impractical or inefficient. This Chapter discusses simple random sampling in more detail, and then considers several other sampling plans widely used in behavioral ecology. We begin by defining several terms used in describing sampling plans.

Survey sampling plans can be classified by the number of stages, by the method of sample selection used in each stage, and by whether each stage is preceded by stratification. A stage, in survey sampling terminology, is any point in the sample selection process at which population units or groups of units are selected. The selection may be carried out using simple random sampling or various other sample selection methods described in the next Section. Sampling plans with more than one stage are called multistage sampling. Stratification means that prior to selection we divide the population, or the portion of it that we are sampling from, into groups of population units, which we call strata, and we then select units or groups of units from *each* stratum. The sampling plan is the entire scheme used to select the sample.

Developing an ability to describe sampling plans accurately is important because different plans have different formulas for calculating estimates and standard errors. On the other hand, as noted above, many real examples involve nonrandom sampling, and deciding which sampling plan provides a suitable approximation for purposes of identifying an appropriate statistical analysis in such cases is often difficult. We approach this subject in three stages. In this Section, we provide several artificial examples, not involving nonrandom sampling or specific biology. These examples are intended to clarify the terms and kinds of sampling plans. Later in the Chapter, in the detailed discussions of each technique (Sections 4.4–4.8), we provide partially realistic examples, generally not involving any nonrandom sampling, intended to illustrate the situations in which behavioral ecologists may find each technique particularly useful. At the end of the Chapter (Section 4.10) we take up fully realistic examples, including the one given at the start of this Chapter.

Suppose we are interested in estimating the density of some biological entity in plots. The study area is square and contains 900 plots. We are content to express density, initially, as mean number per plot. Conversion to other scales will be accomplished later (Section 2.4). Since the population units are plots, the sampling plan is the scheme for deciding how to select the plots for the sample. A simple random sample (without replacement) would be obtained by drawing plots in such a manner that every plot not already in the sample had an equal chance of being selected each time we drew a new unit for the sample. Any spatial arrangement of the plots is possible, but in many samples they would be distributed rather unevenly across the study area. Figure 4.1 shows an example of a simple random sample we selected using a computer. Another approach would be to distribute the plots evenly across the study area (Fig. 4.2). This approach to sample selection is referred to as systematic sampling and is considered in more detail in Section 4.2. A third approach to sample selection is nonrandom selection, a term that refers to any plan in which formal random selection is not employed. Nonrandom sampling is uncommon when selecting

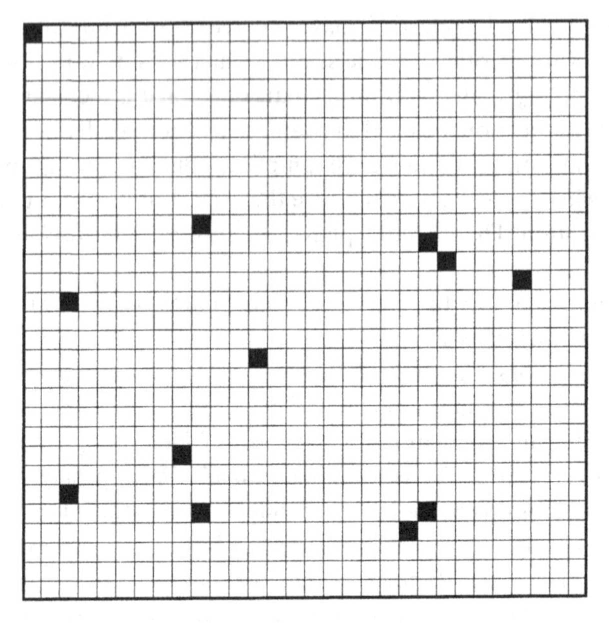

Fig. 4.1. A simple random sample of 12 population units (small squares) selected from a population of 900 units.

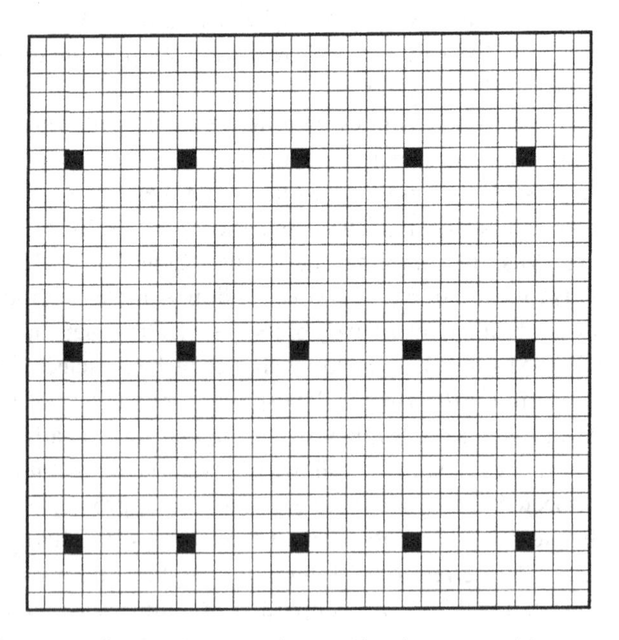

Fig. 4.2. A systematic sample. Note that spacing is not equal between rows and columns.

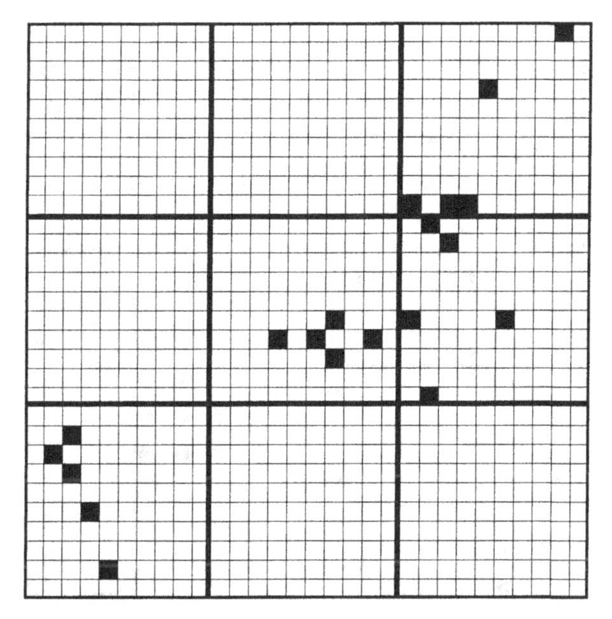

Fig. 4.3. A two-stage sample with four primary units and five secondary units per primary unit. Simple random selection was used at both stages.

objects like plots as in examples such as the one we are considering. Selection of animals, however, often uses nonrandom methods. To select a sample of size 'n', we may just use the first n animals we encounter. Alternatively, we may include all the animals (of a given species) we can see from a blind. Simple random, systematic, and nonrandom selection methods are the most common sample selection methods in behavioral ecology. We discuss each of them in the rest of this Chapter in more detail along with a few other selection methods that are used less often in behavioral ecology.

Now consider a fundamentally different kind of plan for selecting the sample. First we delineate groups of plots. Next we select some of these groups. Finally, we select population units within the chosen groups. Many different sampling plans are possible using this general approach. For example (Fig. 4.3), we might delineate nine groups, each with 100 plots, then select four groups using simple random selection, and then select five plots within each group, again using simple random selection. Sample size is thus 20 population units. Notice, however, that this plan is not equivalent to taking a simple random sample of size 20. A simple random sample of size 20 would tend to cover the population more evenly than a plan in which all units in the sample are in just four of the nine groups.

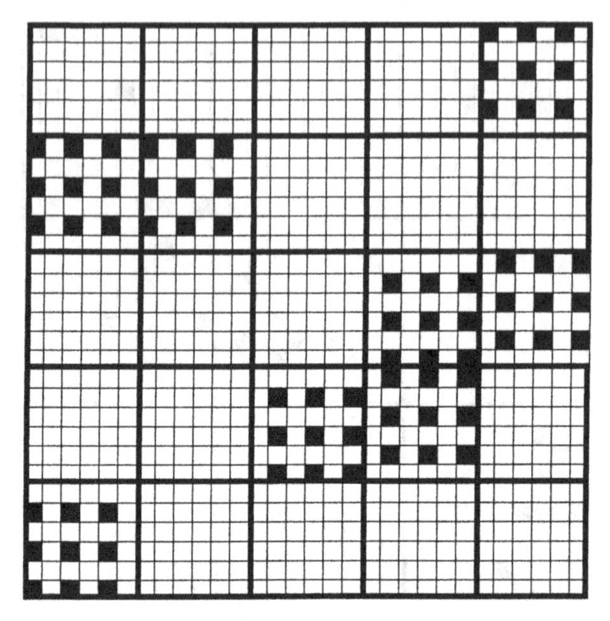

Fig. 4.4. A two-stage sample with eight primary units and nine secondary units per primary unit. Primary units were selected with simple random selection. An independent systematic sample of population units was selected in each primary unit.

This plan illustrates the concept of a 'stage' in sampling. As already noted, the word 'stage', in survey sampling terminology, means selecting groups of population units with the understanding that subsequent sample selection will be restricted to the chosen groups. In the example just given, delineating groups of 100 units each and selecting a sample of these was the first stage in the sampling plan. Subsequently, we selected population units in each group; this was the second, and final, stage, in this 'two-stage' sampling plan.

The groups selected at the first stage are called primary sampling units, or just primary units. In this example, we used simple random selection at stage one to select the primary units and simple random selection again at stage two, to select population units. Other plans, however, may be used. For example (Fig. 4.4), suppose we delineated square primary units with six population units on a side. This produces 25 primary units, each with 36 population units. At the first stage, we select eight of these primary units by simple random selection. At the second stage, we select nine population units in each primary unit using systematic selection. Note that selection in Fig. 4.4 was independent in different primary units: the population units in the sample do not occupy the same position in different primary units.

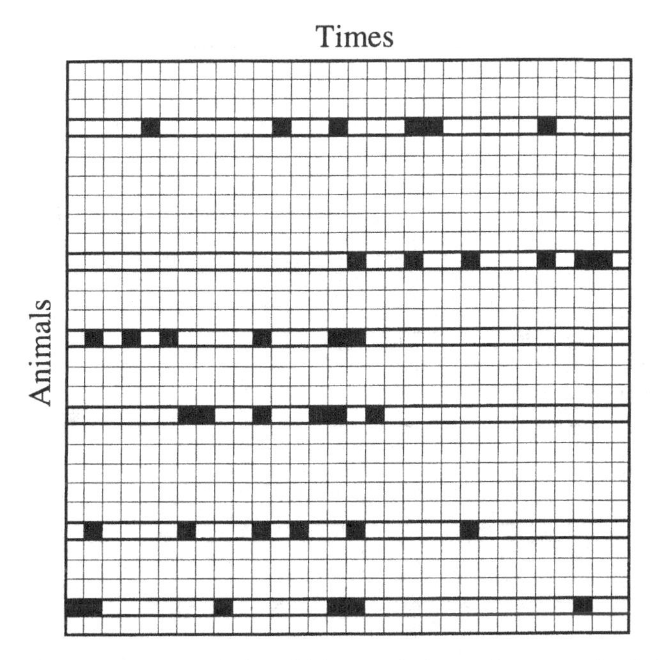

Fig. 4.5. A two-stage sample in which population units are animal-times and primary units are the sets of times at which each animal might be watched. An independent, simple random selection of times was made within animals. Although the plan is simple statistically, it does not distribute intervals across the temporal dimension well and it requires that up to four animals be watched at once.

As noted previously, nonrandom selection is also used frequently in studies of animals. For example (Fig. 4.5), suppose we select the first '*n*' animals encountered, as in the example of one-stage sampling described above, and then randomly select six times to watch each animal. The population unit in this case is one animal watched for one observation period. Two-stage sampling is employed with nonrandom selection in the first stage (selection of animals), and simple random selection at the second stage (selection of observation times). We can represent the population with a two-dimensional array in which each row represents one animal (or the set of times at which it might be watched) and columns represent times (Fig. 4.5). In this case, we may, on the basis of additional assumptions about the process that produced the first n animals encountered, view the animals as a simple random sample, even though they were actually selected nonrandomly. In practice, observation times would be restricted to a certain portion of the day. Statistical inferences extend only to those times, and the columns are thus labelled 'observation periods' not just 'time'.

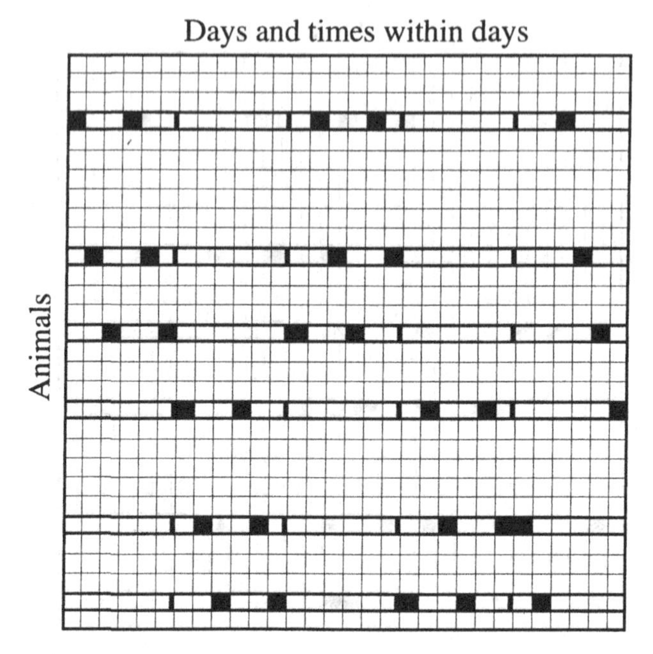

Fig. 4.6. A plan similar to that in Fig. 4.5, but with modifications to accommodate practical constraints. Rows are divided into secondary units of six population units each and three animals are watched each day. Observation periods are divided into five groups of six each. Three animals are watched in periods 1, 2, 3, and 4 with two systematically selected intervals per period. In the last period, each animal is watched once, intervals again being selected systematically.

The concept of multistage sampling can be generalized in three ways, and it is important for the reader to keep these extensions of the basic concept distinct. First, more than two stages can be employed. For example (Fig. 4.6), suppose in the example involving animals that we selected days to watch each animal but also sample within days. The observation period on each day might be 3 h, and we might actually watch animals for two 30-min periods during these 3 h. This would allow us to gather data on up to three animals per day. The population unit in this case is an animal watched for 30 min. We now have three-stage sampling: selection of animals, selection of days (i.e., groups of 30-min intervals), and selection of the population units, 30-min periods. In this example, the days are second-stage sampling units or just secondary units. Selection is nonrandom at the first stage and systematic at the second and final stages. Note that in this example selection of days and time periods within days is not independent. Three animals are always watched on the same days, and when one of these animals is being

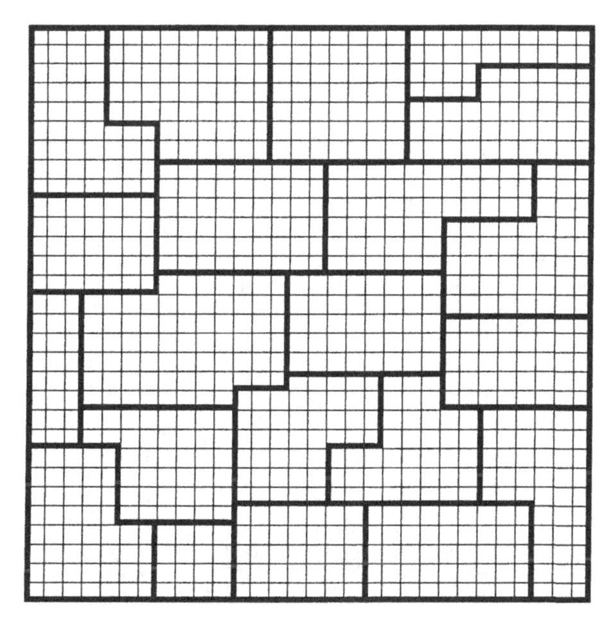

Fig. 4.7. Division of a population into unequal-sized primary units.

watched the other two are not being watched. The issue of independent *versus* nonindependent selection is discussed in Section 4.5.

The second way to extend multistage sampling is by letting the sample size vary within selected groups. For example, we might employ two-stage sampling as illustrated in Figs. 4.3–4.5 but select different numbers of population units within primary units. Unequal samples might be selected within groups at any stage. Figure 4.6 provides an example: we selected two population units on days one to four but only one population unit per animal on day five.

The third way to extend multistage sampling is by letting the group size vary. For example, we might delineate unequal-sized groups as shown in Fig. 4.7 and then select five of these groups using simple random sampling. With three-stage sampling, the second-stage units might vary in size. Letting group size vary is often useful when we wish plots to follow natural boundaries.

Sometimes the investigator has a choice of how to define the population units and sampling plan. Suppose that we select groups of ten small plots, but then count the number of individuals in all ten plots in each selected primary unit. Figure 4.8 portrays this plan with systematic selection of the groups. If we define population units as the small squares in Fig. 4.8 we could describe the plan as involving 100% sampling of the

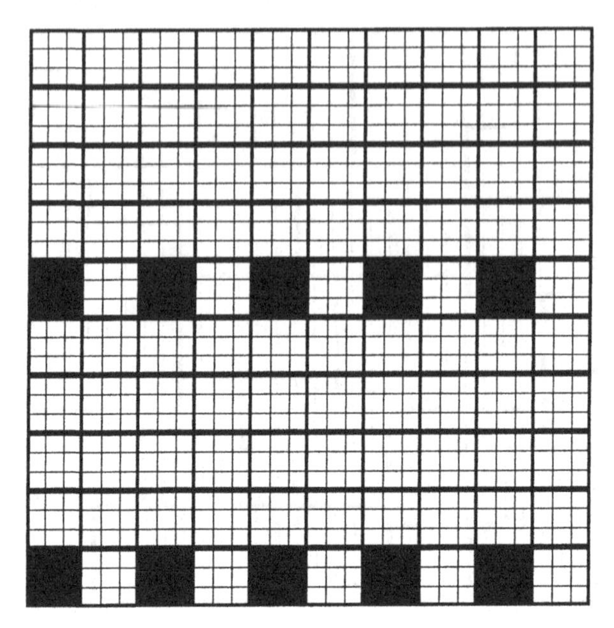

Fig. 4.8. An illustration of 100% sampling within primary units (which were selected systematically).

primary units (i.e., measuring all populations units in each primary unit). This plan is referred to as cluster sampling. Alternatively, we could define the groups of ten small plots as the population units and then describe the plan as one-stage systematic sampling. In either case, results could be expressed on a different scale, for example density per hectare, at later stages in the analysis. Statistical results such as significance levels in tests or whether a confidence interval covers a particular density value are unaffected by how the population unit and sampling plan are defined in this case.

To summarize the discussion so far, one of the fundamental ways in which sampling plans differ is by having different numbers of stages. A stage means selection of groups of population units. At each stage, the groups may be of equal or unequal size, and the sample sizes selected within groups may be equal or unequal. In discussing real examples, it is easy to confuse unequal group size with unequal sample sizes per group. Figures 4.1–4.8 provide several examples illustrating this distinction.

We now turn to another fundamentally different approach in sampling, stratification. Stratification, like multistage sampling, involves subdivision of the population units into groups. In multistage sampling, we select a sample of the groups and confine subsequent selection to units in this

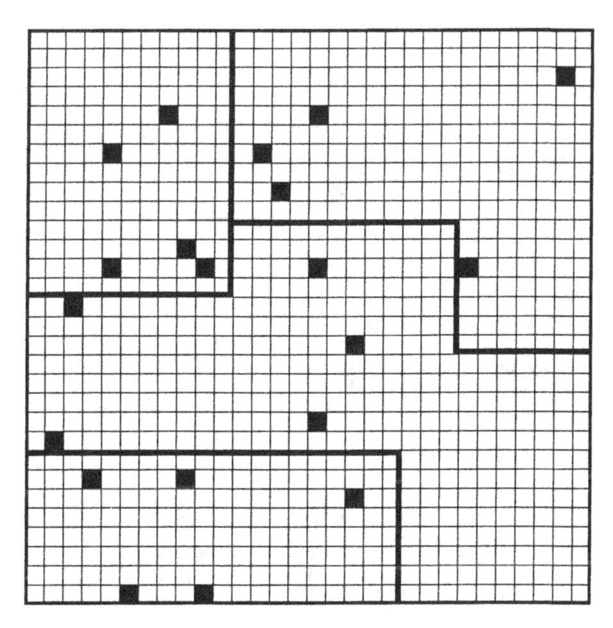

Fig. 4.9. An example of stratified, simple random sampling.

sample. In stratification, in contrast, we subsequently select a sample from every group. The groups in stratification are referred to as strata. Figure 4.9 portrays an example in which we first delineate four strata, and then select five population units within each stratum using simple random selection. Note that this is one-stage sampling because population units are only randomly selected once within strata. This plan is often called stratified, simple random sampling, or more commonly just stratified random sampling. Other sample selection plans can be used. For example, we might use systematic selection within strata in which case the plan would be referred to as stratified, systematic sampling. Different sample selection plans can be used in different strata. For example, simple random selection might be used in two strata and systematic selection in the other two strata. This seems to occur only rarely in behavioral ecology but such plans are permissible (i.e., obtaining unbiased point and interval estimates is possible with such plans).

Stratification can be carried out prior to any stage in sampling. The examples above involve one-stage sampling preceded by stratification. Figure 4.10 portrays an example of two-stage sampling with stratification prior to the second stage. In this example, two strata are delineated in each primary unit and two population units are then selected independently in

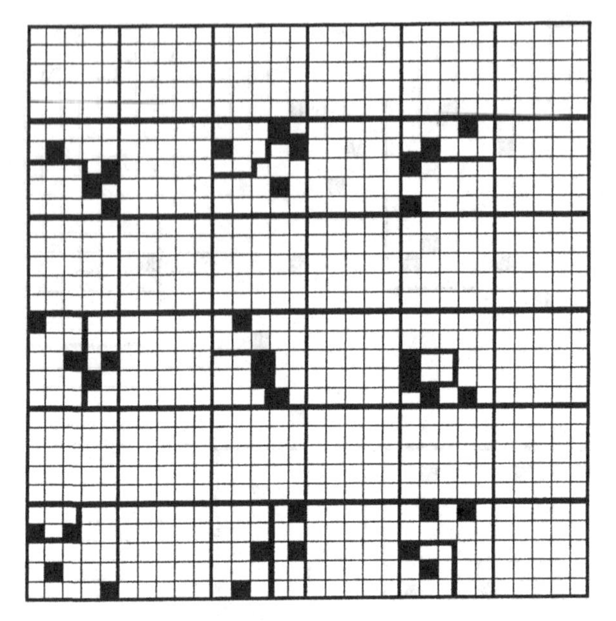

Fig. 4.10. An example of two-stage sampling with systematic selection of primary units, stratification of primary units into two strata, and simple random selection of two population units per stratum.

different strata with simple random sampling. Plans of this sort are sometimes useful when density varies considerably between habitats, and habitats can be mapped within groups but have not been mapped across the entire study area.

The final stage of any sampling plan is selection of the individual population units. If a sampling plan is described solely by the sample selection method, then one may assume that there was only one stage and that stratification was not employed. Thus, the phrases simple random sampling, or systematic sampling, with no additional description, include the unstated qualifiers 'one-stage' and 'without stratification'.

These examples are intended to define and illustrate the sample selection methods, multistage sampling, and stratification and to indicate the multiplicity of sampling plans that can be devised using these techniques. While complex plans can be designed, real sampling plans in behavioral ecology usually are relatively simple. Furthermore, under certain, commonly encountered conditions (explained in Section 4.5) second and subsequent stages can be ignored so one does not have to wrestle with defining stages and the exact plan used at each stage. However, this is not always true; sometimes one does have to consider each stage carefully. Furthermore, the concepts illustrated above must be understood in many real situations, particularly ones involving nonrandom sample selection. We therefore urge readers to study the examples given here until the basic terms are clear.

4.3 The finite population correction

In most introductory textbooks the formula for the standard error of the mean from a simple random sample is given as

$$se(\bar{y}) = \frac{s}{\sqrt{n}}, \tag{4.1}$$

where s is the sample standard deviation and n is the size of the sample. Technically, this formula only applies to infinite populations. In survey sampling, the populations are often finite and sometimes are small enough that the sample includes a substantial fraction of the population. When this is true, an adjustment to Eq. 4.1, known as the finite population correction, is needed. We first explain why the adjustment is needed, then provide the formula for the adjustment in various sampling plans, and then discuss the use of this adjustment in behavioral ecology. We will conclude that the correction is normally not needed, thus this whole issue can usually be ignored by behavioral ecologists. Occasionally, however, use of the correction is advantageous. Furthermore, the adjustment clarifies certain relationships a few of which are noted at the end of the Section.

To see why a correction is needed in Eq. 4.1 when the sample includes a substantial fraction of the population, consider the extreme case of including the entire population in the sample. In this case, repeated sampling would generate exactly the same estimate, and the standard error of the estimate should thus be 0. Expression 4.1, however, would not be 0. Although this is the most extreme case, the same point could be made by imagining that all but one member of the population was included in the sample. In this case estimates would vary slightly from sample to sample according to which member of the population was excluded. But Eq. 4.1 would take no account of the fact that the sample included nearly the whole population, and would again overestimate the true standard error. Readers able to prepare simple computer programs can easily verify that the same principle holds even when smaller fractions (such as 0.5 or 0.25) of the population are included in the population: Eq. 4.1 always overestimates the actual standard error.

The finite population correction is $1 - n/N$ for $v(\bar{y})$ and the square root of $1 - n/N$ for $se(\bar{y})$. Thus, with finite populations, we use

$$se(\bar{y}) = \left(\sqrt{1 - \frac{n}{N}}\right) \frac{s}{\sqrt{n}}. \tag{4.2}$$

The correction is often abbreviated to fpc for finite population correction, and $1-n/N$ is often written $1-f$, meaning one minus the fraction of the population included in the sample. If a small fraction of the population is included in the sample, then including the fpc makes little difference. For example, if the fraction is 5%, then $1-f=0.95$, so $v(\bar{y})$ is only reduced 5% by including the fpc and $se(\bar{y})$ is only reduced 2.5% $(1-\sqrt{0.95})$. Since including the fpc reduces the standard error, ignoring it is conservative. In practice, the fpc is usually ignored if the fraction sampled is less than 5–10%.

In behavioral ecology the population is usually much larger than the sample, so including the fpc makes little difference. Occasionally, the sample does include a large fraction of the population as in studying some endangered species or other small populations. Another case in which including the fpc might seem appropriate arises when all the individuals in a study area have been measured. In such cases, however, investigators nearly always wish to extrapolate their findings to the other animals that might have been there, or other outcomes that might have occurred; for example, by invoking the superpopulation concept (Section 1.2). The fpc is thus not appropriate.

Deciding whether to include the fpc is sometimes difficult. For example, Cochran (1977 pp. 72–6) discusses a hypothetical anthropologist studying the frequency of blood types in a small population. The sample includes a substantial fraction of the population, and Cochran suggests including the fpc since, if we are only interested in estimating the frequency of blood types in this small population, the fpc reduces our estimate of the variability by the proper degree. It might be argued, however, that the anthropologist is probably interested in the underlying gene frequencies, of which the current members in the population are but one of many possible manifestations, hence the actual population of interest is quite large, and the fpc should not be included.

The fpc may be used to clarify the distinction between multistage and stratified sampling. They were treated, above, as completely separate plans, but suppose we delineate ten groups and then sample within eight of them. Technically, this is not stratified sampling which would require selecting population units from all ten groups (in which case they would be called strata). Yet it does not seem too different from stratified sampling. The fpc provides the link between the two plans. When the fpc is included in the formulas for multistage sampling, but samples are taken within all groups, then the formula reduces to the formula for stratified sampling (Appendix One, Box three, expression A1.14).

4.4 Sample selection methods

In this Section we describe the basic sample selection methods used most widely by behavioral ecologists. Each of the methods may be used to select population units, primary sampling units, secondary units, or units at any stage in a multistage design. We emphasize their use in one–stage designs without stratification. Additional comments on their use in multistage and stratified designs are provided in Section. 4.5 and 4.6.

Simple random sampling

The phrase simple random sampling has been used frequently in previous Sections but we have not defined it carefully. Here we provide a formal definition. A simple random sample of size n, selected without replacement, is one in which all possible samples of size n from the population are equally likely to be selected. Alternatively, we could define it as a sample in which the population units are selected one at a time, and all units not in the sample have the same probability of being selected on each draw. Notice that this probability changes between successive draws as the number of remaining units shrinks. Thus, in drawing a simple random sample without replacement from a population of size 100, each unit has a selection probability of 1/100 on the first draw. The 99 units still available for selection after the first draw each have a selection probability of 1/99 on the second draw, and so on.

An important feature of simple random samples drawn without replacement is that the expected value of any statistic, g say, calculated from the sample, such as the sample mean or variance, can be written simply as

$$E(g) = \frac{1}{N^*}\sum_{i=1}^{N^*} g_i, \tag{4.3}$$

where N^* is the number of distinct (equally likely) samples that might be drawn and g_i is the statistic obtained from the ith sample. This fact is used in later Sections of this Chapter. Incidentally, the same statements may be made about sampling with replacement but the set of possible samples (N^*) is larger since each unit can enter the sample more than one time.

Behavioral ecologists sometimes mistakenly assume that sampling plans under which all units have the same selection probabilities automatically produce simple random samples. As an example in which this is not true, suppose that individual animals – males or females – are the population units, but we randomly select pairs of animals for the sample. Each individual does have the same chance of being selected for the sample. Under this plan, however, all possible combinations of units are not equally likely. Many combinations, in fact, are not possible at all. For example no sample including the male from a given pair, but not including the female, is possible. This plan is actually two-stage, random sampling with selection of groups of size two at the first stage, and then selection of all (i.e., both) units in every selected group. The distinction might initially seem unimportant to the reader, but as shown in Section 4.5 this plan has quite different statistical properties from true simple random sampling. In particular, the sample size is best viewed as the number of pairs in this plan, not the number of animals, as would be the case with true simple random selection of animals. Why this so will be discussed in more detail in Section 4.5.

One way to select a simple random sample is by numbering all the units in the population from 1 to N, and then selecting n different numbers from a table of random numbers, or generating the n numbers on a computer using a random number generator. The population units corresponding to the selected numbers comprise the sample. This is equivalent to putting N tickets numbered 1 to N in a hat and 'randomly' removing n of them.

With populations that can be represented in two dimensions, a method similar in principle but easier to apply in practice is to produce a picture of the population overlain by a grid such that each cell of the grid covers one population unit. For example, if the population units were one-hectare plots and we had a map of the study area, then the grid would be scaled so that each cell covered one hectare on the map. To select units for the sample, we would use a table of random numbers to independently choose values along the X-axis, and along the Y-axis. The cell in the grid identified by each X, Y combination would be selected for the sample. Since the sampling is to be done without replacement, if a cell has already been selected, then a new X, Y combination is chosen. If the study area was irregularly shaped, X, Y points could arise which did not correspond to any cell in the study region. If this occurs, these X, Y points would be ignored and we would continue drawing pairs of random numbers until the required sample size had been obtained. In reality, a full grid is not needed, one just has to work out the proper intervals on the X and *Y-axes* so that a unit square covers an area on the ground equal to one hectare.

This procedure is most useful for sampling spatial populations, but in principle it can be used for other populations. For example, one could select times at which animals would be watched by listing animals down the side of a page and potential observation times across the top of the page. This same process could then be followed to select a simple random sample of animal-times. In practice, however, this method may be difficult to implement as our sample of times may require us to sample more animals at a given specific time than we have resources for. More 'balanced' methods of sample selection such as systematic sampling are often used instead.

One-stage, simple random sampling is seldom employed in behavioral ecology for at least two reasons. When the population units are animals, we seldom know the true population size. Even if we did, numbering all the animals so that we could randomly select the ones for our sample is almost never practical. Instead, typically some form of nonrandom selection is used; for example, taking the first n individuals that we encounter. On the other hand, although formal simple random sampling could often be applied to select plots or periods of time, investigators nearly always prefer to use systematic selection of units so that the sample is more evenly distributed across the entire population. This practice is discussed in more detail later (see 'Systematic sampling', p. 102).

Even though simple random sampling is rarely used to select the units in a one-stage sampling plan, when computing standard errors of the estimates practitioners often use the formula appropriate for simple random sampling. The consequences of using the formula appropriate for simple random sampling even when such sampling is not used are discussed in greater detail later in this Chapter.

The estimated mean, with simple random sampling, is just the sample mean. The standard error with simple random sampling is discussed in Section 2.6. The equations for the point and interval estimates are given in Box 2, Appendix One.

Nonrandom sampling

As already noted, selecting units with a well-defined, random sampling plan is often not practical. As a result, nonrandom sampling is widely used in behavioral ecology, especially to select animals. Although nonrandom sampling is often unavoidable, interpretation of the estimates and their standard errors is often difficult. With random selection we obtain an unbiased estimate of the population mean and a nearly unbiased estimate of its standard error. When nonrandom sampling is used, we lose both of these properties. The sampling method may tend to capture animals that are young or nonterritorial, for example, and these individuals may differ from older or territorial individuals with respect to the variable being measured.

If all animals in one area are studied, they may be more like one another than those in a random sample from the region, due to effects of habitat. In this situation, standard errors will be underestimated. Instances in which the standard error will be overestimated can also be imagined. Nonrandom samples should thus be avoided whenever possible. When this is not possible investigators should recognize the limitations of their statistical inferences.

Semantic difficulties occasionally arise in deciding whether a sample selection method should be called random or nonrandom. For example, suppose the population units are plants of a particular species. We select the sample by first selecting a set of points on a map using simple random selection. We then select the plant nearest each point for inclusion in the sample. The question at this point is whether we have selected a simple random sample of plants of the particular species. The plan includes simple random sampling and involves a set of well-defined rules for determining which population units to include in the sample. However, if the plants occur in groups, then plants at the edge of a group have more chance of being included in the sample than plants in the interior of the group. We therefore cannot legitimately claim to have selected a simple random sample of plants. Furthermore, notice that it would be extremely difficult to calculate the selection probabilities for the plants in our sample. In this book, we include under the category nonrandom selection any method in which selection probabilities vary and are unknown.

Systematic sampling

Systematic selection of units is widely used in behavioral ecology. As the name implies, the units in the sample are distributed 'systematically' or evenly across the population rather than randomly. This fact has several consequences, some useful, others troublesome. We first discuss ways of selecting a systematic sample and then describe some of the special statistical issues that arise in using this method. Guidelines are then provided for deciding when and how to use systematic sampling. We use a spatially defined population to illustrate many of the points in this Section but they generalize easily to temporal populations. Consider a small population of just 36 units (Fig. 4.11) and imagine that we want to select nine of these for the sample. In the example, we imagine the units to be population units, but in practice they might also be primary (or other) units in a multistage design. The most common approach used by behavioral ecologists to select the sample is to delineate groups as indicated by the heavy lines. In this example, since we want nine units in the sample and population size is 36,

Fig. 4.11. Artificial population used to illustrate simple random and systematic sampling.

each group has four units. One of the population units is then selected randomly. The sample consists of all the units in the same position as the randomly selected unit. In the figure, these units are indicated by shading. Various methods can be used to select the position. As described above, a single number may be selected from 1 to 36 and then all units in the same position comprise the sample. Instead, positions within a group could be numbered 1 to 4 and one of these values selected with all nine units that occupy this position comprising the sample. Finally, a particular group could be selected and then one of the units in it randomly selected with all units in the same position comprising the sample. In practice, the last method is probably most common, especially with large, spatially defined populations. Investigators can determine group size, randomly select a starting point in the first group, and then follow simple rules to find the rest of the units such as 'travel north placing plots every 30 m until reaching the border of the study area; then go 50 m west and place a plot at that location; then work back to the south placing plots every 30 m' and so on. More complex methods for choosing a systematic sample can be imagined (e.g., employing a 'knight's move' between units) but are seldom used in behavioral ecology.

Biologists sometimes wonder whether the interval between units should be measured exactly. In most situations there is little need for this. It is important, however, that the value of the variable to be measured does not affect the placement of plots. For example, in counting the number of plants in small plots, slight differences in where the plot is placed may have a large effect on the count obtained. Consequently, it is important that we avoid any tendency to place plots in high- or low-density areas. Using sophisticated surveying instruments to measure distances and bearings exactly does result in unbiased selection of plot locations, but is likely to be difficult and time consuming especially in rough terrain. A more practical method is to pace between units and take the last few steps without looking at the ground. The plot is then located exactly at the tip of one's toe or some constant distance from it. If terrain is so rough that even this approach might result in some bias in plot placement, then one might pace to within some short distance – for example 5 m – of each plot and measure the remaining distance.

The point estimate from systematic samples is unbiased as long as the starting point is selected randomly (when this is not done, technically there is no random sampling at all). This feature is relatively easy to see by examining Fig. 4.11. Four samples are possible and between them they include all the population units. The mean of the four sample means is thus exactly equal to the population mean so the estimate is unbiased.

This intuitive reasoning can be made more exact for readers who have been following the use of expected values. Let the population be of size N and be divided into n groups each of size M (i.e., $Mn = N$). There are thus M possible means, $\bar{y}_k$, say, $k = 1,\ldots,M$. The Mn population units may be designated y_{ki} where k indicates the group and i the number within the group. We may thus demonstrate that $E(\bar{y}) = \bar{Y}$ by writing

$$E(\bar{y}) = \frac{1}{M}\sum_{k=1}^{M}\bar{y}_k = \frac{1}{M}\sum_{k=1}^{M}\left(\frac{1}{n}\sum_{i=1}^{n}y_{ki}\right) = \frac{1}{Mn}\sum_{k=1}^{M}\sum_{i=1}^{n}y_{ki} = \bar{Y}. \tag{4.4}$$

The precision and estimated precision of systematic samples are more problematic. From the first Sections of this Chapter, it can be seen that systematic sampling is actually multistage sampling with selection of just one primary unit and then 100% sampling of the selected population units. Thus, in the example given in Fig. 4.11, one of the primary units consists of all the shaded plots. There are three other primary units: the nine plots directly below the shaded ones, the nine plots directly to the left of the shaded ones, and the nine plots to the lower left of the shaded ones. We might not initially recognize this as multistage sampling because we typically think of primary sampling units as groups of adjacent units. Furthermore, we typically envisage sampling within the primary units rather than selecting all population units. Nonetheless, on reflection the reader should be able to see that systematic sampling really is multistage sampling with $n = 1$ primary unit selected. And this fact should immediately make us wonder, 'how are we going to estimate standard errors with $n = 1$?'

The answer is that no general unbiased estimator is known for the variance of the mean from a systematic sample. In practice, investigators almost always use the formula for a simple random sample as an approximation. While the approximation is often quite good, cases also arise in practice when this approach leads to substantial errors. Understanding when systematic sampling does and does not work well requires that we make a clear distinction between actual and estimated precision. We consider these separately in the following text.

With respect to actual precision, systematic samples are usually more precise than simple random samples of the same size. The reason for this is that variables studied by behavioral ecologists tend to exhibit trends across space (or time in sampling from temporal populations) so that adjacent units tend to have similar values. A systematic sample covers the population evenly whereas a random sample has a more patchy distribution. Units are usually concentrated in some areas and sparse in others. We may even draw a sample in which no units are selected from large areas that have particularly high or low values of the variable. These problems are avoided with systematic samples.

Systematic samples, however, are not necessarily more precise than simple random samples. If the value of the variable varies in a cyclic or periodic way, and the period corresponds to the interval used in selecting the systematic sample, then the systematic sample may be much less precise than a random sample. For example, in Fig. 4.11, if the variables in the upper two cells in each box of four cells happened to be large and the two values in the lower part of each box happened to be small, then this particular systematic selection plan would not work well at all. Two of the possible samples would have only large values and two would have only small values. Whichever sample we happened to draw, it would not provide a very good estimate of the population mean. Thus actual standard errors, that is the sample-to-sample variation, would be high. With a random sample we would be likely to obtain a much more representative set of population units. Although it is important to understand the potential for this sort of problem, we have seldom seen it occur in practice. In the great majority of cases, systematic sampling in behavioral ecology yields more precise estimates than simple random sampling.

We now turn to the issue of estimating precision with a systematic sample. As already noted, no unbiased method is known for estimating the variance of the mean of a systematic sample. It can be shown, however, that precision is generally overestimated when the formula for simple random sampling is used. Thus, confidence intervals will tend to be too long. More specifically, it can be shown that the bias in estimating $V(\bar{y})$ is positive (estimates tend to be too high) when the systematic sample is actually more

1	1	4	3	7	7
1	2	4	3	8	8
1	3	6	5	7	7
3	3	5	4	9	8
4	3	7	6	9	10
3	4	7	6	8	8

Fig. 4.12. Artificial population used to compare simple random, systematic, and stratified sampling.

precise than the simple random sample, and negative when the reverse is true.

To illustrate these points, we assigned numbers to the 36 units in Fig. 4.11 and sampled from them using simple random and systematic sampling. Values were assigned to the population units by numbering the large squares sequentially from 1 to 9. Values for individual units were then obtained by randomly adding 1 to these values, not changing them, or subtracting 1 from these values. Thus, the numbers in the first square were 0, 1, or 2; those in the second square were 1, 2, or 3; and so on (Fig. 4.12). This produced a population of numbers with a clear trend but also with local variation. We wrote a short computer program to sample from the populations using simple random and systematic selection. The actual variance and standard error of the simple random sample were calculated with expressions 2.13 and 2.18; the actual values for systematic sampling were calculated from the four possible sample means using this plan. The average estimates were calculated by obtaining 10,000 samples for the simple random method and by calculating the average of the four possible estimates obtainable using systematic sampling.

The means of the four possible systematic samples were quite similar: 5.11, 5.00, 5.33, and 5.11. The means from simple random sampling were much more variable. For example, the first ten values were 4.22, 4.22, 4.89, 4.44, 6.00, 4.00, 5.56, 5.22, 4.89, and 6.33. The results (Table 4.1) illustrate the points made above. For simple random sampling, the estimated variance was unbiased to three decimal places whereas the estimated standard error had slight negative bias as expected (see Section 2.7). Systematic sampling was much more precise than simple random sampling but the esti-

Table 4.1. *Actual and estimated (Ave. est.) precision of simple random and systematic sampling ($n=9$) from the population depicted in Fig. 4.12*

	Variance of $\bar{y}$		Standard error of $\bar{y}$	
Sampling plan	Actual	Ave. est.	Actual	Ave. est.
Simple random	0.558	0.558	0.747	0.738
Systematic	0.015	0.812	0.121	0.899

mators of variance and standard error, based on the formula for simple random sampling, both had substantial positive bias. The effect of this bias is that test results might turn out to be nonsignificant that should have been significant had an unbiased estimate of variance been available.

Although the difference in precision between simple random and systematic sampling, and the bias in systematic sample estimates were considerable, we emphasize that this was purely a result of the way we constructed the population and sampling plan. We encourage readers familiar with computer programming to carry out similar investigations with other hypothetical populations. Doing so will give you a better understanding of when systematic samples perform well and poorly. You might also try to develop an unbiased variance estimate for systematic sampling (even though you probably will not succeed). Some authors discuss taking several systematic samples at random and using these to estimate variance. This approach is occasionally feasible but tends to have low precision because sample size is the number of systematic samples. (see Section 4.5 Multistage sampling).

Selection with unequal probabilities

In this method population units have different probabilities of being selected for the sample rather than equal probabilities, as in simple random or systematic sampling. As an example, suppose we wish to estimate density or total population size of something that lives in woodlots. The population unit is viewed as a unit of area such as a hectare. Woodlots are the groups of population units. Travel time between woodlots is significant, so we want to include more large woodlots in the sample, thereby increasing the total number of individuals detected. We accomplish this by selecting points at random on a topographical map of the study area which indicates woodlots. Points falling outside a woodlot are ignored. We continue selecting points until we have the desired number of woodlots. This plan is not simple random sampling, because larger woodlots have a greater chance of being included in the sample.

Sampling without replacement using unequal selection probabilities raises fairly difficult statistical issues. The basic problem is that selection probabilities change in a way dependent on which units are already in the sample. As a trivial example, suppose the group sizes in the population are 10, 20, 30, and 40 and that selection probabilities are proportional to the sum of the group sizes still in the population. Thus, on the first draw, the probability of selecting the first group is 10/100, the probability of selecting the second group is 20/100 and so on. After we have drawn one group, the selection probabilities will be different depending on which group was selected on the first draw. For example, the probability of obtaining group one on the second draw would be 10/80 if the first draw yielded group two but 10/60 if the first draw yielded group four. With simple random sampling, selection probabilities on any given draw depend only on how many groups have been removed from the population, not on which particular groups have been selected.

Notice that the complexity described above does not occur if we sample with replacement. In our example, the probability of drawing unit one is always 10/100 regardless of which units have already been selected. As a result of this difference, sampling is sometimes carried out with replacement when selection probabilities vary. Alternatively, investigators may assume their population is so much larger than their sample that they can treat the sample as having been selected with replacement even if in fact it was not. With sufficiently large populations this practice causes little error. Formulas for means and standard errors with unequal sampling with replacement are provided in Appendix One (Box 3, Part D). As noted there, the selection probabilities may be replaced by numbers proportional to these probabilities. Thus, in the woodlot example, we may use woodlot size in place of the selection probabilities. This is helpful because we do not then need to measure the size of every group in the population. For a description of sampling without replacement, we recommend Chapter 9A of Cochran (1977) or Chapter 6 of Thompson (1992). The Horvitz–Thompson estimator is probably the most widely used method for sampling without replacement using unequal selection probabilities.

Sampling with unequal probabilities has not been widely used in behavioral ecology, though cases like the woodlot example above occur occasionally. One reason may be that when the groups do vary substantially in size, we are often interested in comparing different size categories. We may therefore prefer to stratify the population, putting units of different size into different strata. We can then compare means from different strata as well as obtain an overall estimate.

Line transect methods sometimes employ unequal probability sampling. For example, suppose we randomly select transects and then walk along each one until we encounter an object such as a bush, tree or pond. We then measure some attribute of the object encountered. If the objects are of variable size, and if we include each object traversed by our transect line, then we have unequal probability sampling. The selection probabilities depend on the maximum width of the object perpendicular to the transect. These widths are therefore proportional to the selection probabilities (and may be used as the z_i, i.e., probability of selecting the ith primary unit, in Appendix One, Box 3, Part D).

Finally, we note that sampling with unequal probabilities sometimes occurs but is unnoticed by investigators. Some biologists might treat the examples above as simple random sampling. However, any time that selection probabilities vary and can be calculated, then the formulas for unequal probability sampling should be used. Failure to do so may lead to seriously biased estimates, a point that can be verified easily by means of a simple computer simulation.

4.5 Multistage sampling

This method is described and illustrated in Section 4.2. Briefly, in multistage sampling the population units are first divided into groups called primary units, and then several primary units are selected using one of the basic selection methods described in Section 4.4. Subsequent sampling is confined to the selected primary units. If population units are selected from these groups then we have two-stage sampling. The 'stages' are: (1) selection of primary units; and (2) selection of population units within the primary units. Alternatively, groups of population units may be selected within primary units. These are then defined as secondary units and subsequent sampling is confined to them. In this case we have three or more stages in the design.

Multistage sampling is widely used in habitat studies, behavior studies, wildlife surveys, and other branches of behavioral ecology. For example, many surveys of animals involve repeated coverage of several randomly selected survey routes. In such studies the population unit is a route-time, and the primary unit is a survey-route, or more specifically the group of times at which the route might be run. In behavior studies with individually recognizable individuals, investigators often select several animals and then make repeated observations on each. The population unit in these cases is an animal-time, and each primary unit consists of the possible times at which the animal might be observed. Most applications of multistage sampling in behavioral ecology involve either (1) selecting 'plots within plots' as illustrated at the beginning of this Chapter, (2) selecting animals or plants and then recording measurements on each at several times, or (3) selecting locations and then recording measurements on each at several times.

Sample selection methods

Primary units are generally selected with nonrandom methods to choose animals or systematic methods to choose plots or periods in time. Simple random sampling is seldom used for the reasons given in discussing one-stage sampling (Section 4.2). The primary units are nearly always treated as though they had been selected with simple random sampling. All of the cautions regarding this practice in one-stage sampling (Section 4.2) apply to multistage sampling as well. Selection of primary units with unequal probabilities is uncommon.

The later stages of a multistage design do sometimes involve simple random sampling. For example, subplots may be selected using simple random sampling within plots or if groups of animals are selected as primary units, then simple random selection of individuals within groups may be made at the second stage of the plan. Systematic selection, however, is much more common than simple random selection, and nonrandom selection at the second and later stages is widespread.

The standard formulas developed for multistage sampling assume that selection of secondary units is independent in different primary units. Independence implies that selection of secondary units in one primary unit has no effect on which secondary units in any other primary unit are selected. Lack of independence does not affect the expectation of standard (linear) point estimates of means and totals. The usual estimates of standard errors, however, are biased when sampling is not independent in different primary units.

The assumption of independent sampling within primary units is often violated in behavioral ecology due to practical constraints. For example, suppose that we want to watch several marked animals from a blind 2 days a week, and we record the number of an activity such as fights or displays. We select Mondays and Thursdays and watch each animal on each day. The animals would probably be viewed as primary units and the days as secondary units. Secondary units, however, are not selected independently in different primary units because all animals are watched on the same set of days. To further illustrate the problem we consider a similar example with actual numbers. To keep the example simple, suppose we watch five animals and the study will last just 4 days. The actual number of fights or displays by each animal on each day is shown below:

	Day			
Animal	1	2	3	4
1	1	3	4	2
2	2	5	3	1
3	1	2	4	0
4	0	4	5	1
5	1	3	3	1

In this example, the counts for all animals are low on days 1 and 4 and higher on days 2 and 3. Now suppose we only take observations on 2 days for each animal. If selection of days was independent within animals, then the sample would probably include both high and low values. On the other hand, if we select 2 days at random and observe all five animals on these same 2 days, then there is a one-third chance that we would select days 1 and 4 (and thus have all low counts) or days 2 and 3 (and thus have all high counts). As already noted, the lack of independence does not cause any bias in the estimate of the mean (i.e., the mean of all possible estimates exactly equals the true mean). Precision and estimated precision, however, are affected. In this example, the plan is less precise than a plan with independent selection of days for each animal. Furthermore, use of a formula that assumes independent sampling would produce biased estimates of the standard error. The direction of the bias depends on whether units in the sample are more similar, or less similar, to each other than would be true with independent sample selection. When the units are more similar to each other – the most common case in behavioral ecology – then the bias is negative. We tend to underestimate standard errors and conclude that we have more precision than we really do. This would be the case in our example.

Lack of independence, however, does not always cause this sort of problem. As a trivial example, suppose the animals were thumbtacks which were tossed once on each of the 4 days. The activity measured was a 1 if the tack landed on its side and a 0 otherwise. Results obviously would not be influenced by adopting a plan such as the one above in which data collection was restricted to certain days. The reason is that the measurements are unaffected by time; that is, we expect the same result regardless of what day we record the data.

Returning to the animal example, the critical issue is thus whether a 'day effect' exists, meaning that the expected outcome is different on different days. If it is, then independent sampling in different primary units is important. Biological considerations may help one decide how serious the day effects are likely to be. In our example, the effects might seem less important if the animals were far apart and thus unlikely to be affected by a given stimulus such as a predator or potential mate, or if they were watched sequentially and for long periods rather than simultaneously.

Correct estimation of standard errors accounting for dependent sampling in different primary units is more complex. The standard survey sampling formulas can be modified by the addition of covariance terms and further assumptions about the nature of the dependence. Another approach is to use multivariate methods which incorporate the correlation structure into the analysis (e.g., Crowder and Hand 1990). We recommend consulting a statistician when dependent sample selection in primary units cannot be avoided for practical reasons and is likely to have large effects on precision. For the rest of this Section, we assume that sampling is independent in different primary units.

Weighting of estimates from primary units

Multistage sampling is generally used to estimate the mean of the results from the primary units, and usually a 'simple' or 'unweighted' mean, $\bar{y}=\frac{1}{n}\sum y_i$, is calculated, where y_i is the estimate from the ith primary unit and n is the number of primary units. Unequal weighting means that we wish to use a different formula. Instead of $\bar{y}\sum\left(\frac{1}{n}\right)y_i$ we wish to use $\bar{y}=\Sigma w_i y_i$, where the w_i do not all equal $\frac{1}{n}$. This situation probably arises most often in behavioral ecology when population units are individuals (plants, animals, other objects) and the first-stage unit is a plot. In such cases, the number of individuals per plot often varies and simple means of the estimates from each primary unit will not, in general, yield an unbiased estimate of the mean of the population units. For example, suppose that we count the number of flowers per plant in each of four plots. The number of plants, flowers, and flowers/plant in each plot are as shown below:

Plot	No. of plants	No. of flowers	Flowers per plot
1	5	20	4.0
2	10	60	6.0
3	2	5	2.5
4	8	40	5.0
Totals	25	125	—

The mean number of flowers per plant is 125/25 = 5.0. Now suppose that we calculated the mean/plant in each plot and then the simple mean of these numbers. The result would be $(4.0+6.0+2.5+5.0)/4=4.4$. Thus, the simple mean per plant, and the simple mean of the means per plot are not equal. In general, with unequal sized groups, the simple mean of the primary unit means is a biased estimator of the mean per population unit. Unequal weighting of the means must be used to obtain an unbiased estimate of the mean per population unit. Procedures for cases like this one are discussed later in this Chapter (see p. 118).

When sample sizes within primary units vary it may be tempting to weight the means per primary unit by sample size (i.e., w_i = number of observations on primary unit i). For example, suppose we are using telemetry methods to study habitat use by some species. The population unit is an 'animal-time', and the variable, with respect to any given habitat, is 'yes' (in the habitat) or 'no'. The sampling plan is multistage with animals (or more precisely the set of possible times for each animal) as primary units. In defining the parameter, we would probably weight animals equally on the basis that each provided an equally useful opportunity to study habitat use. Suppose, however, that some animals are harder to detect than others. For example, we might get 60 locations on some individuals but only 10 on others. We get more precise estimates from the first set of animals so it is tempting to give them more weight in the analysis. Weighting estimates by sample size would accomplish this. Unfortunately, however, weighting by sample size tends to yield biased estimates. The dangers of this approach can perhaps best be seen with an example. For simplicity, suppose we have just two animals with actual and sample values as shown below:

Animal	Proportion of time spent in habitat 1	No. of attempted locations	No. of times found	Expected no. of times found in habitat 1	Expected proportion of times found in habitat 1
1	0.6	100	100	60	0.6
2	0.2	100	50	10	0.2

With just two animals the value of the parameter, assuming they are weighted equally, is $(0.6+0.2)/2=0.4$. The expected value of the estimate from each animal is equal to the true mean for that animal and thus the expected value of the estimate is unbiased. That is, using the rules for expectation in Section 2.4

$$E[(p_1+p_2)/2]=0.5[E(p_1)+E(p_2)] \tag{4.5}$$

$$=0.5[0.6+0.2]=0.4.$$

Suppose, however, we weight the estimates by sample size, that is the number of times each animals was detected. The estimate is [100(0.6) + (50)(0.2)]/150 = 0.47, rather than 0.40. Thus, the estimate is biased. The magnitude of the bias depends on several factors including the difference in actual values among primary units, the difference in sample size, and (especially) the correlation between the value of the parameter for the primary unit and sample size in the primary unit. In the example just given, the actual values varied substantially (threefold for true values, sixfold for sample size) and the correlation was high (1.0 actually), and even so the bias was relatively small. This is often the case with simulated, but plausible, data (which we encourage readers to verify with simulation programs). Nonetheless, using a weighted average, the estimate is biased for the simple mean of the values of the population units. We therefore generally recommend against weighting estimates by sample size unless there are convincing reasons for believing that the correlation between the true value of the primary unit and sample size is close to zero. A statistician may be able to provide advice on these issues in specific applications. Incidentally note that this case provides an example of statistical bias (Section 2.2) – bias results solely from the way the estimates are calculated.

Equally weighted primary units

In this Section, we first discuss the definition of the parameter to be estimated and the form of the point estimate. We then discuss the actual variance of the estimate and how it is affected by allocation of effort between primary and secondary units. The formulas for the estimated variance and standard error are then discussed. The formulas have different forms so the reader needs to be clear about which quantity – true variance or estimated variance – is being discussed.

We assume in this Section that the parameter is defined as the simple mean of the actual values from the primary units

$$\overline{Y} = \frac{1}{N}\sum_{i=1}^{N} Y_i, \tag{4.6}$$

where N is the number of primary units in the population and Y_i is the parameter from the ith primary unit. In most cases of practical interest to behavioral ecologists, Y_i is a mean, for example the mean per plot, or mean per animal. Proportions are also frequently of interest (e.g., the proportion of animals surviving, producing offspring, etc.) but as noted in Section 2.2, they are treated throughout this book as a mean. Cases do arise, however, in which the quantity of interest is more complex such as a diversity index or other measure of community structure. These cases can usually be analyzed using the methods of this Section as long as we are interested in the mean of these quantities over the primary units. The appropriate estimate of $\bar{Y}$, in this case, is simply the sample analogue

$$\bar{y} = \frac{1}{n}\sum_{i=1}^{n} y_i, \tag{4.7}$$

where n is the number of primary units and y_i is the unbiased estimate of Y_i from the ith primary unit in the sample.

The actual variance of $\bar{y}$ depends on the variation between and within the primary unit means and on the sampling plan. With two-stage sampling to estimate the population mean per unit ($\bar{Y}$) and primary units of equal size, M,

$$V(\bar{y}) = \frac{1-f_1}{n} S_1^2 + \frac{1-f_2}{nm} \overline{S_2^2}, \tag{4.8}$$

where n = the number of primary units in the sample, m = the number of secondary units measured in each primary unit, N = the number of primary units in the population, $f_1 = n/N$ and $f_2 = m/M$. S_1^2 is the variance (with $N-1$ weighting) among true values per primary unit,

$$S_1^2 = \frac{\sum_{i=1}^{N}(\bar{Y}_i - \bar{Y})^2}{N-1},$$

where $\bar{Y}_i$ = the true mean for primary unit i and $\overline{S_2^2}$ is the average variance within the primary units

$$\overline{S_2^2} = \frac{1}{N}\sum_{i=1}^{N} S_{2i}^2$$

with

$$S_{2i}^2 = \frac{\sum_{j=1}^{M}(y_{ij} - \bar{Y}_i)^2}{M-1},$$

where y_{ij} is the value for population unit j in primary unit i. As usual, the standard error of the estimate is the square root of $V(\bar{y})$.

As noted in Section 4.3, the first-stage finite population correction, f_1, is usually small. If, in addition, f_2 is small, then $V(\bar{y})$ is approximately

$$V(\bar{y}) = \frac{1}{n}\left(S_1^2 + \frac{\overline{S_2^2}}{m}\right). \tag{4.9}$$

It can be seen from Eqs. 4.8 and 4.9 that increasing either n or m decreases $V(\bar{y})$, but that the law of diminishing returns affects m. Eventually, as m

increases, a point is reached at which $V(\bar{y})$ is essentially S_1^2/n, and further increase in m has little affect on $V(\bar{y})$.

We now turn to formulas for estimating $V(\bar{y})$. Notice that the parameter, the true variance, is a nonlinear transformation of the population units. It should thus not be completely surprising that the sample analogue of Eq. 4.8 above would be a biased estimate of $V(\bar{y})$. A different formula is thus needed. The usual estimator can be written in several ways. The most common form is

$$v(\bar{y}) = \frac{1-f_1}{n} s_1^2 + \frac{f_1(1-f_2)}{nm} \overline{s_2^2} \tag{4.10}$$

where

$$s_1^2 = \frac{\sum_1^n (\bar{y}_i - \bar{y})^2}{n-1},$$

$\bar{y}_i$ = the mean of the sample from primary unit i, $\bar{y} = \Sigma\, y_i/n$, and

$$\overline{s_2^2} = \frac{1}{n}\sum_{i=1}^{n} s_{2i}^2 = \frac{1}{n}\sum_{i=1}^{n}\left(\frac{\sum_{j=1}^{m}(y_{ij} - \bar{y}_i)^2}{m-1}\right)$$

An alternative form, obtained from 4.10 by substituting n/N for f_1, is

$$v(\bar{y}) = \frac{s_1^2}{n} - \frac{1}{N}\left(s_1^2 - \frac{(1-f_2)\overline{s_2^2}}{m}\right) \tag{4.11}$$

As noted many times previously, in behavioral ecology N is usually large so the right-hand term drops out and we obtain the very simple formula

$$v(\bar{y}) = \frac{s_1^2}{n} \tag{4.12}$$

Furthermore, it can be shown (e.g., Cochran 1978 p. 279) that use of Eq. 4.12 tends to overestimate the actual variance, the bias being S_1^2/N. Thus, use of Eq. 4.12 is conservative. Virtually all studies in behavioral ecology that use multistage sampling employ Eq. 4.12 to obtain the standard errors, i.e,

$$se(\bar{y}) = \frac{s_1}{\sqrt{n}}. \tag{4.13}$$

If N is not large relative to $s_1^2 - (1-f_2)\,\overline{s_2^2}/\mathrm{m}$ then use of the complete expression 4.10 may be preferable.

The fact that $V(\bar{y})$ can usually be estimated solely from the estimates obtained from each primary unit has several consequences. First, if primary units are selected using simple random sampling, then the variance is unbiased even if systematic sampling (with an independent random start in each primary unit) is used within primary units. Thus the problems caused by systematic selection with one-stage samples generally do not occur with multistage sampling. This is why, in discussing systematic sampling earlier (Section 4.2), we stressed that the discussion applied only to one-stage systematic sampling, not to all uses of the method.

Second, the estimate for each primary unit does not have to be a mean, it can be any quantity as long as the goal is to estimate the grand mean of the estimates from each primary unit. For example, in studying vegetation along transects, a measure of foliage height diversity or horizontal patchiness might be calculated for each transect. These quantities are often nonlinear. Working out the formula for their standard errors would often require use of the Taylor series expansion (Section 2.11) and would be quite complex. If the investigator is interested in the mean of the estimates, however, and f_1 is small then the standard error of the mean is calculated quite simply using Eq. 4.13.

Third, when f_1 is small, second-stage finite population corrections have minor effects on the standard error even if they are large; thus, they can be ignored. The same is true of all subsequent finite population corrections in plans with more than two stages.

If sample sizes vary among primary units but the estimated population mean is calculated as an unweighted mean (as suggested above – See 'Weighting of estimates from primary units') then no change is needed in the formula for the standard error as long as the simple version (Eq. 4.13) is used. If the more complex form is used, then Eq. 4.11 must be modified to acknowledge the variation in sample size. With two-stage sampling we use

$$v(\bar{y}) = \frac{s_1^2}{n} - \frac{1}{N}\left(s_1^2 - \frac{1}{n}\sum_{i=1}^{n}\frac{(1-f_{2i})s_{2i}^2}{m_i}\right) \qquad 4.14$$

where m_i = sample size in primary unit i, $f_{2i} = m_i / M$, and

$$s_{2i}^2 = \frac{\sum_{i=1}^{m_i}(y_{ij} - \bar{y}_i)^2}{m_i - 1}$$

Constructing primary units

In many applications, practical constraints leave the investigator little choice in how to construct primary units. For example, primary units may be animals (e.g., the set of times at which each could be observed), survey routes, or days, and no way to alter the sampling plan may be feasible. In other cases, however, choices do exist. For example, when data are collected using transects, investigators may have a choice of whether to orient the transects north–south or east–west. The general principle to follow, in such cases, is that each primary unit should be as representative of the entire population as possible. This results in each primary unit mean having about the same value, and, as a result, the variance of the primary unit means is small. Thus, suppose density varied from north to south, across a population, but was about the same east to west. Other factors being equal, we would orient transects north to south and thus capture as much of the population variability as possible within each transect.

Unequally weighted primary units

Procedures for unequally weighted primary units are complex and the case seldom arises in behavioral ecology so we discuss it only briefly. Readers are referred to Cochran (1977) or Thompson (1992) for a more extensive discussion. In this Section we assume two-stage sampling to estimate the mean per population unit, $\overline{Y}$.

Terminology can be confusing when primary units, and perhaps sample sizes within primary units, vary. By way of review, N and n = the number of primary units in the population and sample, M_i = the size of (i.e., number of population units in) primary unit i, and m_i = the sample size (i.e., number of population units measured) in primary unit i.

When primary units vary in size, the population mean can be expressed in several ways

$$\overline{Y} = \frac{\sum_i^N \sum_j^{M_i} y_{ij}}{\sum_i^N M_i} = \frac{\sum_i^N M_i \overline{Y}_i}{N\overline{M}} = \frac{1}{N}\sum_i^N \left(\frac{M_i}{\overline{M}}\right)\overline{Y}_i, \tag{4.15}$$

where $\overline{M}$ is the mean of the M_i, y_{ij} is the value of the variable for the jth population unit in the ith primary unit, and $\overline{Y}_i$ is the true mean from the ith primary unit. Each term in the expression above can be used to make a different point. The left-hand expression for $\overline{Y}$ (the one with a double sum) is the simplest definition of the population mean. It is the sum of the values of all the population units divided by the total number of population units. It is often called the 'simple mean per unit'. The middle expression shows how the parameter can be expressed as a function of means. Both numerator and denominator use the fact that a sum may be written as the number of items times their mean. The right-hand expression indicates how the parameter may be viewed as a 'weighted' mean of the primary unit means. The weights are $M_i/\overline{M}$, the ratio of the size of the primary unit i to the mean primary unit size. When all primary units are the same size, then all the weights equal 1.0 and they drop out reducing the expression to Eq. 4.6 for equal-sized primary units.

The weights can be expressed in other ways. For example, from the right-hand expression, we could write

$$\overline{Y} = \sum_i^N W_i \overline{Y}_i \tag{4.16}$$

where $W_i = M_i/N\overline{M}$. We prefer defining weights as $M_i/\overline{M}$ because this definition makes the relation to sampling with equal-sized primary units clearer.

The formula for the actual variance of the population mean may be written as

$$V(\overline{y}_w) = \frac{(1-f_1)}{n} \frac{\sum_i^N W_i^2(\overline{Y}_i - \overline{Y})^2}{N-1} + \frac{1}{nN}\sum_i^N \frac{W_i^2(1-f_{2i})S_{2i}^2}{m_i}, \tag{4.17}$$

where the subscript w indicates 'weighted' and $W_i = M_i/\overline{M}$. The resemblance of this case to the case of equal-sized primary units and sample sizes is worth noting. Expression 4.8, with the subscript u added to indicate 'unweighted', may be written

$$V(\overline{y}_u) = \frac{(1-f_1)}{n} \frac{\sum_i^N (\overline{Y}_i - \overline{Y})^2}{N-1} + \frac{1}{nN}\sum_i^N \frac{(1-f_2)S_{2i}^2}{m}.$$

Thus with unequally weighted primary units, the only difference from Eq. 4.8 is that we add the weights, W_i. Expression 4.17 is more general in the sense that if all primary units are the same size, and all samples within primary units (m_i) are the same size, then Eq. 4.17 reduces to Eq. 4.8. Expression 4.17 also applies when sample sizes within primary units are constant (in that case simply replace m_i with m).

An approximate formula for the estimated variance is

$$v(\overline{y}_w) = \frac{(1-f_1)}{n} \frac{\sum_i^n w_i^2(\overline{y}_i - \overline{y})^2}{n-1} + \frac{1}{nN}\sum_i^n \frac{w_i^2(1-f_{2i})s_{2i}^2}{m_i}, \tag{4.18}$$

where $w_i = M_i/(\Sigma M_i/m)$. This expression can be re-written as

$$v(\bar{y}) = \frac{1}{n}\frac{\sum_1^n w_i^2(\bar{y}_i - \bar{y})^2}{n-1} - \frac{1}{N}\left(\sum_1^n \frac{w_i^2(\bar{y}_i - \bar{y})^2}{n-1} - \frac{1}{n}\sum_1^n \frac{w_i^2(1-f_{2i})s_{2i}^2}{m_i}\right), \tag{4.19}$$

which closely resembles Eq. 4.14 for equal-sized primary units except for the weights, w_i, which make the expression more complicated. If all primary units are the same size and all sample sizes are equal then Eq. 4.19 reduces to Eq. 4.14. As with Eq. 4.14, the right-hand term in Eq. 4.19 usually can be omitted because N is large. In this case, we have

$$v(\bar{y}) = \frac{1}{n}\frac{\sum_i^n w_i^2(\bar{y}_i - \bar{y})^2}{n-1}. \tag{4.20}$$

Use of Eq. 4.20 produces a conservative (i.e., 'safe') estimate of $v(\bar{y})$. Expression 4.20 also applies to sampling plans with three or more stages as long as N is large; $\bar{y}_i$ is the estimate obtained from primary unit i and, as usual, must be an unbiased estimate of the primary unit mean. If N is not large relative to the difference between the two terms in Eq. 4.19 then use of Eq. 4.19 may be preferable.

Cases occasionally arise in behavioral ecology in which the population mean of the primary unit sizes, $\Sigma M_i/N$, is known. For example, imagine flying transects across an irregularly shaped study area to estimate the density of an animal. The transects are primary units and will often vary in size. In this case, ΣM_i is just the sum of the areas covered by all of the transects, which is the size of the study area, and N is the number of transects. When $\Sigma M_i/N$ is known, one has a choice of using it, rather than the sample mean, $\Sigma M_i/n$, in defining the w_i. Appendix One, Box 3 provides guidance on which approach to follow and formulas for $v(\bar{y})$ using the population mean in defining weights.

Problems caused by having too few primary units

If only a few primary units are measured, and if their means are rather different from each other, then the standard error will be large. As a result, there will be little power to detect differences, and confidence intervals will be large. In other words, the study is likely to be inconclusive. These statements unfortunately are true even if a large sample is collected within primary units because standard errors in such cases are dominated by the variances of the primary unit means.

This point is quite important in behavioral ecology because obtaining

Table 4.2. *Effect of few primary units on precision and power*[1]

Primary units (n)	Secondary units (m)	Population units (nm)	$SE(\bar{y})$	95% CI on $\bar{y}$	Power
10	5	50	1.9	4.3	0.89
8	8	64	2.1	4.9	0.83
6	12	72	2.3	5.9	0.70
5	18	90	2.5	6.9	0.59
4	25	100	2.8	8.9	0.43
3	40	120	3.2	13.8	0.21
2	75	150	3.9	50.0	0.00

Note:
[1] Power for testing $H_0: \bar{Y}_1 = \bar{Y}_2$ when actually $\bar{Y}_1 = 45$ and $\bar{Y}_2 = 54$g. In each population, $S_1^2 = 0.003$, $\bar{S}_2^2 = 0.003$. Confidence interval, CI $= t_{0.025,\mathrm{df}}\, SE(\bar{y})$; power obtained using methods in Appendix One with two-tailed test and level of significance $= 5\%$.

enough primary units is often difficult. The following example may help show the importance of the issue. Suppose we are studying the 'inside strength' of several species' eggs to compare hatching difficulty among species. The trait may vary across each species' range so eggs are obtained from more than one locale. Suppose that the actual puncture resistance is 45g and that the way 'locales' are defined the variance of locale-specific means (S_1^2) is 30 and the average within-locale variance ($\bar{S}_2^2$) is also 30 (these are realistic values). Finally, suppose another species has a mean puncture resistance of 54g (20% more than the first species) and the same variances. Table 4.2 shows several hypothetical allocations of effort between primary and secondary units. The total sample size increases as the number of primary units decreases to reflect decreasing travel costs.

The right-hand columns of the table provide several measures of how efficiency is affected by reducing the number of primary units. With ten primary units all the measures of efficiency indicate that samples sizes are adequate to estimate parameters with considerable precision and to determine which population mean is larger with acceptable power (89%). As the number of primary units decreases, however, the situation deteriorates. Standard errors increase moderately as the number of primary units decreases. The t-values also increase, especially as the number of primary units drops below four or five. The width of the 95% confidence interval increases even faster because it is the product of the standard error and t-value. Power also decreases steadily but most dramatically when the number of primary units falls below four or five. Notice that these trends occur even though the *total* sample size is much larger with few primary units. This situation, while certainly not universal in behavioral ecology, is quite common. The major point of the example is that many studies which *do* have enough data if the effort is distributed among seven or more primary units *do not* have enough data when the sampling is restricted to just a few primary units. Readers who know a programming language may find it interesting to carry out similar simulations using realistic values and sampling plans for their own studies.

As a practical matter, one should usually avoid multistage sampling with fewer than five to seven primary units. Sometimes it is nearly as easy to select several primary units (and measure relatively few population units in each) as to select only a few primary units and measure many more population units in each. When this is true, we recommend selecting eight to ten (or more) primary units.

Alternative definitions of the population

In some studies, logistic constraints preclude having more than a few primary units in the sample. When this is true, it is sometimes possible to restrict the population to which statistical inferences are made and thereby maintain adequate power. In this Section we discuss three general approaches using the following example. Suppose we are studying some aspect of breeding behavior or ecology of males such as the incidence of cuckoldry, level of parental care, or influence of territory quality on reproductive success. Due to practical constraints we can carry out the work in one to three, but not more, study areas. Studies such as this one often involve nonrandom selection, and this raises several conceptual issues. These will be discussed in Section 4.10. For the present we will assume that the study areas really could be selected randomly and we could thus legitimately make statistical inferences to males in a larger region. As noted above, however, our ability to reach any conclusions with so few primary units is extremely small so we may prefer another approach.

The three strategies to be discussed all entail restricting the spatial extent of the target population to the area(s) in which we actually collect data. The 'size' of the sampled population – and thus the population about which we can make statistical inferences – is thus much smaller. As will be seen, however, these methods often provide a means of obtaining much higher power for statistical tests. Another advantage of these approaches is that the statistical analysis does assume that the study areas were randomly selected.

First, we might carry out the study on only one area. Sample size would then be the number of males. If all males in this area were studied, then we would probably argue that these individuals could be viewed as a sample from a hypothetical superpopulation (Section 1.2). The population would thus be limited spatially to the single study area but would include all males that might have been present in it during the study or all outcomes that might have occurred with the males. As noted in Section 1.2, this approach may be advantageous if the parameter or process under study varies spatially. A possible disadvantage is that even the largest of the available areas might not contain enough males for the study.

A second option is to carry out studies in two or three areas but treat results from each area as a separate study. Results would then take the form, 'in each of two areas, a statistically significant relationship…'. Statistical inferences still extend only to the studied areas. This approach is sound statistically but like the first approach requires that enough males be present in each study area to provide satisfactory precision. Also, the total number of males that must be studied is much higher than if they could all be treated as a single sample.

The third option has been used less often in behavioral ecology but seems legitimate to us. Viewing the study areas as sampling units is natural because they are separated in space. From a statistical point of view, however, there is no necessity for doing this. Suppose instead that we define the spatial dimension of the sampled population as including all three study areas. We collect data from all three of them. If we actually measure all of the males in these areas, then we invoke the superpopulation concept and argue that the males or results may be regarded, for statistical analysis, as a random sample from a much larger, hypothetical population. Alternatively, we randomly select which males to study. In either case, we have a single, random sample of males and carry out our analysis accordingly. We do not treat the study areas as primary units and thus cannot make statistical inferences to other areas; such inferences are based on nonstatistical reasoning (Section 1.4). Sample size, however, is the number of males not the number of areas and precision thus may be much higher.

There is, of course, an issue in this third option as to whether the mean from a potentially internally variable population is interesting on biological grounds. If the means in 2 years, for example, are sufficiently different biologists might feel that little can be learned from the combined mean. An example involving the spatial dimension was provided in Section 1.2. As noted there, however, the issue is a purely biological one.

Each of the examples above involves a trade-off between the 'size' of the sampled population and the precision or power of the statistical analyses. They show that, with a given data set, we may be able to identify different populations about which inferences may be made. The sampling plan may thus be different, and, as a result, analytical methods and precision may differ. The concept of alternate populations about which inferences may be made with a given data set will be critical in Section 4.10 on nonrandom sampling and in Chapter Six on pseudoreplication. The conceptual issues in those contexts are less clear cut, so we encourage readers to study the example in this Section as a prelude to the later discussion.

Investigators sometimes wonder whether certain variables should be used to subdivide the population. One of the most common examples is whether data collected in different years can be treated as a single sample or not. Most behavioral ecologists are not familiar with survey sampling terminology and thus do not express the question as whether they should define years as primary units, but that, essentially, is the question. In nearly all cases, the answer is that years should not be defined as primary sampling units. The years are sequential and thus are not selected randomly from a larger set of years. Furthermore, in many studies consecutive years are likely to have similar values for the parameters being estimated so they are more alike than two randomly selected years. Also, and of more importance, even if we could define years as primary units we would not want to, since this would generally cause us to have a small sample size. Investigators generally recognize this but often feel for some reason that they 'have to' separate years if the mean value of the variable differs significantly between years. There is no statistical reason for this, however, any more than one has to separate males from females in estimating average weight just because the two groups' mean weight is different. In an example involving males and females, one may want to separate them (e.g., by stratification) to increase precision, but there is no necessity for doing so.

4.6 Stratified sampling

In stratified sampling the population units are divided into groups so that each unit is in exactly one group and units are then selected from *each* group according to some sampling plan. In multistage sampling, we also begin by dividing the population units into groups. The distinction is that in multistage sampling we then select a random sample of the groups, and confine subsequent sampling to these groups, whereas with stratification we select units from each group. We call the groups strata to distinguish them from primary sampling units. As with multistage sampling, sample selection must be independent in different strata.

Stratified sampling is often used in wildlife surveys that cover large areas. Use of stratification ensures that the sample is distributed throughout the area, permits heavier sampling in some areas, and lets the investigator obtain separate estimates for each portion of the study area (i.e., for individual strata). The Breeding Bird Survey, coordinated by the U.S. Fish and Wildlife Service, provides a typical example of stratified sampling. The survey area (much of North America) was first subdivided into 1° latitude–longitude blocks (the strata). Roadside survey routes were then drawn randomly from within these strata. Waterfowl surveys to estimate abundance and production on the breeding grounds are designed in the same way except that the strata are larger, and survey lines are located systematically within each stratum. The distance between survey lines varies from stratum to stratum, thus the sample is not equivalent to a single, systematic sample covering the entire study area. National surveys of woodcock and mourning doves, and numerous state and provincial surveys also use stratified sampling to select routes.

Estimation

We follow the same procedure as in previous Sections, first defining the parameter, then discussing the usual estimate of it. We next discuss the true variance, and finally present the estimated variance and standard error. We first consider formulas to estimate the population mean with one-stage sampling in each stratum. The results are then generalized to include estimating other quantities and employing multistage sampling in some or all first-stage strata. In multistage sampling, the case of equal-sized groups was discussed first. In stratified sampling, strata are seldom of equal size, and analytical methods are not much simpler even if they are, so we proceed immediately to the more general case in which stratum size varies.

Stratified sampling is almost always used to estimate means, including proportions, or functions of means such as densities and totals. The parameter can be defined with the same notation used for multistage sampling. With one-stage sampling in strata, the population mean is

$$\overline{Y} = \frac{\sum_i^M \sum_j^{M_i} Y_{ij}}{\sum_i^N M_i} = \frac{\sum_i^N M_i \overline{Y}_i}{N\overline{M}} = \frac{1}{N}\sum_i^N \left(\frac{M_i}{\overline{M}}\right)\overline{Y}_i, \tag{4.21}$$

where N is now the number of strata in the population, M_i is the size of (i.e., number of population units in) the ith stratum, $\overline{M}$, is the mean of the M_i, y_{ij} is the value of the variable for the jth population unit in the ith stratum, and $\overline{Y}_i$ is the true mean from the ith stratum.

This parameter may also be written as

$$\overline{Y} = \sum W_i \overline{Y}_i \tag{4.22}$$

where $W_i = M_i / N\overline{M}$ and is the proportion of the population covered by stratum i. Thus, 'stratum size' does not have to mean the actual number of population units; it is sufficient to know what proportion of the population falls in each stratum. This point will be important later when we discuss methods based on *estimated* stratum sizes.

The point estimate of the population mean is just the sample analogue of Eq. 4.21

$$\overline{y} = \frac{1}{N} \sum_{i}^{N} \left(\frac{M_i}{\overline{M}} \right) \overline{y}_i \tag{4.23}$$

where $\overline{y}_i$ is the estimate of $\overline{Y}_i$ and is assumed to be unbiased. Readers who have been following the use of expectation will be able to see that $\overline{y}$ is an unbiased estimate of $\overline{Y}$. Thus, by successive application of the rules regarding the expected value of a constant times a random variable and the expected value of the sum of random variables (Section 2.4), we may write

$$E\left[\frac{1}{N}\sum\left(\frac{M_i}{\overline{M}}\right)\overline{y}_i\right] = \frac{1}{N} E\left[\sum\left(\frac{M_i}{\overline{M}}\right)\overline{y}_i\right] = \frac{1}{N}\sum E\left[\left(\frac{M_i}{\overline{M}}\right)\overline{y}_i\right] \tag{4.24}$$

$$= \frac{1}{N}\sum\left(\frac{M_i}{\overline{M}}\right) E(\overline{y}) = \frac{1}{N}\sum\left(\frac{M_i}{\overline{M}}\right)\overline{Y}_i$$

$$= \overline{Y}.$$

With a little practice, one quickly develops the ability to skip most of the steps in the equations above. The point estimate is a linear combination of the variables in the sample, so we expect that an unbiased estimate of the parameter will be provided by the sample analogue.

Notice that as with multistage sampling and primary units of unequal size, we would not wish to use the simple mean of the units in the sample as our estimate (Section 4.5). They must be combined in a way that properly reflects variation in stratum size and sample size within strata.

The actual variance of the point estimate, $\overline{y}$, is

$$V(\overline{y}) = \frac{1}{N^2} \sum_{i=1}^{N} \left(\frac{M_i}{\overline{M}} \right)^2 V(\overline{y}_i). \tag{4.25}$$

Thus, the actual variance of the estimated mean in each stratum is calculated using whatever methods are appropriate, and then these variances are combined in a weighted average, with weights equal to the square of relative stratum sizes, to obtain the overall variance.

The estimated variance is just the sample analogue

$$v(\bar{y}) = \frac{1}{N^2}\sum_{i=1}^{N}\left(\frac{M_i}{\bar{\bar{M}}}\right)^2 v(\bar{y}_i). \tag{4.26}$$

Thus, to obtain the estimate, we first obtain an unbiased estimate of the variance of the point estimate, $\bar{y}_i$, within each stratum, $v(\bar{y}_i)$. These are then combined in a weighted average with weights equal to the square of $(M_i/\bar{\bar{M}})$ to obtain the overall variance. The standard error of the estimate, as usual, is just the square route of $v(\bar{y})$

$$se(\bar{y}) = \sqrt{\frac{1}{N^2}\sum_{i=1}^{N}\left(\frac{M_i}{\bar{\bar{M}}}\right)^2 v(\bar{y}_i)}. \tag{4.27}$$

The rules of expectation in Section 2.9 can also be used to show that the estimated variance is unbiased. Note that we have defined $v(\bar{y})$ as an unbiased estimate of $V(\bar{y})$. This being true, we may write

$$E\,[v(\bar{y})] = E\left[\frac{1}{N^2}\sum\left(\frac{M_i}{\bar{\bar{M}}}\right)^2 v(\bar{y}_i)\right] = \frac{1}{N^2}\sum\left(\frac{M_i}{\bar{\bar{M}}}\right)^2 E\,[v(\bar{y})] \tag{4.28}$$

$$= \frac{1}{N^2}\sum\left(\frac{M_i}{\bar{\bar{M}}}\right)^2 V(\bar{y}_i) = V(\bar{y})$$

As with multistage sampling, the point estimate (in repeated sampling) does not have a t distribution so, strictly speaking, t tables are not appropriate. An approximate method has been worked out, however, under which one calculates an 'effective number' of degrees of freedom. A t-value is then obtained from a t-table using this number of degrees of freedom and then standard procedures based on the t-table are used to construct confidence intervals and carry out tests. With one-stage sampling in strata, the formula for effective degrees of freedom is usually calculated using Satterthwaite's approximation (Cochran 1977 p. 96) which is

$$df = \frac{(\Sigma g_i)^2}{\Sigma[g_i^2/(n_i-1)]}, \tag{4.29}$$

where $g_i = W_i^2\, v(\bar{y}_i)$ and n_i = sample size from the ith stratum.

Construction of strata

In constructing strata, one tries to form groups so that the units within each group arc similar. This reduces standard errors within strata and thus the overall standard error is small. As an admittedly trivial example, suppose that the entire population consisted of just six population units and the values for them were 1, 2, 3, 8, 9, 10. We wish to estimate the population mean using a sample of size two. Notice that there are three small values (1, 2, 3) and three large values (8, 9, 10). With simple random sampling there is a 40% chance that both units in our sample would come from the group of small values or the group of large values. In either case, the point estimates would be far from the true mean. Thus, the precision of a simple random sample, in this case, is fairly low. Now suppose that we could assign the first three units to one stratum and the second three to a second stratum. We could then select one unit from each group thus avoiding any possibility of the very poor estimates that might be obtained with simple random sampling. This example illustrates the point that stratification works well when one can assign similar units to the same strata. Note, too, that the method works especially well when the stratum means are very different. Thus, stratification in our example would pay even bigger dividends compared to simple random sampling if the second group of numbers was 98, 99, 100.

Readers may be interested to note the difference in the principles used to construct the groups for multistage sampling and for stratified sampling. In multistage sampling, we sample the groups and estimate precision from the variation among group means. We therefore try to keep this variation as small as possible by including as much of the population variability within each group (i.e., primary unit). In stratified sampling, we measure all groups; sampling occurs only within groups and precision is estimated by the sample-to-sample variation within groups. We therefore try to keep this variation small by making the population units within each group as uniform as possible.

Use in behavioral ecology

Although most texts on sampling methods stress the utility of stratification, the method is seldom used by behavioral ecologists. We suspect there are at least three reasons for this. First, it is clear from the above formulas that stratum sizes must be known or at least estimated (see later), and this may take time away from the collection of data. Second, if it is anticipated that the means will differ greatly between strata (the case in which stratification is most effective), the biologist may prefer to think of the strata as separate populations and make comparisons between them. Third, many biologists like the way that systematic samples cover the population uniformly, whereas this is not guaranteed with stratified random sampling unless the population is partitioned into a large number of small strata.

Despite these points, stratification is sometimes useful in behavioral ecology. For example, as discussed previously, the approach is widely used in wildlife surveys so that sampling effort can be concentrated in certain areas. Furthermore, while systematic samples have small actual variance, no way is known to obtain unbiased estimates of this variance, and the usual estimates (based on simple random sampling) often suggest that precision is much lower than it really is. Stratified sampling, on the other hand, avoids this problem. The estimate of variance is unbiased. Here is an example showing the potential utility of stratification. We once encountered a biologist studying habitat preferences of radio-collared gray fox. He wanted to monitor their use of habitat throughout the 24-h day. During the night he checked each animal once every 4 h, and felt comfortable with the resulting data. During the day, however, the sampling plan seemed too labor intensive because the fox seldom moved during the day and used the same resting area day after day. The biologist believed, however, that he should standardize procedure by sampling uniformly throughout the day, so he continued to record locations every 4 h day and night. If he had defined daytime and nighttime strata, he could have greatly reduced sampling intensity during the daytime period because there was virtually no variation in habitat use during the day. His standard error for $\bar{y}_{\text{day}}$ would have been small despite the smaller sample size. Furthermore, if habitat use during the day and night differed, then the standard error of his overall estimate would probably have been even smaller than the standard error he actually obtained because, with stratified sampling, the overall standard error comes solely from the variation within strata.

Poststratification

In typical stratified sampling, one selects units from the first stratum, then the second stratum, and so on. In poststratification, one selects units from throughout the population and then assigns them to the appropriate stratum after selection. The main difference is that with poststratification sample sizes within strata are not predetermined. As an example of this technique, suppose we are estimating the average value of a quantity which differs between males and females. It might be weight, for example, or time spent in a given habitat or behavior. Our objective is to estimate the mean for the population (we may also wish to estimate means for each sex, but that is a different problem). Suppose we select a sample of 15 individuals, and that by chance we obtain 4 males and 11 females despite the fact that the sex ratio is known to be 50:50. The simple average of the 15 measurements is not a very satisfactory estimate because females dominate the sample. As an alternative, we could calculate the mean for the males and the mean for the females and then use the simple mean of these two values as our estimate. This amounts to viewing the plan as one-stage sampling with poststratification into two strata, males and females. Stratum sizes are known to be equal under the assumption of a 50:50 sex ratio.

The true variance of the estimate with poststratification can be written in two ways. The most obvious approach is to acknowledge that the sample sizes within strata would vary from sample to sample. Some authors (e.g., Schaeffer *et al.* 1979) adopt this approach which then also has implications for how the variance is estimated. Another approach (e.g., Cochran 1977) involves a rationale under which variation in sample size per stratum does not have to be recognized. The rationale is a little complex but is worth understanding because it may occasionally be invoked in other applications in behavioral ecology.

Suppose that sampling has been completed and the application involved one-stage, simple random sampling within strata. Let the number of units that we obtained in the ith stratum be denoted as m_i, $i=1,2,\ldots,N$, N=number of strata. We now wish to calculate the actual variance, an unbiased estimate of it, and, from this, the estimated standard error. The variance of $\bar{y}$, as always, refers to the sample-to-sample variation in the point estimate, $\bar{y}$. Normally, in calculating the value of $V(\bar{y})$ we would consider all possible samples. Suppose, however, that we restrict our attention to those samples with the same sample size in each stratum as the sample that we actually obtained. This set includes all possible ways to select m_1 units from the first stratum, and the mean of these samples is thus exactly equal to the true mean for that stratum. The same can be said of each stratum. As a result, the overall mean of the estimates from this restricted set of samples is exactly equal to the true population mean. We can thus substitute this restricted set of samples for the full set (in which sample sizes per stratum vary) because the two groups have the same parameter of interest. In the restricted set, however, the sample sizes within strata do not vary, and we thus treat them as constants. We therefore use the usual formula for $v(\bar{y})$ (Eq. 4.26) in which sample sizes really are predetermined. This type of rationale is often described by saying that we calculate variances 'conditioned on' the values of the sample sizes in our particular sample.

As noted above, this rationale, while complex, is sometimes useful to behavioral ecologists in other contexts. To be sure it is clear, consider why the rationale would not apply to multistage sampling with unequal-sized primary units. The cases are somewhat similar in that both seem to involve size as a random variable: sample size per stratum in the case of poststratification, and size of primary units in the case of multistage sampling. The question thus arises, 'with multistage sampling, could we condition on primary unit size and thereby treat the mean of the primary unit sizes in our sample ($\overline{m}$ in Section 4.5) as a constant?'. In this case, however, we cannot condition on the particular set of primary units we happened to select because their mean, in general, is not equal to the population mean. Thus the rationale does not work. This entire issue is a particularly good one to work through using a computer simulation, though it would be a little more difficult to program than some of the previous examples.

Estimated stratum sizes

In some applications, stratum sizes are not known exactly but must be estimated. For example, suppose that we are handling or examining a large number of individuals (plants or animals) and that we classify them into cohorts using age, sex, or other attributes. A subsample of each cohort is selected and some relatively expensive measurement (e.g., involving analysis of a blood sample) is recorded. One of the goals is to estimate the population mean for this variable. We would generally not want to use the sample mean as our estimate. Instead, we would weight the cohort means by the relative abundance of each cohort. In other words, we would view the data as a stratified sample, with each cohort comprising one stratum. A difficulty now arises, however. We do not know stratum sizes, we only have estimates of them since we did not examine all members of the population.

Examples such as this one may be viewed as involving two samples, one to estimate stratum sizes, and one to estimate means (or other quantities) for each stratum. The plan is referred to as 'double sampling' for stratification in recognition of the fact that two samples were selected. The additional uncertainty caused by estimating stratum sizes is incorporated into the analysis by the addition of an extra term in the formula for the standard error (see Appendix One, Box 3, Part C). The effect of this additional term depends on how well stratum sizes are estimated. In the limiting case of measuring all members of the population in the first sample, the formula reduces to the equation for stratification with known stratum sizes.

4.7 Comparison of the methods

Comparing the precision of different estimators using computer simulations and a small population is an effective way to gain additional insights

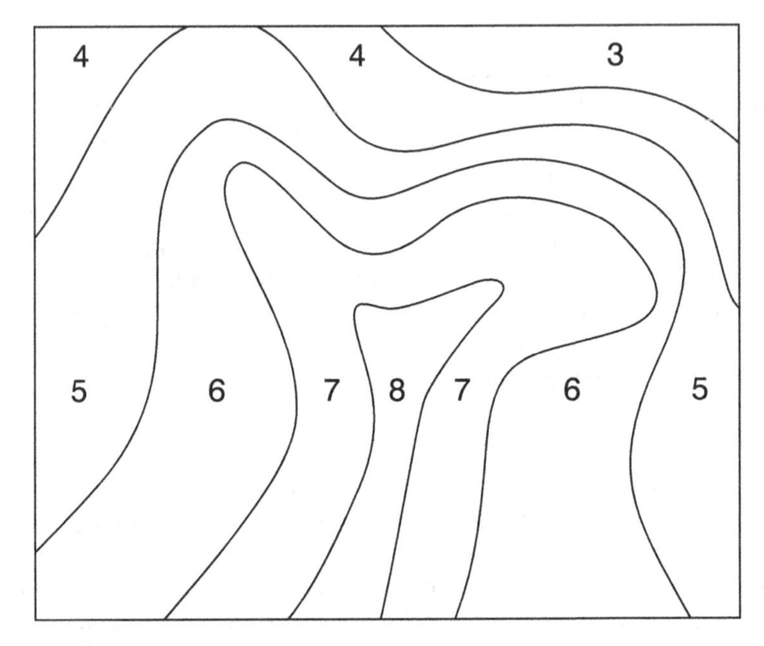

Fig. 4.13. Hypothetical population from which the simulation results in Table 4.3 were obtained. The variable had the values 3 to 8, as indicated

into the strengths and weakness of the various sampling methods. In this Section, we illustrate the technique and provide results from one such simulation. We urge readers to carry out similar analyses on their own.

We compared precision and bias of simple random, systematic, and stratified sampling in an artificial population (Fig. 4.13). We used sample sizes of 16, 36, and 64 (Table 4.3). Simple random sampling was least precise, followed by systematic sampling, followed by stratified sampling. The estimators for standard errors with systematic samples had a positive bias of 10–20% (compare columns c and e in Table 4.3), a much smaller bias than in our example with systematic sampling in Section 4.4. As noted there, results such as these depend heavily on fine details of the population. Stratified sampling with only a few units per stratum is seldom used in behavioral ecology but it would be interesting to see this sampling plan evaluated in more field situations.

4.8 Additional methods

This Section considers a final few methods from survey sampling theory which have been little used in behavioral ecology but which we have occasionally found useful.

Table 4.3. *Precision of different sampling plans used to estimate the mean of the population illustrated in Fig. 4.13*

	Actual standard errors			
Sample size (a)	Simple random samples (b)	Systematic samples (c)	Stratified samples[a] (d)	Ave. est'd *se* ($\bar{y}$) among the syst. samples[b] (e)
16	0.30	0.25	0.21	0.30
36	0.19	0.15	0.09	0.19
64	0.14	0.13	0.05	0.14

Notes:
[a] Two units per stratum.
[b] Values were slightly larger than those in column b.

Ratio and regression estimators

We have already considered the estimation of ratios, such as population size in two different years. We pointed out that when differences are expressed on a proportional basis (e.g., 'older females had 20% more young than younger females'), the estimate is a ratio and this needs to be recognized in the analytical methods. In this Section, we consider methods for estimating means or functions of them using an approach that involves a ratio. The same methods must be used for the analysis as were presented earlier, but the conceptual basis for estimating means using ratio methods has not been previously described. Thus we concentrate in this Section on the rationale of this approach.

Suppose that we are interested in estimating the mean value of a trait that is rather difficult to measure exactly, but that is closely correlated with some other variable which can be measured quickly and easily. For example, suppose we are measuring the age of trees by coring and counting rings. Age in some stands is fairly closely correlated with tree diameter and we will assume that this is true for our population. Below, we develop a method to improve our estimate of mean age by using information about diameters.

Suppose that we measured diameters of the trees in our sample as well as age. We denote the mean age, $\bar{y}$, and the mean diameter, $\bar{x}$. In repeated sampling we would obtain pairs of values ($\bar{y}$,$\bar{x}$). If diameter is closely related to age, then a plot of these pairs might look something like Fig. 4.14. With most real populations the set of all samples would be too large for a scatterplot to be useful, but we can imagine that the plotted dots indicate the distribution of the complete set of samples. We have also

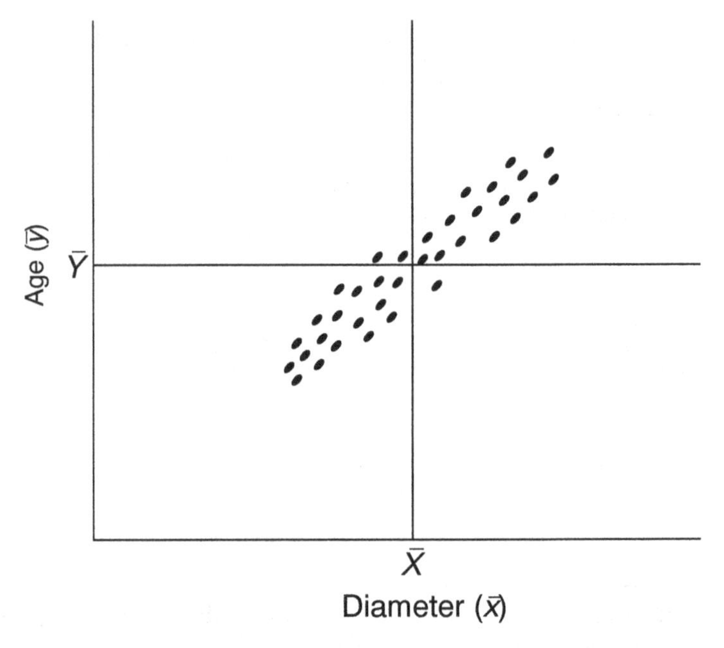

Fig. 4.14. Artificial distribution of sample means when two correlated variables, age and diameter, are measured on each unit. Each dot represents one sample.

plotted the population means of the diameters and ages on the figure, and we have drawn vertical and horizontal lines through these values. Obviously, with only one real sample (one point on the plot), we do not have the ability to draw these lines. Suppose, however, that we somehow knew that our particular sample had a mean diameter less than the population mean. By inspecting the figure, one can see that most samples with $\bar{x}_i < \bar{X}$ also have $\bar{y}_i < \bar{Y}$. This is a result of our assumption that diameter is correlated with age. Thus, if we did know that in our single sample $\bar{x}$ was less than $\bar{X}$, we would naturally wish to adjust our estimate of the mean age upwards.

Ratio and regression estimators provide a method for making this sort of adjustment and thereby increasing the precision of the estimator. In their simplest form, they require that the population mean of the x_i to be known. For the ratio estimator $\bar{y}_r$, the adjustment is

$$\bar{y}_r = \left(\frac{\bar{X}}{\bar{x}}\right)\bar{y}, \tag{4.30}$$

where $\bar{X}$ is the population mean and $\bar{x}$ is the sample mean. This approach is used when the slope of a regression line through the sample means is believed to pass through the origin. When this is not true, then a regression estimator can be used which accounts for a nonzero intercept (see Cochran 1977, Chapter Six).

The standard error of the ratio estimate is easily obtained since the estimate is a constant ($\overline{X}$) times a random variable ($\overline{y}/\overline{x}$). As a result, we may write

$$se\left(\overline{X}\frac{\overline{y}}{\overline{x}}\right)=\overline{X}se\left(\frac{\overline{y}}{\overline{x}}\right), \tag{4.31}$$

(see p. 32) and $se(\overline{y}/\overline{x})$ may be calculated using the equations introduced on p. 41 and summarized in Appendix One, Box 9. As already noted, ratios, in general, are biased estimators of the corresponding parameters, but the bias is generally negligible. We therefore refer to ratios as essentially unbiased.

An important point in understanding ratio and regression estimators is that they are essentially unbiased regardless of the true relationship between $\overline{y}$ and $\overline{x}$. Thus, even if the relationship between the two variables is much weaker than that in Fig. 4.14, the point estimates are still valid. In such cases, the use of the auxiliary variable may not increase precision, and in some cases their use may even decrease precision. Both point and interval estimates, however, remain essentially unbiased.

In the example of the trees, we first suggested that all trees in the stand must be measured to use the ratio estimator. A natural response might be that measuring all trees would often be impractical and that a large sample should be sufficient to estimate the population mean, $\overline{X}$. This is a reasonable approach but the additional uncertainty caused by not knowing $\overline{X}$ exactly must be incorporated into the formula for the standard error. The sampling plan when $\overline{X}$ is estimated is similar to double sampling for stratification in which a large initial sample is selected to estimate stratum sizes. With the ratio estimator, the sample is selected to estimate the population mean of the auxiliary variable. The plan is referred to as double sampling for ratio estimation. Thompson (1992, Chapter 14) provides formulas for the point and interval estimates.

Although ratio and regression estimators are rather complex and have seldom been used in behavioral ecology, they are sometimes extremely useful. We suggest Chapters 7, 8, and 14 in Thompson (1992) for readers interested in learning more about them.

Visual estimates

Some variables of interest in behavioral ecology can be measured quite accurately and quickly by eye, but are much harder to measure exactly. Examples include estimates of vegetation cover (e.g., against the sky, ground or a density board), abundance (e.g., of seeds in small plots), and flock or herd size. In cases such as these, accurate measurement may require methods such as taking photographs, enlarging them, and then making time-consuming counts or measurements. On the other hand, visual estimates alone may be unsatisfactory because of potential bias. Double sampling provides an excellent method for combining the visual estimates with the exact measurements. The initial sample of visual estimates may be used either in a stratified estimate or a ratio estimate. Below, we provide an example of using visual estimates for a stratified estimate and then discuss how the same data could be used in a ratio estimate.

Suppose we are counting something such as the number of seeds per plot, eggs or larvae per nest, or invertebrates per water sample. The number per population unit is often large so complete counts are time consuming. Surveyors, however, can quickly assign population units to broad categories, on the basis of visual inspection. We use the categories 'few', 'medium', 'many', and 'very many'. To gather the data, an initial large sample of units is examined, and the surveyor assigns each unit to one of these categories by visual inspection. This is the first sample in double sampling. A random number is then obtained from a hand calculator and if the value is equal to or less than some threshold (e.g., 0.20), then the unit is measured exactly. This gives us a second random sample from each stratum.

In this plan, a stratum consists of all the plots that would be assigned to a given category such as 'few'. Stratum size is thus the number of units of this type. Recall from Section 4.6, expression 4.22 (p. 126), that we do not need actual stratum sizes, the proportion of the population units that would be assigned to a given stratum is all we need, and our first sample gives us this estimate. The methods of double sampling for stratification thus are applicable. Incidentally, the reason for deciding which units to measure exactly after assigning them to categories is to ensure that these assignments are not made with greater care than other assignments (i.e., the second sample must be a random subsample of the first sample). As pointed out earlier, this approach produces unbiased estimates of means even if the visual estimates bear little relationship to actual abundance, in which case the stratum means would differ little. In this case, the stratification would provide little or no gain in precision over just using the exact measurements, but the estimates are still valid.

We now turn to using this general approach for a ratio estimator. Suppose that the surveyor made an estimate of actual abundance in each unit rather than simply assigning them to broad categories such as 'few' or 'many'. We could then view these estimates as auxiliary information and use them in a ratio or regression estimator. In this event, we would have an estimate of the population mean of the visual estimates, not its true value which would have required that we visually inspect all possible units. Formulas for double sampling would therefore be used. If the visual estimates were closely correlated with actual abundance, this estimator might have substantially greater precision than use of just the exact measures. As was the case when using this approach for stratification, no assumption is made that the visual estimates are accurate; the ratio or regression estimate remains essentially unbiased even if the visual estimates are poorly correlated with the actual values.

4.9 Notation for complex designs

With certain multistage or stratified designs, calculation of the means per group (primary units or strata) is a little complex. This happens when the means should *not* be calculated as the simple average of all the observations in the primary unit or stratum. For example, suppose we are estimating some quantity using three-stage sampling with plots, subplots, and quadrats as the primary, secondary, and tertiary units. We stratify subplots (using habitats as strata for example) and sample some habitats more intensively than others. Plots are all the same size, as are subplots. How should we calculate the estimated mean per plot? We should not use the simple mean of the quadrats because we stratified the subplots and sampled more intensively in some strata than others. Similar problems occur often in behavior studies. For example, suppose we divide the nesting attempt of a bird into incubation and nestling periods and then record some aspect of the male's behavior each time it visits the nest. The number of visits will vary between observation intervals, and we may collect more data during some periods of the attempt than others. How should we calculate the mean per visit so that it will be unbiased?

Problems such as these are basically issues of how to weight the observations. In simple cases, the correct approach is often obvious. In more complex cases, it may be helpful to have formal guidelines and to write out a formula that will be used to calculate the means per primary unit. The formula can then be inspected one part at a time to make sure that no errors have crept in. The biggest problem in deriving such a formula is developing a workable notation. In this Section we explain one procedure for deriving a formula to calculate unbiased means per primary unit. We use the example above in which we selected plots, subplots, stratified the subplots, and then selected quadrats within each stratum. Plots and subplots were each the same size, sampling intensity varied between strata, and strata varied in size.

The first step is deciding how many subscripts will be needed and what each should mean. In most studies, the number of subscripts equals the number of stages in the design plus the number of times we stratify. Here we have three stages and stratification prior to selection of quadrats. We will use i, j, k, and l as subscripts referring to plots, subplots, strata, and quadrats respectively. Thus, y_{ijkl} is the measurement from the lth quadrat in stratum k in subplot j of plot i. Next, we need symbols for the sample sizes for each subscript except the last one – plots, subplots/plot, and quadrats/stratum in this case. We will use n, m, and g respectively. Thus, n is the number of plots, m_i is the number of subplots in plot i, and g_{ijk} is the number of quadrats in stratum k in subplot j of plot i. Finally, we need subscripts to indicate stratum sizes and a symbol for the number of different strata. We will use H for the number of strata and W to indicate stratum size. W_{ijk} = the proportion of the jth subplot in plot i that is covered by stratum (i.e., habitat) k.

Summarizing,

y_{ijkl} = measurement on the lth quadrat in the kth stratum in subplot j of plot i.
g_{ijk} = number of quadrats in habitat k in subplot j of plot i.
W_{ijk} = proportion of subplot j in plot i that is covered by stratum k.
m_i = number of subplots in plot i.
n = number of plots.

The next step is to decide how we should combine observations indicated by the *last* subscript (quadrats within a given habitat, subplot, and plot in this case). The basic principle is to combine observations in such a way that we obtain an unbiased estimate of the true mean for the group. We have one-stage sampling within habitats, so we use the simple average of the observations as the estimate. Thus, within a given habitat and subplot, our estimate of the mean/population unit will be

$$\bar{y}_{ijk} = \frac{1}{g_{ijk}} \sum_{l=1}^{g_{ijk}} y_{ijkl}. \tag{4.32}$$

Next we consider how to combine these estimates to obtain an unbiased estimate of the mean per subplot. Since we have stratified sampling within subplots, we need to weight the means/habitat by the proportion of the subplot covered by each habitat. The estimate of the mean/population unit within the jth subplot (of plot i) is therefore

$$\bar{y}_{ij} = \sum_{k=1}^{H} W_{ijk} \bar{y}_{ijk}. \tag{4.33}$$

The last step, combining subplots to obtain means/plot is relatively simple because subplots were all the same size. The formula is thus

$$\bar{y}_{\mathrm{i}} = \frac{1}{m_{\mathrm{i}}} \sum_{\mathrm{j}=1}^{m_i} \bar{y}_{\mathrm{ij}}. \tag{4.34}$$

These $\bar{y}_i$ are then used (since plots were all the same size) to estimate the population mean as $\bar{y} = \Sigma \bar{y}_i / n$. A single equation for the $\bar{y}_i$ may be obtained by combining the equations above

$$\bar{y}_i = \frac{1}{m_i} \sum_{j=1}^{m_j} \sum_{k=1}^{H} W_{ijk} \frac{1}{g_{ijk}} \sum_{l=1}^{g_{ijk}} y_{ijkl}. \tag{4.35}$$

Since the means per primary unit ($\bar{y}_i$) are combined using equal weighting ($\bar{y} = \Sigma \bar{y}_i / n$), the $se(\bar{y})$ is just $se(\bar{y}_i)/\sqrt{n}$ (see expression 4.13, p. 116 or Appendix One, Box 2).

Note that if we had a different number of stages, if the number of subplots/plot had varied, or if stratification had been used prior to other stages, then a different set of symbols and equations would have been needed. The number of different formulas that may be needed in different studies is thus very large.

4.10 Nonrandom sampling in complex designs

In Section 4.4 (pp. 101–2), we discussed the fact that nonrandom sample selection is often unavoidable in behavioral ecology, emphasizing the difficulties that may result. Section 4.4, however, was restricted to one-stage sampling. We now extend this discussion to designs that involve multistage sampling, stratification, or both.

In this Section we use the phrase 'nonrandom sampling' broadly including cases in which selection is supposed to be independent (e.g., in different primary units) but in fact is not independent.

Analysis of samples collected with nonrandom selection involves assuming that the sampling plan actually followed produces estimates with essentially the same statistical properties as some other sampling plan which is to be used as the basis for point and interval estimation. This assumption is usually based largely on biological, rather than statistical, considerations. For example, with one-stage, systematic sampling, we use the formulas for simple random sampling and hope that any errors are not too serious. We have shown, in that case, that the point estimate remains unbiased but that standard errors are usually overestimated, sometimes substantially (Section 4.4, pp. 102–7).

With one-stage sampling there is generally little choice in how to analyze the data: the formulas for simple random sampling are used. With more complex designs, however, more than one alternative may exist, and different investigators (biologists or statisticians) may not agree on the best course. While this potential for disagreement must be recognized, it is still helpful in such cases to clarify the assumptions inherent in each approach and to reach agreement that if those assumptions – wise or not – are adopted, then the rest of the analysis is appropriate. This process at least reduces the scope of the disagreement to whether the assumptions are reasonable, a topic that is often evaluated on the basis of biological evidence rather than statistical analysis. As noted in the first Section of this Chapter, biologists can then argue biology, not statistics, which is presumably preferred by all participants in the discussion.

A common example of nonrandom sampling in behavioral ecology is systematic sampling in two 'dimensions', where dimension refers to a spatial or temporal variable or to individuals such as animals or plants. First, consider an example of sampling solely in space. Suppose we lay out n transects spaced across a study area at equal intervals, and then place m plots along each transect using the same spacing as between transects. This produces a regular grid of plots (Fig. 4.15a). We would probably treat this as a one-stage systematic sample with sample size equal to nm. Now suppose that the distance between transects was greater than the distance between plots along transects as in Fig. 4.15b. This design begins to look like two-stage sampling with transects as primary units and plots as secondary units. We might thus be tempted to analyze the data this way, calculating means per transect and thereby obtaining a sample of size n rather than nm. As noted in Section 4.5, if the number of transects is small, this approach may result in substantially lower power than viewing the data as a one-stage sample of size nm.

Viewing the sample as two-stage is probably more conservative, which may be desirable, but in that case we should recognize two additional problems. First, analysis of multistage samples assumes simple random selection of the primary units (transects in this case) but in our example they were selected systematically. Second, independent selection of the secondary units in different primary units is also assumed, but this was not the case; secondary units are lined up in rows. Finally, having recognized that this systematically selected sample may be better viewed as a multistage sample, one might wonder whether the same is true of a regular grid pattern as in Fig. 4.15a. Thus, we might select either rows or columns and designate them as primary units. It is easy to show that with real data sets of this sort,

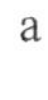

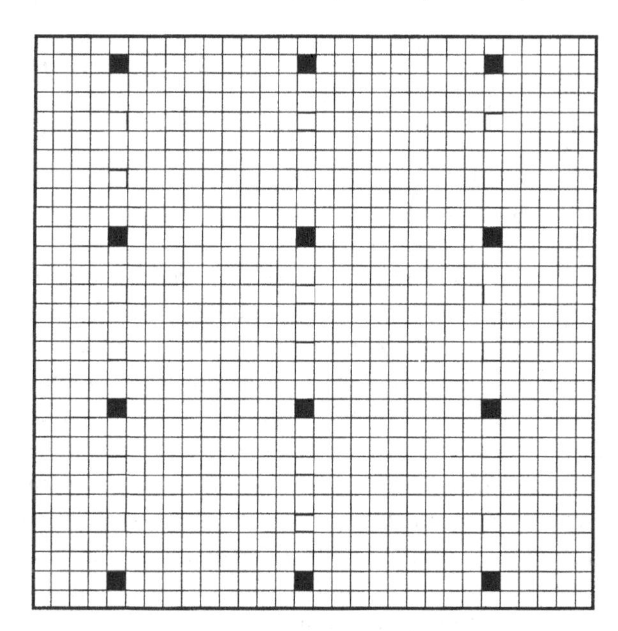

b

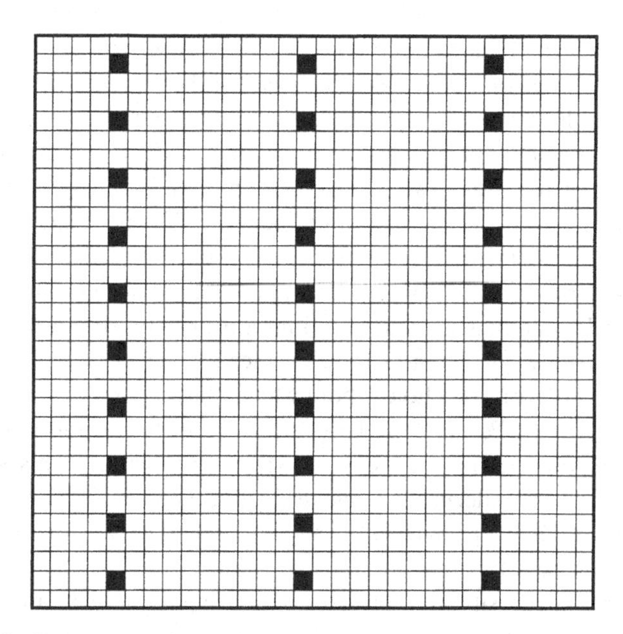

Fig. 4.15a,b. Two arrangements of sample stations along transects.

estimated standard errors and power may vary substantially according to which of these approaches one follows. Obviously, choosing the approach on the basis of which one happens to produce the best results is not legitimate. The decision should based on considerations such as those in Section 4.5 on evaluating nonindependence.

Other cases of sampling in two dimensions generally involve time as one of the dimensions. Thus, we may visit each of several routes, plots or locations at regularly spaced intervals, or we may record measurements on a given sample of plants, animals, or other objects at regularly spaced intervals. In such cases, it is customary to view time as the second dimension, thus considering the plot or individual as the primary sampling unit. For example, if the number of animals seen, heard, or captured on several, regularly spaced occasions at each of several fixed locations is recorded then the natural tendency among behavioral ecologists seems to be to regard the number of locations as 'the sample size', which, by implication, establishes locations as primary units. If we make repeated measurements on plants or animals, then they are generally thought of as the basic units for analysis, that is, as the primary units.

These customs are probably useful because a detailed quantitative evaluation of the issue is quite complex and will usually require making difficult estimates of quantities that behavioral ecologists are unfamiliar with. Note, however, that the guideline may be rather arbitrary. For example, if ten plots are each visited three times, then we obtain an effective sample size of ten. But if we selected ten times and visited the three plots each time, then our sample size would be three. The population diagrams would appear identical (except for being rotated 90°) but, as noted in Section 4.5 (pp. 120–122), the second approach would probably have far less power. In fact, in many real situations a study with such a small sample size would not be worth conducting. This seems a heavy price for what is in reality a largely arbitrary choice about which dimension is used to designate the primary units. Perhaps the best lesson from this example is to avoid systematic sampling in two dimensions whenever possible. If this is not possible then we advise deciding, while designing the study, which dimension will be used to define primary units and ensuring that the plan provides an adequate number of primary units.

Another difficult situation involves measuring all units in a given study area. We begin with a case that is fairly easy to resolve, and proceed to more difficult cases. Suppose we record repeated measurements on all the individuals of a mobile species in a study area. We have already noted (Section 1.2) that the individuals can usually be viewed as having been (self-)selected from a larger superpopulation. With repeated measures, the sampling plan

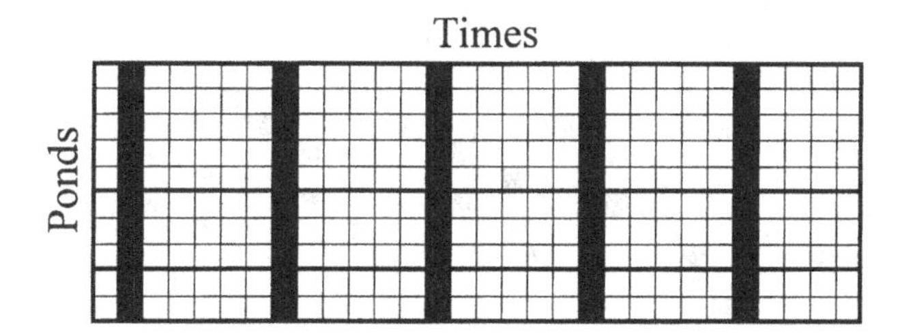

Fig. 4.16. A series of ponds repeatedly measured on the same days.

is viewed as multistage, and the individuals (or more precisely the set of measurements that might be recorded on them) become primary units. With animals (or other objects) that move around, it is often not too difficult to imagine the existence of a superpopulation and to feel that the statistical analysis is useful. If test results are significant we will generally feel that the sample results can be extrapolated to a larger population, though the spatial or temporal limits of the population may be difficult to establish. On the other hand, if the result is nonsignificant, then we have good reason for not generalizing the results at all. The analysis thus serves the usual purpose: it tells us whether the study has revealed which population parameter is larger or was inconclusive.

Now, however, suppose we are studying long-lived objects that do not move around much, such as trees or ponds, and suppose further that the study area is fairly distinctive or unusual, rather than being one example of a widespread type. In these cases, extrapolation to larger populations may be more tenuous and less interesting. If all of the subjects were measured, and there is no temporal variation, then no statistical analysis is needed or appropriate because the data set does not represent a random sample from a larger population of interest. In many studies, however, sampling still occurs in time, and our interest lies in means across the entire study period. For example, we may need to estimate average water temperature during July in a series of ponds but lack a continuous record. The population unit is then a 'pond-time'. Even though all ponds were measured, sampling still occurs in time, and thus statistical analysis is needed to estimate one mean or to compare means.

In this case, as usual, the formulas depend on the sampling plan. If all units were measured on the same days, then a population diagram would look like that in Fig. 4.16 and the sample should probably be viewed as one-stage systematic with sample size equal to the number of sampling occasions (five in the example). Most behavioral ecologists would probably view ponds as determining the sample size (because the importance of independent selection in different primary units is generally not recognized),

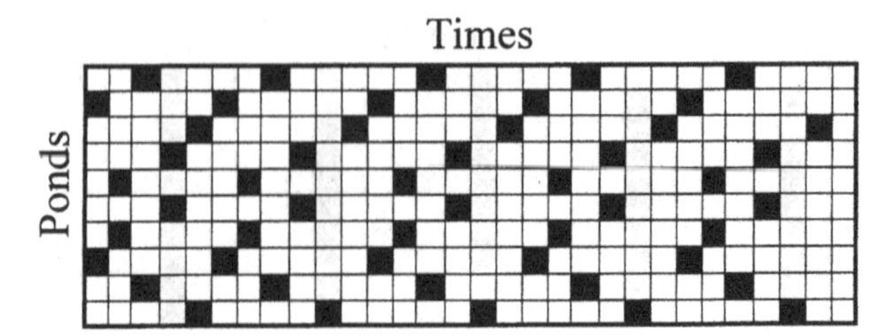

Fig. 4.17. The same population as in Fig. 4.16 but measurements are made on different days.

but the population diagram makes it clear that viewing the plan as one-stage systematic is more appropriate. Now suppose that an independent systematic sample was taken from each pond with the constraint that no more than two units were measured on a single day. An example of this plan is shown in Fig. 4.17. This plan is actually stratified systematic sampling (ponds are strata) with the qualification that starting times were not selected completely independently. Most analysts would probably feel that the dispersion of units (pond-times) across the population was sufficiently even that the formulas for simple random sampling (Appendix One, Box 1) could be used. Other approaches might be imagined; for example, the number of measurements might differ between ponds in which case the formulas for stratified sampling (Appendix One, Box 3) might be preferable.

The approach described here seems reasonable on the basis that the superpopulation concept is not invoked when doing so is questionable or unsatisfying. Furthermore, the population is well defined and each alternative above represents application of well-accepted sampling theory. It must be acknowledged, however, that editors and referees, not used to the implied flexibility, may object to this type of rationale which must therefore be explained with particular care. In many real cases, unfortunately, authors may be better served by invoking the superpopulation concept and thus treating individuals as primary units, even if the rationale for this approach is questionable, rather than risking rejection of their work simply because the approach above is less familiar.

At this point in the discussion, readers may be growing weary of the need for close reasoning and tenuous assumptions and the danger that such efforts may not be viewed favorably by others. We completely agree. Nonrandom sampling, particularly in two dimensions or with multistage designs, causes numerous problems that are difficult or impossible to resolve in a satisfactory manner. The real point of this Section is thus to stress the importance of delineating populations large enough to be interesting, and to develop well-defined sampling plans that avoid nonrandom

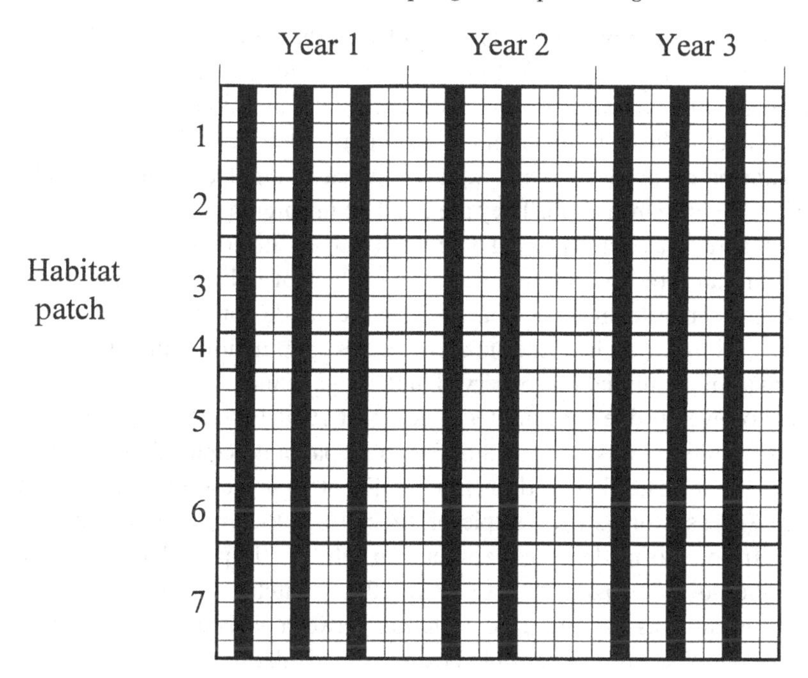

Fig. 4.18. Unequal-sized habitat patches censused on several occasions.

and nonindependent selection. In particular, decide how the population and sampling plans will be defined *before* data collection begins, or at least be sure that this process will not lead to unpleasant surprises if it is carried out after the data collection.

To end this Section on a more optimistic note we return to the example posed at the start of the Chapter (p. 85). As noted there, the investigator probably had few options in this case and, even though it raises difficult issues, a meaningful analysis should nonetheless be possible. We therefore outline how the analysis might be carried out.

The example involved studying habitat preferences by flying aerial surveys across most or all of a study area several times and recording which water bodies had animals near to them and what habitat they were in. Recording water bodies suggests that the investigator wanted to view them as part of the definition of the population unit (and this was in fact true). On reflection, however, it may be best to ignore water bodies and simply view the habitat patches as sampling units. Since they are unequal in size, they must be viewed as groups of arbitrarily defined (but smaller) units, such as hectares. In this particular case, the study area was distinctive, and viewing the sampled patches as coming from a superpopulation was not appealing. Statistical inferences were thus restricted to the study area (during the period of study). The population diagram for a single habitat type would thus appear as in Fig. 4.18 in which rows represent the spatial dimension and columns represent the temporal dimension. The finer, horizontal lines indicate a single unit of area (e.g., a hectare), the heavier lines indicate patch borders. Black squares indicate population units (patch-times) that were measured. They are lined up in columns because they were all measured on each survey. There would be eight such pictures, one for each habitat type, to fully represent the study. Normally, with this type of diagram, separate measurements would be made for each unit in the sample. Thus for each of the smaller squares (population units) we would record a 1 if an animal was observed in the square and a 0 otherwise. In this case, however, it does not matter which particular hectare, within a given patch, an animal was in; recording in which patch it occurred provides all the information needed for the estimates. Fig. 4.18 suggests that the data set be viewed as a one-stage sample of times from the entire 3-year period study. In comparing densities between two habitats, the paired nature of the data should be recognized since each habitat patch was observed at the same time and, as a result, the correlation between observed densities must be taken into account. This could be done by calculating the differences, for each day, between densities and using these differences as the measurements (e.g., in testing whether the sample mean of the differences was significantly different from 0.0). For example, to compare densities in habitats 1 and 2, we could define $d_{12i} = y_{1i} - y_{2i}$, where y_{1i} and y_{2i} are the densities in habitats 1 and 2 on survey i. We would then calculate $\bar{d}_{12} = \Sigma d_{12i}/8$ and $se(\bar{d}_{12}) = sd_{12i}/\sqrt{8}$ and use these quantities in a t-test to determine whether the observed difference in densities was statistically significant.

4.11 Summary

Survey sampling theory provides alternatives to one-stage, simple random sampling. Choosing a sampling plan, or identifying the plan after data have been collected, is often difficult but may be assisted by a 'population diagram' which portrays population units and units in the sample. Several examples are provided in this Chapter. Systematic selection of units often provides better coverage of the population than simple random selection. Unfortunately, no unbiased estimate of variance is known and estimates based on the formula for simple random sampling usually have positive bias (i.e., tend to overestimate standard errors). Nonrandom selection is also common in behavioral ecology studies and, while frequently unavoidable, usually causes uncertainty over what population is really being sampled. Multistage sampling and stratification both involve partitioning the population into groups. In multistage sampling some of the groups are selected randomly and subsequent selections are made only in the selected groups, whereas in stratified sampling population units are selected from every group. Multistage sampling is widely used in behavioral ecology. In nearly all cases, point and interval estimates are based on the estimates from primary units. As a result, systematic sampling within primary units does not cause any difficulty in error estimation. In most studies, the estimates from primary units are weighted equally, and calculations are quite simple. Occasionally, however, the estimates need to be weighted by primary unit size in which case investigators have some options and calculations may be somewhat more complex. Stratification has not been widely used in behavioral ecology, except in wildlife monitoring studies, perhaps because some of the same benefits may be obtained using systematic sampling. Unbiased estimation of standard errors, however, is possible with stratified designs that closely resemble systematic samples; thus, this approach may also be useful to behavioral ecologists. Other designs such as double sampling, poststratification, ratio and regression estimators, and sampling with unequal selection probabilities are discussed briefly. Multistage sampling with nonrandom selection poses particularly difficult problems because estimated precision may depend strongly on subjective decisions about what assumptions to adopt. Examples include systematic sampling in two dimensions and multistage sampling when all individuals in a study area are sampled. While these problems are often unavoidable in practice, the difficulty of treating them rigorously indicates the great benefit of using well-developed sampling plans whenever possible. This Chapter describes many opportunities for achieving this goal, some of which have not been widely used by behavioral ecologists.

5
Regression

5.1 Introduction

This Chapter reviews methods for studying the relationship between two or more variables. We begin with a brief description of scatterplots and simple summary statistics commonly calculated from them. Emphasis is given to examining the effect of outliers and influential points and to recognizing that measures of association such as the correlation coefficient only describe the linear (i.e., straight line) relationship. Simple linear regression is then described including the meaning of the slope, the basic assumptions needed, and the effects of violating the assumptions. Multiple regression is then introduced, again with emphasis on quantities of direct value to behavioral ecologists rather than on the analytical methods used to obtain these results.

We make a slight notational change in this Chapter. In prior Chapters, we have used lower-case letters for quantities associated with the sample and upper-case letters for quantities associated with the population. In discussions of regression, upper-case letters are generally used for individual values and their means regardless of whether they are associated with the sample or population. Regression coefficients for the population are generally identified by the symbols β (e.g., β_o, β_1 ...) and the corresponding sample estimates are denoted b_o, b_1 and so on. These practices are so well established in the statistical literature that we follow them in this Chapter even though it introduces some inconsistency with other Chapters.

5.2 Scatterplots and correlation

Among the simplest and most commonly used methods for studying the relationship between two quantitative variables are scatterplots and correlations. A scatterplot is produced by plotting the pair of measurements on each sample unit as a point on a graph with the vertical axis

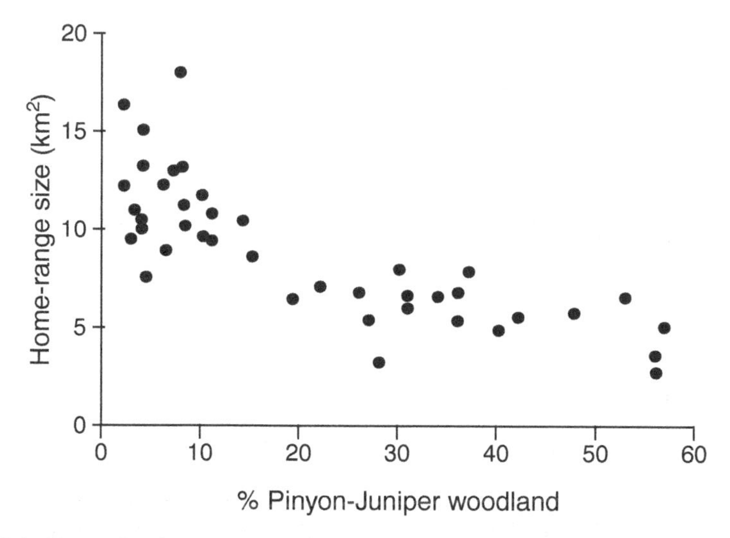

Fig. 5.1. Example of a scatterplot showing coyote home-range sizes *versus* per cent available pinyon-juniper habitat. (From Gese *et al.* 1988.)

representing one of the variables (the dependent variable if one of the variables is so designated) and the horizontal axis the other (the independent variable if one is so designated). The result is a plot that looks like a scattering of points. For example, Gese *et al.* (1988) used a scatterplot (Fig. 5.1) to describe the relationship between coyote home-range size (the dependent variable) and the percentage of available pinyon-juniper habitat (the independent variable). The scatterplot suggests that home range size tended to decrease as percentage pinyon-juniper habitat increased.

When a categorical variable is used to explain changes in a quantitative variable a scatterplot can also be constructed. The categories of the categorical variable are represented as equally spaced points on the horizontal axis and the resulting scatterplot looks like a series of vertical bars above these points. One may wish to display the mean or median of the quantitative variable at each value of the qualitative variable on the plot. For example, Diernfeld *et al.* (1989) plotted the mean of plasma vitamin E (the dependent variable) from peregrine falcons *versus* the categorical independent variable month (Fig. 5.2). One could suppress the individual data points and show only the mean and a bar representing the standard deviation of the data points for a given month to keep the plot from appearing overly cluttered. Among other things, we see that in July mean plasma vitamin E is highest.

It is customary to attempt to quantify any relationships that appear in a scatterplot. The simplest possible relation is when the points in a

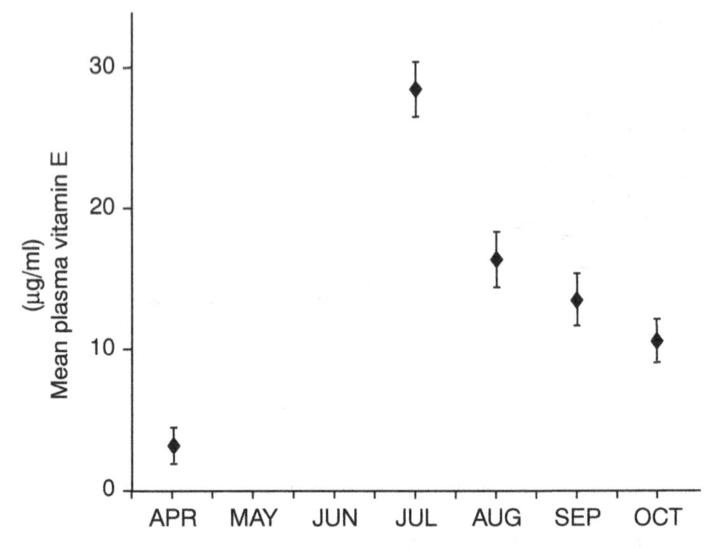

Fig. 5.2. Example of a scatterplot involving a categorical independent variable. The plot displays both the mean of plasma vitamin E per month and a bar representing the standard error of the means. (From Dierenfeld *et al.* 1989.)

scatterplot appear to be centered along a straight line. When above average values of one variable tend to accompany above average values of the other variable we say that the variables are positively associated. When above average values of one variable tend to accompany below average values of the other the variables are said to be negatively associated. The more tightly the points appear to be clustered about a straight line, the more highly associated they are. A numerical measure of this degree of association is the (Pearson product moment) correlation coefficient. Suppose we have a sample of n observations on two variables X and Y denoted

$$(X_1, Y_1), (X_2, Y_2), \ldots, (X_n, Y_n)$$

The sample correlation coefficient, r, may be expressed in several ways

$$r = \frac{\operatorname{cov}(X_i, Y_i)}{\operatorname{sd}(X_i)\operatorname{sd}(Y_i)} \tag{5.1}$$

$$= \frac{1}{n-1}\sum_{i=1}^{n}\left(\frac{X_i - \bar{X}}{\operatorname{sd}(X_i)}\right)\left(\frac{Y_i - \bar{Y}}{\operatorname{sd}(Y_i)}\right)$$

$$= \frac{\sum (X_i - \bar{X})(Y_i - \bar{Y})}{\sqrt{\sum (X_i - \bar{X})^2 \sum (Y_i - \bar{Y})^2}}.$$

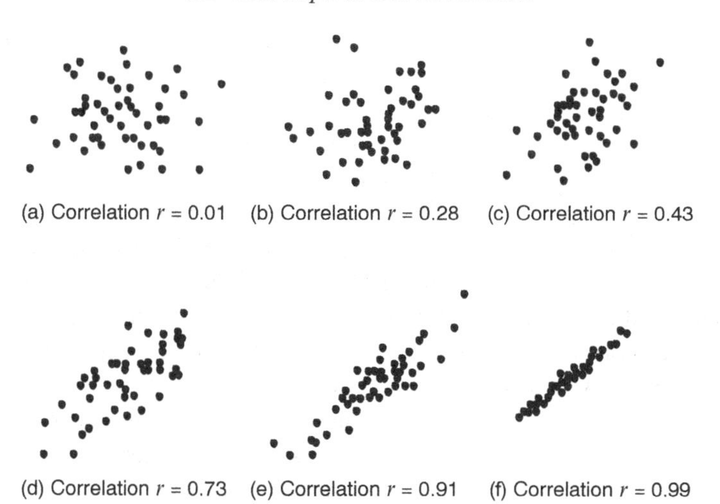

Fig. 5.3. Examples of scatterplots and their associated correlations. (From Moore and McCabe 1979.)

The first expression describes r as the ratio of the covariance of X_i and Y_i to the product of their standard deviations. The second versions expresses r as a function of the products of deviations, $X_i - \overline{X}$ and $Y_i - \overline{Y}$, each standardized (divided by) its standard deviation. The third expression shows that r may be calculated solely from the appropriate sums of squares rather than mean squares.

If whenever X_i takes on a value above its mean Y_i tends also to be above its mean, then the products of the deviations in the summation tend to be positive and r is positive. Likewise, if whenever X_i takes on a value above its mean the corresponding value of Y_i tends to be below its mean, then the products of the terms in parentheses in the summation tend to be negative and r is negative. Positively associated variables will therefore have a positive correlation and negatively associated variables will tend to have a negative correlation. It can be shown that the correlation coefficient must take on a value between -1 and $+1$, achieving the values ± 1 only if all the observations fall exactly on a straight line. Data that fall exactly on a horizontal line are defined as having a correlation coefficient of 0.0 (Y_i has constant value independent of the value of X_i). The correlation coefficient is undefined for data that fall perfectly on a vertical line (X_i has constant value). In these latter examples, either X_i or Y_i is constant and hence questions concerning how changes in one variable relate to changes in the other cannot be answered, because one of the variables does not change. Some sample scatterplots and the associated value of r are given in Fig. 5.3.

Two subtle points about the relation between scatterplots and the correlation coefficient are worth mentioning. First, the correlation coefficient depends on the *vertical* distances between the observations and the line rather than on the perpendicular distances to the line. The differences between these two distances can be large if the line is very steep. Thus, two scatterplots may appear to be equally tightly clustered (in the sense of perpendicular distance) about a line, yet have quite different values of r because the lines have quite different slopes. Second, a scatterplot can show a very distinct trend or pattern and yet the correlation can be 0. This is because the correlation coefficient indicates only whether there is a *straight line* relation between two variables. When a scatterplot displays a relationship other than a straight line, the 'strength' of the relationship can be measured through other measures such as the coefficient of multiple determination using a multiple regression model (Section 5.4) or by investigating whether a straight line relation exists between transformations (functions) of the two variables. An example of this latter method is given in Section 5.3 on simple linear regression.

A few words of caution concerning correlation may be useful. The presence of correlation between two variables, even a substantial correlation (near $+1$ or -1), does not imply that a cause and effect relationship exists between the two variables. The correlation may be present because both variables are responding to changes in some third variable. Correlation may also be present but impossible to interpret when a cause and effect relation exists between two variables, but this effect is 'mixed up' or 'confounded' with the fact that changes in several other factors are also causing changes in the two variables.

Additional features of scatterplots that one should be aware of are *outliers* and *influential* points. An outlier is a point that lies well above or below the 'band' or 'cloud' formed by the rest of the points. An influential point is one that has a strong effect on the impression that a trend is present in the data, i.e., removal of this point would have a significant effect on our impression of the trend present. Isolated points at the margins of a scatterplot are often influential.

In the scatterplot in Fig. 5.4 from Fryxell *et al.* (1988), the circled point would be considered an outlier as it lies well above the 'band' formed by the other points. This point is also influential. In the scatterplot in Fig. 5.5, adapted from Renecker and Hudson (1989), the circled point is influential. The trend suggested by the plot with the point present is much steeper than that when the point is removed.

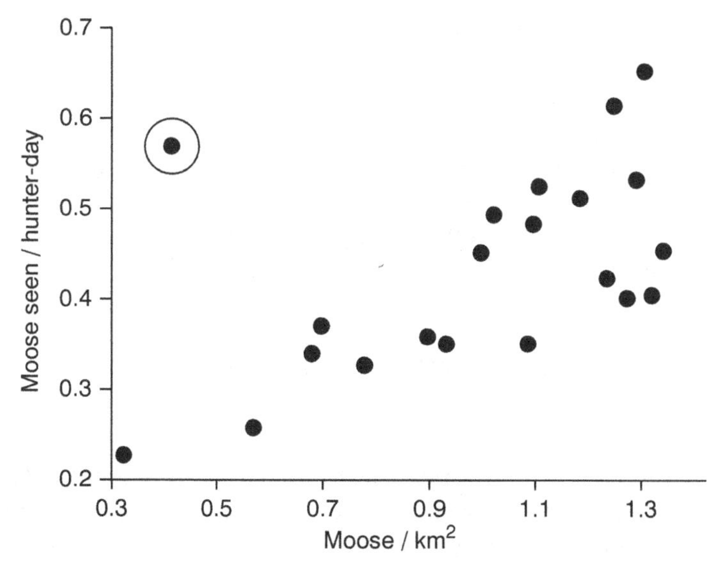

Fig. 5.4. Example of an outlier (circled) in a scatterplot. The plot displays moose density (per km^2) *versus* moose seen per hunter-day. (From Fryxell *et al.* 1988.)

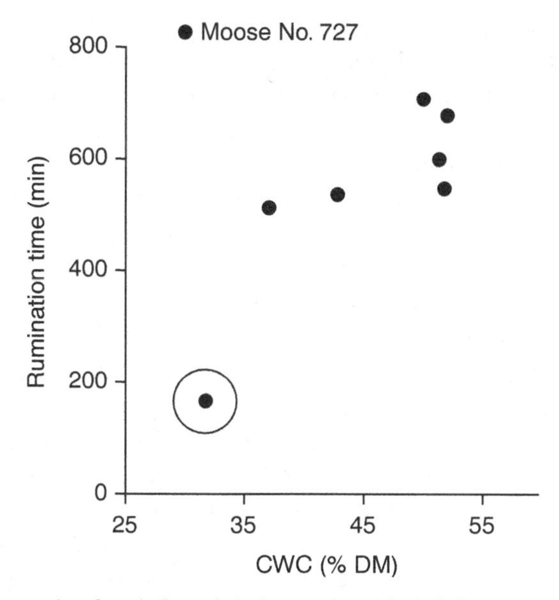

Fig. 5.5. Example of an influential observation (circled) in a scatterplot. The plot displays the time a female moose spent ruminating cell wall constituents (CWC) in the diet dry matter (DM) for selected times from December 1982 to January 1984 at Ministik Wildlife Research Station, Alberta, Canada. (From Renecker and Hudson 1989.)

When outliers are present, it is worth investigating whether the observation corresponding to the point is special in some way. If it is, analyzing it separately from the remaining data may be worthwhile. When a point is influential, the data should be analyzed twice, once with the point present and once with the point absent. Thus, conclusions based on analysis with the point present must be regarded with caution if they differ from conclusions based on analysis with the point absent. Conclusions whose validity rests on a single observation cannot be made with confidence.

5.3 Simple linear regression

Formal models

If the scatterplot or correlation coefficient suggests that the relation between *bivariate data* (i.e., data in which two variables are measured on each sample unit) is a straight line trend (recall that the equation of a straight line is $Y = b_0 + b_1 X$, where b_0 is the Y-intercept and b_1 is the slope), often one will want to explore this apparent relation further. The method for such exploration is called *simple linear regression*. The word 'regression' was coined by the English scientist Sir Francis Galton (1822–1911) and was based on his study of the relation between the heights of fathers and their sons. He found that tall parents tended to produce offspring that were taller than average, but not as tall as their parents and called this phenomenon 'regression toward mediocrity'.

In behavioral ecology, the question most commonly of interest is whether the apparent straight line trend indicates that a true straight line trend exists in the population from which the data were drawn. To answer this question, one must think about the population from which the sample (X_1, Y_1), $(X_2, Y_2), \ldots, (X_n, Y_n)$ comes. If one thinks of the X_i as independent variables and the Y_i as dependent variables, a formal regression analysis assumes that the distribution of all Y_i in the population, having a given value on the X-axis (X), is normal with mean $\beta_0 + \beta_1 X$ and variance σ^2. Thus the population units are scattered around the line $Y = \beta_0 + \beta_1 X$, and σ^2 describes how tightly the points cluster around the line. In particular, the proportion of population units within a given vertical band about the line $Y = \beta_0 + \beta_1 X$ is determined by the normal distribution. Notice that the variance σ^2 of the population of y-values for a given X is independent of X. This is referred to as the assumption of *homogeneity of variance*. One makes no assumptions about the distribution of the X_i in the population. If neither the X_i nor Y_i values are regarded as independent variables and both are regarded as random, the *bivariate normal* distribution may provide an adequate description of their joint distribution. The bivariate normal distribution has the property that for a given value on the X-axis (X) the distribution of the Y_i for i is such that $X_i = X$ is normal with mean $\beta_0 + \beta_1 X$ and variance σ^2. In addition, the X_i are also assumed to be normally distributed in the population with mean μ_x and variance σ^2, and in the population of all units the correlation coefficient between the X_i and Y_i is assumed to be ρ. Discussion of the bivariate normal distribution is beyond the scope of this Chapter. The interested reader is referred to a text on regression or multivariate analysis (e.g., Neter *et al.* 1983). Although this description of the population is rather complicated, the validity of inferences depends on the extent to which this description holds.

Inference

In behavioral ecology, researchers usually wish to make inferences about the slope β_1, generally testing whether the slope is 0 (which under the assumption of simple linear regression is interpreted as equivalent to testing whether or not there is a relation between Y and X). One may also wish to make inferences about the intercept β_0, the variance σ^2, the correlation ρ (when the population is bivariate normal), and predictions of future values of Y for a given X based on the line $Y = \beta_0 + \beta_1 X$. The first step is to obtain estimates of the slope β_1 and intercept β_0. This is generally done by the method of least squares. This method finds the equation of the straight line (called the least-squares regression line) with the property that it minimizes the sum of the squares of the vertical distances of the individual data points from the line. If we have n observations on two variables X and Y, denoted $(X_1, Y_1), (X_2, Y_2), \ldots, (X_n, Y_n)$, using calculus one can show that the least-squares line has the equation $Y = b_0 + b_1 X$ where

$$b_i = \frac{\sum_{i=1}^{n} (X_i - \bar{X})(Y_i - \bar{Y})}{\sum_{i=1}^{n} (X_i - \bar{X})^2} \tag{5.2}$$

$$b_0 = \bar{Y} - b_1 \bar{X}.$$

Notice in the denominator of b_1 that if all the X_i are equal the denominator will be 0 in which case β_1 is undefined. Hence we must take observations at two or more different values of X_i (and hence at least two observations) to estimate b_0 and b_1. It also turns out that at least three observations involving at least two different values of the X_i are necessary to estimate variances and make statistical inferences. If the pairs (X_1, Y_1), $(X_2, Y_2), \ldots, (X_n, Y_n)$ in the sample are independent (this will be true if sample units were selected by simple random sampling), one can show that unbiased estimates of β_0 and β_1 are given by the least-squares estimates b_0 and b_1.

An unbiased estimate of σ^2 is given by the mean square error (MSE)

$$\mathrm{MSE} = \frac{1}{n-2} \sum_{i=1}^{n} (Y_i - b_0 - b_1 X_i)^2 \tag{5.3}$$

and, if appropriate, ρ is estimated by the correlation coefficient, r, for the sample data. In addition b_0 and b_1 have normal distributions with means β_0 and β_1 and actual standard errors

$$\sigma(b_0) = \sigma \sqrt{\frac{1}{n} + \frac{\overline{X}^2}{\sum_{i=1}^{n} (X_i - \overline{X})^2}} \tag{5.4}$$

$$\sigma(b_1) = \frac{\sigma}{\sqrt{\sum_{i=1}^{n} (X_i - \overline{X})^2}}$$

respectively. Estimates of these standard errors, denoted $se(b_0)$ and $se(b_1)$, are obtained by replacing σ by its estimate $\sqrt{\mathrm{MSE}}$. MSE, $se(b_0)$, and $se(b_1)$ all have $n-2$ degrees of freedom. Confidence intervals and hypothesis tests for β_0 and β_1 based on normal theory are applicable and the general procedures discussed in Sections 3.2 and 3.3 can be used. For example, a $(1-\alpha) \times 100\%$ confidence interval for the true slope β_1 of the regression line is

$$b_1 \pm t_{n-2,\alpha/2} \mathrm{se}(\mathrm{b}_1), \tag{5.5}$$

and an α-level test of the hypotheses $H_0: \beta_1 = 0$ *versus* $H_1: \beta_1 \neq 0$, i.e., a test of whether the slope differs from 0, is to reject H_0 if

$$|b_1| > t_{n-2,\alpha/2} \mathrm{se}(b_1). \tag{5.6}$$

Rejection of H_0 implies there is evidence that a straight line relation between X and Y explains some of the variability in Y and hence that the correlation, ρ, is nonzero. In fact one can show that the test of $H_0: \beta_1 = 0$ *versus* $H_1: \beta_1 \neq 0$ is equivalent to testing $H_0: \rho = 0$ *versus* $H_1: \rho \neq 0$.

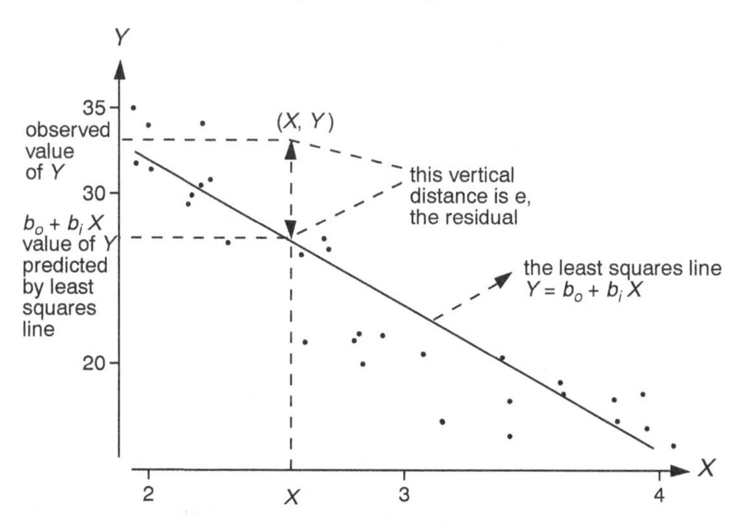

Fig. 5.6. The least-squares line superimposed on a scatterplot. Also shown for a particular point are the observed value of the Y, the predicted value of Y, and the residual as the vertical distance from a point to the regression line.

Examining assumptions

The validity of any inference one makes depends on the extent to which our formal model describes the population from which the data were drawn. Thus, a complete regression analysis should include examination of the assumption that the data follow a normal distribution, have homogeneity of variance, follow a straight line, and are independent. 'Unusual' observations, such as outliers or influential points, should also be identified and their effect on conclusions should be investigated. Examination of assumptions is often carried out by examining the residuals $e_i = Y_i - b_0 - b_1 X_i$. Notice that the residual is simply the difference between the value of Y actually observed and the value Y would have if it fell exactly on the least-squares line (Fig. 5.6).

Two ways in which the residuals are used to examine assumptions are as follows. First, recall that the scatter of the population units (of which our data are a sample) about a line is determined by the normal distribution. In particular, the proportion within a given vertical band about this line is determined by the normal distribution. One consequence of this is that the residuals, which measure how far a particular observation is from the least-squares line, should behave approximately as though they have a normal distribution with mean 0. If one calculates all the residuals (many statistical software packages that do regression will calculate residuals), one can use

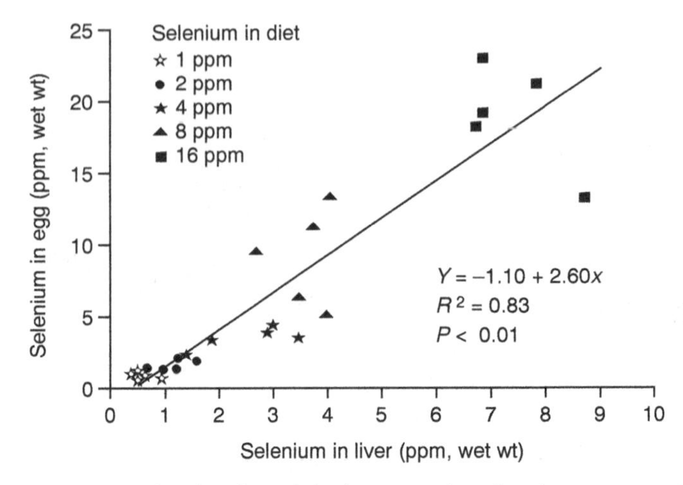

Fig. 5.7. Example of a violation of the homogeneity of variance assumption. The scatterplot displays the relationship between the concentration of selenium in the liver of female mallards and the concentration in the eighth eggs for females fed diets containing 1, 2, 4, 8, or 16 ppm selenium as selenomethionine. (From Heinz *et al.* 1989.)

statistical procedures to investigate whether the residuals appear to have a normal distribution. Second, the homogeneity of variance assumption implies that the population units should display the same magnitude of variability about the regression line for any point on the x-axis. As a consequence, the magnitude of the residuals should not display any tendency to increase or decrease as the associated value of x increases or decreases. Such tendencies may indicate a violation of the homogeneity of variance assumption.

Corrective action may be necessary if the assumptions are not met. For example, if the data do not appear normal or do not satisfy the homogeneity of variance assumption (violations of these two assumptions often occur together), one may try replacing (transforming) the values of $Y_1,\ldots,Y_n$ by some function of these values, i.e., by $f(Y_1),\ldots,f(Y_n)$. Common functions, f, are the logarithm, square root, reciprocal, or arcsin. Often, the transformed Y-values will satisfy the assumptions of normality or homogeneity of variance and regression can proceed on the pairs $[X_1,f(Y_1)],\ldots,[X_n,f(Y_n)]$. One must proceed with caution, however, because conclusions will now refer to the transformed data. For example, if a reciprocal transformation is used and one determines that the relation between $1/Y$ and X is $1/Y = -2X$, it is incorrect to conclude that increases in X are associated with decreases in

Y. In fact, the relation between Y and X is $Y = -1/(2X)$ and Y increases (becomes less negative) as X increases.

Although a formal discussion concerning the effects of violations of assumptions on the resulting inference is quite mathematical, a few comments can be made. If the expected value of Y at a given X is in fact $\beta_0 + \beta_1 X$ (i.e., on average the value of Y at a given X is $\beta_0 + \beta_1 X$), then the least-squares estimates of β_0 and β_1 are unbiased even if the assumptions of homogeneity of variance, normality, and independence do not hold. If the expected value of Y at a given X is in fact $\beta_0 + \beta_1 X$ and the assumptions of homogeneity of variance and independence hold, then MSE is an unbiased estimate of the variance even if the assumption of normality does not hold. For testing hypotheses or constructing confidence intervals all assumptions must hold, although these procedures are felt to be 'robust' to departures from the assumption of normality, i.e., these inferences are still valid even if the assumption of normality is somewhat suspect.

The above discussion of simple linear regression is rather sketchy. The important thing to remember is that simple linear regression is a statistical tool for studying possible straight line relationships between pairs of variables. A thorough discussion of simple linear regression, including checking of assumptions, can be found in any book on regression analysis (e.g., Neter *et al.* 1983). Many introductory texts on statistics also contain discussion of simple linear regression (e.g., Moore and McCabe 1993)

5.4 Multiple regression

Formal models

The techniques used in simple linear regression can be extended to provide methods for examining relationships other than straight line relationships between two variables. The general method is called multiple regression analysis and applies to 'linear models'. Suppose a sample of n units is selected from some population, and for each of these units one records a dependent variable Y and p independent variables $X_1, \ldots, X_p$. One allows some of the X-variables to be functions of the others; for example, one might allow X_2 to be the square of X_1. Let Y_i and $X_{1i}, \ldots, X_{pi}$ be the values of the variables for unit i. One is said to have a linear model if the population from which the units are drawn and the method of selecting units are such that the relationship between Y and the X values can be written as

$$Y_i = \beta_0 + \beta_1 X_{1i} + \beta_2 X_{2i} + \ldots + \beta_p X_{pi} + \varepsilon_i. \tag{5.7}$$

The ε_i represent the effects of measurement error and independent variables not included in the model whose individual effects are considered to be small relative to those of the X values. The Y_i are assumed to be independent (this is reasonable if units are selected by simple random sampling). The Y_i of units having a given X value, $X_1,\ldots,X_p$, are assumed to be normally distributed with mean $\beta_0+\beta_1X_{1i}+\beta_2X_{2i}+\ldots+\beta_pX_{pi}$ and constant variance denoted by σ^2 (this is called the homogeneity of variance assumption because this variance is the same regardless of the values of the X-variables). These last three assumptions concern the population from which the data are drawn. β_0, $\beta_1,\ldots,\beta_p$ are unknown constants (parameters) that one wishes to make inferences about. The assumptions imply that the ε_i are independent. Each ε_i has mean 0, variance σ^2, and is normally distributed. Thus, a given value, Y_i , of the dependent variable will generally not equal $\beta_0+\beta_1X_{1i}+\beta_2X_{2i}+\ldots+\beta_pX_{pi}$ exactly, but discrepancies average to 0, i.e., will yield 0 when averaged over all units in the population for which the X-variables$=X_{1i},\ldots,X_{pi}$. In mathematical language, one says the expected value of Y_i is $\beta_0+\beta_1X_{1i}+\beta_2X_{2i}+\ldots+\beta_pX_{pi}$. The quantity $\beta_0+\beta_1X_{1i}+\beta_2X_{2i}+\ldots+\beta_pX_{pi}$ is called the multiple regression function. As with simple linear regression, the validity of inferences in multiple regression depends on the extent to which the above assumptions hold. Thus, checking whether the assumptions hold reasonably well is an important component of any multiple regression analysis.

The β_k are often interpreted as the 'effects' of the X-variables in the sense that a unit change in X_k, holding the other X-variables fixed, is accompanied by a change of size β_k in Y, on average. In practice, the observed changes due to X_k (as reflected by the b_k, the least-squares estimate of β_k) may be directly caused by X_k or may occur because a change in X_k causes changes in other variables, which in turn cause a change in Y. If β_k is 0, changes in X_k, with the other X-variables fixed, produce no change in Y and so X_k has no 'effect' on Y. Testing the hypothesis that $\beta_k=0$ is one way to examine whether X_k has an 'effect' on Y. In practice, however, one may encounter the following problem. In many experiments, the researcher has little or no control over the values of the X-variables. This occurs in observational studies, for example, when units are selected from some population by simple random sampling and the X-variables are characteristics of the unit (i.e., age, sex, mass). In such cases, the X-variables are likely to be correlated and so a change in X_k is associated with changes in some of the other X-variables , which in turn affect Y through the other β_s. One never observes a set of units in which only X_k varies and all the other X-variables are constant. Roughly, this means that we get no 'direct' information about β_k, only 'indirect' information that is subject to the additional variation in the other X-variables. If the correlation among the X-variables is large, this additional variation is large and one's uncertainty about β_k is increased. This state of affairs is called multicollinearity, and can lead to uncertain inferences about β_k, i.e., estimates will have large standard errors.

The term 'linear' in linear model is borrowed from a branch of mathematics called linear algebra, and refers to the fact that such an equation says that the expected value of Y is a linear function of the β_i. This means that the model is a sum of terms of the form

(parameter)$\times$(some function of the independent variables).

It does not mean that Y is a straight line function of the X_i. For example, if only a single independent variable X is measured, one can define $X_j = X^j$, so that the linear model is

$$Y_i = \beta_0 + \beta_1 X_i + \beta_2 X_i^2 + \ldots + \beta_p X_i^p + \varepsilon_i, \tag{5.8}$$

i.e., Y is a polynomial function of X. Models that are not linear functions of β_i, such as

$$Y_i = \beta_0 + \beta_1 X_i^{\beta_2} \tag{5.9}$$

are called nonlinear models.

Inference and interpretation

Analysis of a linear model proceeds in a manner analogous to that used for simple linear regression. The method of least squares can be used to obtain estimates, b_i, of the β_i. A word of caution is needed here. Recall from the discussion of simple linear regression that one needed at least two different values of the independent variables to estimate the two parameters β_0 and β_1. In multiple regression a similar problem arises. One needs at least $p+1$ different sets of values of the independent variables (and hence at least $p+1$ observations) to be able to estimate the $p+1$ parameters $\beta_0, \beta_1, \ldots, \beta_p$. If, in addition, one wants to get estimates of variances one needs at least $p+2$ observations on at least $p+1$ different sets of values of the independent variables. Furthermore, to evaluate the fit of the model, via a so-called lack of fit test (see Neter *et al.* 1983 for details), one needs at least one repeat observation at a fixed set of values of the independent variables. Of course, the more observations taken, the better, and, for purposes of inference, the wider the range of values of the independent variables at which observations are obtained (with several repeat observations at several sets of values of the independent variables), the better. Unfortunately, the cost or difficulty of obtaining observations may place severe limits on the number of observations one can obtain. It is therefore difficult to give guidelines concerning how many observations to take, and perhaps the best advice is to consult a statistician knowledgeable in design of experiments.

If one can estimate the β_i and if the errors, ε_i, are assumed to be normal, formulas for the standard errors, $se(b_i)$, and of the estimates, b_i, can be derived and used to construct confidence intervals and test hypotheses with the normal theory methods discussed in Chapter Three. The formula for degrees of freedom is $n-p-1$. Unfortunately, these formulas are rather complex and a knowledge of matrix algebra is necessary for their derivation (see, for example, Neter *et al.* 1983 for details). In practice, these estimates and standard errors are obtained with statistical software.

Interpretation of results can be complicated. The following example illustrates this. Bergerud and Ballard (1988) used multiple regression to study the effect of snow depth (mean depth in cm over an 8-month winter period), wolf numbers (in winter after birth), and total caribou numbers as an index of caribou recruitment in south central Alaska. The index was defined as the percentage of 2.5-year-old caribou among all caribou ≥ 2.5 years old. Several multiple regression models were run. One result was

$$\text{recruitment} = 20.980 + 0.128 \text{ snow depth} - 0.064 \text{ wolves}.$$

This model would predict, for example, that for a mean snow depth of 50 cm and 200 wolves in the winter after birth, recruitment would be $20.980 + 0.128 \times 50 - 50.064 \times 200 = 14.58\%$. Notice that this model has a positive coefficient for the snow depth term, which would seem to suggest that increased snow depth (indicating a more severe winter) increases recruitment. This is counterintuitive and is undoubtedly due to multicollinearity, i.e., number of wolves and snow depth may be correlated and so the coefficients are somewhat difficult to interpret. The presence of multicollinearity is further indicated by the fact that a simple linear regression of recruitment on snow depth yielded the model

$$\text{recruitment} = 23.261 - 0.166 \text{ snow depth}.$$

In this model, the coefficient for snow is negative, i.e., less snow yields higher recruitment, which would seem more plausible. When the coefficient of a term in a multiple regression model changes sign or changes in size dramatically when other independent variables are added to the model, multicollinearity is often present, and interpretation of individual coefficients must be done with care, if at all.

In a multiple regression analysis one often reports more than simply least-squares estimates and standard errors of parameters. Several measures of how well the model fits the data and generalizations of the correlation coefficient discussed in simple linear regression may also be reported. For the model

$$Y_i = \beta_0 + \beta_1 X_{1i} + \beta_2 X_{2i} + \ldots + \beta_p X_{pi} + \varepsilon_i, \tag{5.10}$$

these include the following:

1. *The sum of squares total* (SSTO). This measures the total variation in the dependent variable Y and is given by the formula

$$\text{SSTO} = \Sigma(Y_i - \overline{Y})^2. \tag{5.11}$$

2. *The sum of squares for error* (SSE). This measures how much the values of the dependent variable vary about the fitted multiple regression model $b_0 + b_1 X_1 + \ldots + \mathrm{b}_p X_p$, where b_k is the least-squares estimate of β_k. The formula for SSE is

$$\text{SSE} = \Sigma(Y_i - b_0 - b_1 X_{1I} - \ldots - b_p X_{pi})^2. \tag{5.12}$$

   Additionally, one can define the mean sum of squares for error (MSE) to be SSE/$(n-p-1)$, where n is the number of observations. MSE is an unbiased estimate of σ^2, the variance of the errors e_i, and provides a measure of how well the model fits the data; the smaller the value of MSE the better the fit. As already pointed out, the number of observations (n) must exceed $p+1$ for MSE to be defined. If n is less than $p+1$ then one can show that SSE is 0. In addition, the denominator of MSE will be ≤ 0.
3. *The sum of squares for regression* (SSR). This measures how much of the variation in the dependent variable is accounted for by the multiple regression model and is given by the formula

$$\text{SSR} = \text{SSTO} - \text{SSE}. \tag{5.13}$$

4. The *coefficient of multiple determination*, denoted R^2. This measures the fraction of the variation in the dependent variable accounted for by the multiple regression model and is given by the formula

$$R^2 = \text{SSR}/\text{SSTO}. \tag{5.14}$$

   R^2 is always between 0 and 1 and is similar to SSR in interpretation. One shortcoming of SSR as a measure of how well a multiple regression model fits a set of data is that whether or not SSR is sufficiently large depends on how large SSTO is, i.e., SSR must be interpreted relative to SSTO. R^2 accomplishes this automatically by taking the ratio of SSR and SSTO; as such, it is a unitless quantity. In simple linear regression, one can compute R^2 and it turns out to equal the square of the usual correlation coefficient.

For this reason R, the positive square root of R^2, is often perceived as the generalization of the correlation coefficient from simple linear regression to multiple regression and is called the *multiple correlation coefficient*.

These four measures are routinely reported by statistical software for regression and form the basis for comparing various multiple regression models. Such comparisons are formally conducted as follows. To determine whether the 'full' regression model

$$Y_i = \beta_0 + \beta_1 X_{1i} + \beta_2 X_{2i} + \ldots + \beta_p X_{pi} + \varepsilon_i \tag{5.15}$$

is necessary to explain the variation in the dependent variable Y, or if the 'reduced' model

$$Y_i = \beta_0 + \beta_1 X_{1i} + \beta_2 X_{2i} + \ldots + \beta_q X_{qi} + \varepsilon_i, \tag{5.16}$$

involving only the independent variables $X_1, X_2, \ldots, X_q$ $(q<p)$, which are a subset of $X_1, X_2, \ldots, X_p$, is adequate to explain the variation in the dependent variable, calculate SSR and SSE for the full model and for the reduced model. Let $\mathrm{SSR}(X_1,\ldots,X_p)$ and $\mathrm{SSE}(X_1,\ldots,X_p)$ denote SSR and SSE for the full model and $\mathrm{SSR}(X_1,\ldots,X_q)$ and $\mathrm{SSE}(X_1,\ldots,X_q)$ denote SSR and SSE for the reduced model. The quantity

$$\mathrm{SSR}(X_{q+1},\ldots,X_p \mid X_1,\ldots,X_q) = \mathrm{SSR}(X_1,\ldots,X_p) - \mathrm{SSR}(X_1,\ldots,X_q) \tag{5.17}$$

is called the extra sum of squares and measures how much better the full model fits the dependent variable compared with the reduced model. If this is sufficiently large, more precisely if $[\mathrm{SSR}(X_{q+1},\ldots,X_p\ldots \mid X_1,\ldots,X_q)/(p-q)]$ $/[\mathrm{SSE}(X_1,\ldots,X_p)/(n-p)]$ exceeds the appropriate critical value of the F statistic with $p-q$ numerator and $n-p$ denominator degrees of freedom, one decides that the full model is necessary. Otherwise one decides that the reduced model is adequate. Formally this tests the hypothesis of whether the independent variables $X_{q+1},\ldots,X_p$ have a significant effect on the dependent variable after accounting for the effects of the independent variables $X_1,\ldots,X_q$. This method is purely statistical and does not take into account the scientific 'reasonableness' of the full or reduced model. The statistical decision must be modified by scientific considerations in any final decision concerning an appropriate model.

An informal test of the above hypothesis is often carried out by simply comparing the R^2 values for the full and reduced models and selecting the full model if the R^2 is appreciably higher, although how much higher is 'appreciably higher' is rather subjective. The formal hypothesis test is probably the better way to make comparisons.

In Bergerud and Ballard (1988), the two multiple regression models mentioned on p. 162 i.e.,

$$\text{recruitment} = 23.261 - 0.166 \text{ snow depth},$$

and

$$\text{recruitment} = 20.980 + 0.128 \text{ snow depth} - 0.064 \text{ wolves},$$

have R^2 values of 0.10 and 0.79, respectively. These values suggest that snow depth is not a particularly significant predictor of recruitment, but wolf numbers, when added to a model containing snow depth, is a significant predictor of recruitment. The authors also fit a model using only wolf numbers as an independent variable and obtained

$$\text{recruitment} = 24.379 - 0.057 \text{ wolves},$$

with $R^2 = 0.75$. This suggests that wolf numbers are a significant predictor of recruitment but that the addition of snow depth to a model containing wolf numbers is not particularly significant (R^2 increases only to 0.79). Unfortunately, no information about formal tests of hypotheses are mentioned in the paper so these conclusions are somewhat subjective.

Several general observations can be made from this example. First, the value of R^2 increased in the above models when an additional independent variable was added. This always occurs in multiple regression, i.e., the addition of an independent variable will always cause R^2 to increase (or at worst stay the same). This is intuitively plausible, because the addition of independent variables provides extra information and cannot detract from our predictive ability. One could always ignore the extra information. Because R^2 can be inflated by adding independent variables, one must be careful to avoid adding extra independent variables simply to get a large R^2. A balance between reasonable R^2 and relatively few independent variables (simplicity of the model and hence ease in interpretation) is the goal. This is called parsimony.

Second, notice that interpretations were a bit awkward. For example, in comparing the model with snow depth and wolf numbers as independent variables to the model with only snow as an independent variable, we concluded that wolf numbers added predictive power to a model already containing snow depth as an independent variable. This 'conditional' interpretation is a bit different than merely saying that wolf numbers are a significant predictor of recruitment. This illustrates the care needed to interpret the results of a multiple regression analysis.

Third, we mentioned that the model with only snow depth as an independent variable had an R^2 of 0.10, which did not seem to be particularly significant. Actually it is possible in multiple regression to have a very low R^2 (any value >0, even 0.000001) and yet have statistical significance in a formal hypothesis test. Conversely, it is possible to have a large value of R^2 and not have statistical significance. For this reason it is good practice to conduct formal tests of hypotheses in addition to reporting R^2 values.

Fourth, again examining the model with only snow depth as an independent variable, we were tempted to conclude that snow depth was not useful as a predictor of recruitment. Technically one can conclude only that a straight line relationship between snow depth and recruitment has not been demonstrated. One might find that a multiple regression model, such as

$$\text{recruitment} = \text{constant} + b_1 \times \text{snow} + b_2 \times \text{snow}^2 + b_3 \times \text{snow}^3$$

has a fairly high R^2 and is statistically significant, indicating that snow depth is useful for predicting recruitment, but the prediction relation is more complicated (here a cubic polynomial) than a simple straight line relation. Bergerud and Ballard (1988) reported that a three-way analysis of variance (ANOVA) was conducted and neither the main effect of snow depth nor its interaction with other variables was significant. This kind of analysis does suggest that snow depth is not useful as a predictor of recruitment (although the authors do not make it clear exactly how the variable snow depth was categorized so as to make it a classification variable suitable for ANOVA). In general, multiple regression tends to provide information about the specific way in which an independent variable is useful for predicting a dependent variable. ANOVA (or regression with indicator variables – see later) is more suitable for determining whether an independent variable is useful in some way (no specific functional form specified) for prediction.

Fifth, notice that the dependent variable, being a percentage, is constrained to lie between 0% and 100%. For the model with snow depth as the only independent variable, a snow depth of 150 cm would predict recruitment at –1.639% which is, of course, nonsense. Examination of the authors' data shows that actual snow depth never exceeded 75 cm. Substituting a value of 150 cm, therefore, involves extrapolating to data outside the range used to estimate the multiple regression model. Such extrapolation should be avoided and multiple regression models should be considered valid only for the range of data used to establish the model.

Partial correlation

The coefficient of partial correlation is often reported in multiple regression analyses. Consider once again the multiple regression model

$$Y_i = \beta_0 + \beta_1 X_{1i} + \beta_2 X_{2i} + \ldots + \beta_p X_{pi} + \varepsilon_i. \quad (5.18)$$

The amount of additional variability explained by adding X_j to a model already containing the r variables $X_{k1},\ldots,X_{kr}$ is called the coefficient of partial determination between Y and X_j given $X_{k1},\ldots,X_{kr}$ and is defined to be

$$r^2_{j\cdot k1,\ldots,kr} = \mathrm{SSR}(X_j|X_{k1},\ldots,X_{kr})/\mathrm{SSE}(X_{k1},\ldots,X_{kr}),$$

where $\mathrm{SSR}(X_j|X_{k1},\ldots,X_{kr})$ is defined as in Eq. 5.17. The corresponding coefficient of partial correlation is the square root of $r^2_{j\cdot k1,\ldots,kr}$ with the sign equal to that of b_j in the fitted model

$$Y = b_0 + b_jX_j + b_{k1}X_{k1} + \ldots + b_{kr}X_{kr}. \quad (5.19)$$

The relationship between the coefficient of partial determination and the coefficient of partial correlation is analogous to that between the coefficient of multiple determination (R^2) and the correlation coefficient (r) in regression. In particular, the coefficient of partial determination is easier to interpret than the coefficient of partial correlation. Compton *et al.* (1988) fitted a multiple regression model, with number of deer (ND) observed at various locations along the lower Yellowstone River as the dependent variable. Amount of riparian cover in hectares (RC) and amount of riparian cover with cattle in hectares (GR) were the independent variables. The fitted model was

$$\mathrm{ND} = -3.69 + 0.92\mathrm{RC} - 0.50\mathrm{GR},$$

with an R^2 of 0.57. The coefficient of partial correlation of GR for a model already containing RC was -0.53. Notice the sign matches that of the coefficient of GR in the fitted model. The coefficient of partial determination is $(-0.53)^2 = 0.28$. We conclude that the addition of GR to a model already containing RC accounts for an additional 28% of the variance [SSE(RC)] still remaining.

Examining assumptions

In any multiple regression one should thoroughly check whether the model assumptions seem reasonable, i.e., whether the errors are normally distributed with mean 0 and constant variance σ^2. For example, in Bergerud and Ballard (1988) the plot of the observed and calculated (from the fitted model) values of the dependent variable shows that the early data tend to have observed values above those predicted by the model, whereas in later years the observed values are below the predicted values (Fig. 5.8). This suggests that the errors do not have mean 0, but a mean dependent on time. Time should probably be included in the model as an additional independent variable. This is good practice for any data collected over time and may require the use of time series analysis for a thorough statistical investigation.

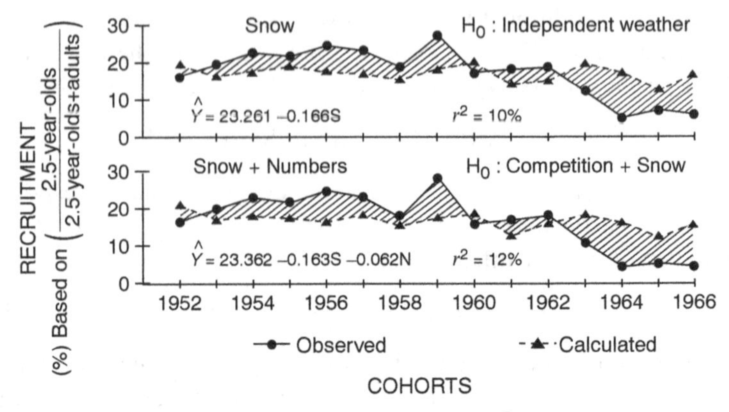

Fig. 5.8. Example of a possible violation of the assumptions of multiple regression. The plots display actual recruitment at 2.5 years of age in the Nelchina caribou herd, south-central Alaska, versus predicted recruitment using first snow depth and then both snow depth and caribou numbers as independent variables, for the years 1952–66. (From Bergerund and Ballard 1988.)

Because the dependent variable in the Bergerud and Ballard (1988) models is constrained to lie between 0% and 100% it cannot technically be considered normally distributed. This problem may not be serious if the values of the dependent variable do not tend to cluster near the extremes of 0% or 100% [they do not seem to cluster near the extremes in the example given by Bergerud and Ballard (1988)] and appear approximately normal over the range of values observed. In such a situation, the multiple regression analysis is probably satisfactory.

A multiple regression analysis may be statistically valid in the sense that all assumptions seem reasonable and the calculations are done properly, but it may be criticized on other grounds. For example, Van Ballenberghe (1989) criticized the multiple regressions of Bergerud and Ballard on the grounds that wolf numbers were obtained artificially and the apparent relation between recruitment and wolf numbers might have been partly due to something in the artificial method of estimating wolf numbers rather than actual wolf numbers, which were not measured.

Categorical variables

Categorical variables can be incorporated into multiple regression models in a number of ways. To illustrate this, suppose one records eye colors of human subjects as brown, blue, or other. Eye color is thus a categorical variable with three categories. One way to quantify this variable might be to denote it by the letter Z and write $Z=1$ if eye color is brown, $Z=2$ if eye color is blue, and $Z=3$ if eye color is other. Suppose we now proceed to use multiple regression to determine the relation between eye color and blood pressure (Y). Treating eye color, Z, as the independent variable and blood pressure, Y, as the dependent variable, we would get a regression equation of the form

$$Y=b_0+b_1Z. \tag{5.20}$$

Unfortunately, this equation predicts that blood pressure for brown-eyed people is b_0+b_1, that blood pressure for blue-eyed people is b_0+2b_1, and that blood pressure for those with other eye colors is $b_0+3\,b_1$. Regardless of the values of b_0 and b_1, our coding scheme used to define Z forces the predicted value of blood pressure, Y, for blue-eyed individuals, as given by the regression equation, to take on a value between that for brown-eyed individuals and that for individuals with other eye colors, even if the data indicate otherwise. Furthermore, the difference in predicted blood pressure, based on the regression equation, between brown- and blue-eyed individuals is automatically the same as that between blue-eyed individuals and those with other eye colors. The way in which Z was defined automatically imposes these relations (possibly incorrect) between eye color and blood pressure as predicted by the regression equation. The above way of quantifying eye color leads to inappropriate models in multiple regression.

A better way to quantify eye color in the example is to define two indicator variables, Z_1 and Z_2, as follows. Let

$Z_1=1$ if the subject has brown eyes
$\quad\;=0$ if the subject does not have brown eyes

and let

$Z_2=1$ if the subject has blue eyes
$\quad\;=0$ if the subject does not have blue eyes.

A variable such as Z_1 or Z_2 which takes on only the values 0 and 1, 1 if a certain characteristic is present and 0 if the characteristic is not present, is called an indicator variable. Notice that for a brown-eyed subject $Z_1=1$ and $Z_2=0$, for a blue-eyed subject $Z_1=0$ and $Z_2=1$, and for a subject with some other eye color $Z_1=0$ and $Z_2=0$. There is thus a unique pair of values for each eye color and hence no need to define a third variable, Z_3. If one fits a multiple regression model as before, one will obtain an equation of the form

$$Y=b_0+b_1\,Z_1+b_2Z_2. \tag{5.21}$$

If a subject has brown eyes, the regression equation predicts a blood pressure of b_0+b_1. If the subject has blue eyes, the regression equation predicts a blood pressure of b_0+b_2. For subjects with other eye colors, the regression equation predicts a blood pressure of b_0. Notice that b_1 and b_2, the coefficients of Z_1 and Z_2 respectively, represent the difference in the effects of brown and blue eyes, respectively, from the effect of other eye colors on blood pressure; thus, b_0, the effect of other eye colors, becomes a sort of reference value. Because b_0, b_1, and b_2 can take on any values, the equation has the flexibility to predict any blood pressures for the different eye colors.

The example above indicates that one must exercise care in quantifying the values of a categorical variable. The second method indicated is the best way to proceed. In general, if a categorical variable is an independent variable, one quantifies it for use in multiple regression by means of indicator or zero-one variables. If a categorical variable can take on c possible values, one defines the $c-1$ indicator variables

$Z_i = 1$ if the categorical variable has the ith possible value
$\quad = 0$ otherwise

for $i = 1,\ldots,c-1$. If $Z_1,\ldots,Z_{c-1}$ are all 0 then obviously the categorical variable has the value c. There is no need to define Z_c because it is redundant. Notice the ith indicator variable 'indicates' whether or not the categorical variable takes on the ith value. The $c-1$ indicator variables are all added to the multiple regression equation to represent the (main) effects of the categorical variable. If the coefficient of any of these indicator variables in the fitted multiple regression model is found, in a hypothesis test, to be significantly different from 0, the effect of that value of the categorical variable differs significantly from that of the cth value. The cth value becomes the reference value. By clever use of indicator variables and their cross products one can represent ANOVA models as multiple regression models and test all the standard hypotheses of ANOVA. Mixing quantitative independent variables with indicator variables allows one to represent analysis of covariance models as multiple regression models. Additional discussion of the regression approach to ANOVA and analysis of covariance can be found in Neter and Wasserman (1974). Use of indicator variables makes multiple regression models more general than might first appear and illustrates the fact that regression, ANOVA, and analysis of covariance have much in common. In fact they are all special cases of general linear models for which an extensive theory exists.

If a categorical variable is the dependent variable, the assumption of normally distributed errors is clearly violated and methods other than multiple regression are needed. Categorical dependent variables arise, for example, in problems of classification such as estimating the sex of an animal based on a set of morphological measurements. Special sets of procedures exist for classification, including discriminant analysis, which we will not discuss, and logistic or loglinear models, which we now discuss briefly.

If the dependent variable takes on only two values, the natural tendency is to treat it as a binomial (Bernoulli) random variable. For such binomial dependent variables, logistic regression or loglinear models are typically used to analyze the data. The basic idea is not to fit a regression model to the dependent variables directly, but instead assume that a regression model exists relating the independent variables to the probability, p, of observing a particular value of the dependent variable {or more precisely to some function of p such as $\log[p/(1-p)]$, which is called the logit of p}. To be more explicit, one might assume that: (1) each observation, Y, is a Bernoulli random variable with a probability of success p; and (2) the probability of success p satisfies

$$\log[p/(1-p)] = \beta_0 + \beta_1 X_1 + \beta_2 X_2 + \ldots + \beta_p X_p, \tag{5.22}$$

where $X_1, \ldots, X_p$ are the independent variables.

The method of maximum likelihood is used to estimate the parameters $\beta_0, \beta_1, \ldots, \beta_p$. Formulas for these estimates cannot be obtained in closed form (i.e., they are an infinite series), so numerical methods are used. Elementary discussion of these topics can be found in Neter *et al.* (1983), whereas more thorough treatment can be found in Cox and Snell (1989). Many computer packages have subroutines to analyze such data. If the dependent variable takes on more than two values, one again fits regression models to functions of the probability that the dependent variable will take on a particular value, using multinomial logit models. Computer packages that handle these models are somewhat less common, but a package called GLIM enables one to analyze such models. In fact, GLIM analyzes so-called generalized linear models of which linear models, loglinear models, and multinomial logit models are special cases. Situations for which the dependent variable follows a Poisson distribution are also covered. For more information see McCullagh and Nelder (1989) or Aitkin *et al.* (1989).

Holm *et al.* (1988) studied the effectiveness of certain chemicals in discouraging deer mice from feeding on corn. These repellents were applied to corn kernels that were then offered to deer mice for a number of days. Each kernel was categorized as having or not having sustained damage. This categorical variable was the dependent variable. The different chemicals were categorical independent variables. As part of the analysis the authors also recorded when a kernel sustained damage, and a loglinear model was used to study how the probability of sustaining damage changed over time.

Stepwise regression

Often in multiple regression many independent variables are measured. Some of these variables may be significantly correlated with each other, and part of the goal of the analysis is to produce a model that makes scientific sense and fits the data well (has a high R^2 or small value of MSE for example) while retaining only a relatively small number of independent variables. One way to find such a model is simply to fit every possible regression model with some or all of the independent variables and to choose the one that strikes the desired balance between scientific sense, good fit, and small number of independent variables. Several rules of thumb are available for deciding what constitutes a desirable balance [see Neter *et al.* (1983) for a discussion of the balance between good fit and small number of independent variables] but ultimately the choice is somewhat subjective. For example, Nixon *et al.* (1988) wished to study the effect of 24 habitat variables (the independent variables) on the presence or absence of deer (the dependent variable). After examining all possible regression models on the basis of R^2, the authors decided that a model involving only five of the independent variables was satisfactory. Notice here that because the dependent variable was categorical with two values a logistic regression might have been more appropriate.

If one has p independent variables there are $2^p - 1$ possible models involving at least one independent variable, so the number of models gets large very rapidly. For example, with $p = 24$, as in Nixon *et al.* (1988), one must examine $2^{24} - 1$, or 16,777,215 models. Even on modern computers examining this many models is time consuming. For large p, therefore, algorithms have been developed that 'cleverly' search for models with good fit while examining only a fraction of the possible models. These algorithms have been implemented in computer packages and are called stepwise regressions. The forward stepwise regression algorithm starts by trying all models with a single independent variable and selecting the one with highest R^2 or highest value of the F-statistic for testing whether the model fits. If this highest R^2 or value of F exceeds a prespecified cut-off; the algorithm accepts this model and continues. It now adds the independent variables not currently in the multiple regression equation to the one it has just accepted and finds the variable that increases R^2 the most or has highest value of F. If this exceeds the cut-off the algorithm accepts this model and proceeds. The algorithm continues to add variables until there is inadequate improvement, at which point the computer stops and prints out the final model accepted as best. Changing the user-specified cut-off values can change the final model produced by the algorithm.

Backward stepwise regression works just the reverse of forward stepwise regression. It begins with all variables in the model and determines which one decreases R^2 or F the least. If this decrease does not exceed a user-specified cut-off, the variable is dropped from the model and the algorithm is repeated. This process continues until no more variables can be removed, at which point it ceases and prints out the final model. The model resulting from a backward stepwise regression may vary as one changes the cut-off values and it need not agree with the model produced by a forward stepwise regression.

The most popular stepwise procedure is the full stepwise regression, which alternates between a forward and a backward stepwise approach. Variables added at a given stage may be removed at a later stage and those removed at a given stage may be replaced later. The user must supply two cut-off values (one for the forward part and one for the backward part) and the choice will affect the final result. The result of a full stepwise regression need not agree with either a forward or a backward stepwise regression. Johnson *et al.* (1989) used stepwise regression (presumably the full stepwise algorithm) to examine the effects of 15 land-use variables on a variable measuring bird damage to grapefruits in southern Texas. The final model involved only three of the independent variables.

Although a stepwise regression will generally lead to a model with reasonably good fit, some words of caution are in order. These algorithms do not examine all possible models so they may miss models with better fit and possibly fewer variables than that produced by the stepwise procedure. Models produced by these algorithms need not make scientific sense nor need they satisfy our regression assumptions. Any model produced by a stepwise procedure should therefore be investigated further. In addition to checking model assumptions, one may wish to add or delete variables to produce a model that achieves a better balance between scientific sense, good fit, and small number of independent variables. One may also wish to compare the model produced by a stepwise procedure to other models. Use of a stepwise procedure does not eliminate the need for additional investigation before deciding on a final regression model. Examination of all possible models is therefore recommended when feasible, i.e., when the number of independent variables is not too large. Stepwise procedures should only be used when this is not the case. For more information on stepwise procedures see Neter *et al.* (1983).

5.5 Regression with multistage sampling

As noted in Section 5.3, standard regression analysis is applicable to data collected only by simple random sampling. Yet as emphasized in Chapter Four, behavioral ecologists often collect data using multistage designs. Two common cases may be distinguished. The first is easy to handle; the second, unfortunately, is not. Both involve two-stage sampling with simple random selection of primary and secondary units.

As an example of the first case, suppose we are studying some behavior or other attribute of individuals as a function of the period during the nesting attempt. In collecting the sample, we distribute effort more or less evenly across the temporal dimension, and an independent sample of individuals is collected in each time period. Multiple measurements are made on each individual within each time period. For example, in studying the frequency of display songs as a function of period, we may record number of display songs in several periods for each of several individuals within a given stage. We thus have two-stage sampling within any given period with selection of birds at the first stage and selection of listening periods at the second stage. This case involves error in the measurement of the $\bar{Y}_i$ because we do not record songs for the ith bird throughout the period of interest. Such cases, however, arise routinely in regression analysis. The analysis must be based on the sample means per bird, $\bar{Y}_i$, but no other special steps are needed. Failure to base analysis on the means amounts to pseudoreplication and tends to cause standard errors to be underestimated, sometimes severely. Cases in which the X_i are only determined after sample selection, and cases in which the X_i are also subject to uncertainty may be analyzed in the same way.

As an example of the second case, suppose we are making repeated surveys along the same routes to estimate temporal trends. The spatial dimension of the population is the set of all routes that might be selected, and the temporal dimension is the set of times at which counts might be made. A simple random sample (in theory) of routes is selected for the first period, but the same set of routes is then surveyed repeatedly. Since the observations at different times are not independent, standard regression does not apply. If the counts on the same route at two different times are about as different as counts on two different routes would be, then little error is caused by ignoring the lack of independence. In reality, however, most surveys have a strong 'place effect', and substantial error may result if standard regression analysis is used.

This problem arises any time that repeated measurements are made on individuals for the purpose of studying relationships between two or more of the variables measured. For example, repeated measurements of marked individuals to study the rate of displays in relation to weather, foraging success in relation to group size, residence time in patches in relation to patch size, and so on may raise these problems. In most cases, we cannot simply calculate means for the variables for each individual because the variation we are trying to study would disappear. For example, if each individual is watched several times in a study of foraging height *versus* temperature, and we calculate mean height and mean temperature, then the means would be very similar for different individuals so we would be left with little variation to study.

Two general approaches exist for dealing with problems of this sort. First, a regression can be carried out for each individual, and tests can then be made on the set of resulting regression coefficients. This approach has been developed extensively in the study of trend data from repeatedly surveyed routes. Regressions are carried out for each route and statistical tests to determine whether population size is increasing or decreasing are based on whether the mean of the regression coefficients is significantly different from 0.0. This approach is discussed more in Chapter Eight. In a study of foraging height in relation to temperature, the same approach could be used. A separate slope would be calculated for each bird and their mean and standard error would be used to describe the relationship. This approach is valid as long as the initial sample may be regarded as a simple random sample. The process, however, does not estimate the same parameter as the slope of the means per primary unit regressed against the independent variable. Thus, the mean of slopes, even if all members of the population were measured, does not necessarily equal the slope of the means (per survey route, per individual, etc.) regressed against time, and the two parameters may differ substantially. This rather complex point is discussed more in Chapter Eight, and examples are provided there. Another problem with this approach is that it may be difficult to apply when more than one regression coefficient is estimated and when sample sizes per individual vary substantially.

The second approach is to carry out the analysis recognizing the dependence in the data set. This is basically a repeated-measures analysis. It is more complex and is available in many statistical packages (SAS, SPSS, BMDP, S^+). Since this type of data is collected quite frequently by behavioral ecologists we describe one approach. In many cases, the X-values are fixed constants as occurs, for example, in surveys (the X-values are years and they would be the same in repeated sampling). The estimated slope – using the standard regression formula – is unbiased in this case. The $se(b)$ is biased, often severely. An essentially unbiased estimate may sometimes be obtained by estimating the covariance terms that acknowledge the repeated sampling of the routes. Statistical inferences apply only to the set of X-values in the sample. Repeated sampling of locations across time is probably the most common example of this design in behavioral ecology. Repeated measurements on the same individuals may be analyzed using the same approach but inferences then extend only to the individuals in the study, not to other individuals in the population. We advise consulting a statistician for assistance and emphasize the importance of doing so before collecting the data. The third approach is to use a bootstrap.

5.6 Summary

Scatterplots and summary statistics provide an excellent way to begin investigating the relationship between variables. Identifying outliers and influential points is particularly important. Simple linear regression provides a quantitative measure of how – and how closely – two variables are associated. Statistical tests require assumptions about the variables that often are not met in behavioral ecology, but moderate departures from these assumptions are not serious. Multiple regression is a natural extension of simple regression, permitting the effects, in combination or alone, of several variables on a dependent variable to be studied. Proper analysis with multistage sampling, unfortunately, may be considerably more complex and consultation with a statistician before collecting the data is recommended.

6

Pseudoreplication

6.1 Introduction

The term pseudoreplication was introduced by Hurlbert (1984) to describe analyses in which 'treatments are not replicated (though samples may be) or replicates are not statistically independent'. In survey sampling terms, the problem arises when a multistage design is employed but the data are treated as a one-stage sample in the analysis. For example suppose m secondary units are selected in each of n primary units. In most cases the nm observations do not provide as much information as nm observations selected by simple random sampling. The correct approach is to base the analysis on the means per primary unit.

The initial point made by Hurlbert and soon thereafter by Machlis *et al.* (1985) was incontrovertible. When multistage sampling is employed, ignoring the sampling plan and treating the data set as though it is a simple random sample can lead to gross errors during interval estimation and testing. This point is emphasized in survey sampling books, and calling attention to the error, which was quite common at the time, was a service to the discipline. Subsequently, however, cases began to appear in which the proper analytical approach was more difficult to agree on. These cases often did not involve large, clearly described populations, and well-defined sampling plans. Instead, they usually involved some combination of incompletely specified populations, nonrandom sampling, nonindependent selection, and small sample sizes. As noted in Chapter Four these issues are inherently difficult and full agreement on the best analytical approach and most appropriate interpretation cannot always be attained. Nonetheless, we believe that progress may be attainable in some cases by applying the ideas developed in Chapter Four. In this Chapter we examine a few of the published debates about pseudoreplication that have seemed particularly difficult to resolve.

6.2 Power *versus* generality

We begin with the general conceptual point, made several times in previous sections, that a given data set may often be viewed as being a sample from more than one population. For example, suppose we have randomly selected several areas and measured several individuals in each. We assume independent selection within areas and that sample size per area varies. It is natural to view this as a two-stage sample, to define the areas as primary units, and to proceed with the analysis accordingly. Under this approach, the statistical population is the set of individuals in areas from which our primary units were selected. The population diagram for an example with seven primary units and four to eight secondary units/primary unit is shown in Fig. 6.1a. As noted in Section 4.5, (pp. 122–124), however, another approach is also legitimate from a statistical point of view, i.e., to define the statistical population as being restricted to the areas that we studied. The sampling plan is then one-stage, preceded by stratification. Each area is a stratum (Fig. 6.1b). We would calculate means per area and use these in the formula for stratified estimates, weighting the study areas equally or perhaps using some measure of their size as the weighting factor. In this case, the statistical population is limited to the particular set of areas we studied. Extrapolation beyond the borders of those areas must be based on nonstatistical rationales. It is important to realize, however, that from a statistical point of view there is nothing improper about this approach. The biologist has clearly delineated a population, carried out a well-defined sampling plan, and analyzed the data in a manner consistent with it. Although this example used areas as strata, the same points could be made if plants, animals, or any other entities were selected rather than areas.

The fundamental point here is that, from a statistical point of view, investigators often have an opportunity to trade statistical power for generality. If they delineate a large population, then power is reduced but their inferences are more general. Alternatively, if they delineate a smaller population, then power is increased but their inferences are less general.

Hurlbert (1984) and others (e.g., McGregor *et al.* 1992) have described pseudoreplication as an error involving statistical analyses inappropriate 'for the hypothesis being tested'. From our perspective, it might be clearer to substitute the phrase 'for the population about which statistical inferences are made'. Thus, the null hypothesis involving a mean $\overline{Y}$, might be

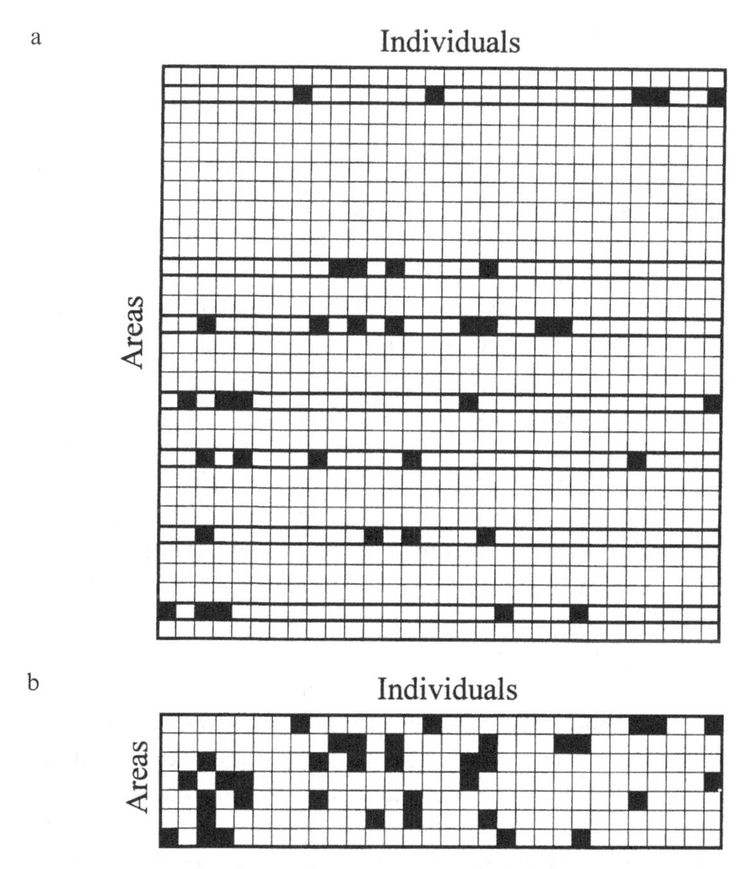

Fig. 6.1. Two ways of defining the statistical population. A. Seven areas were selected randomly and then four to eight individuals were selected in each. The formulas for multistage sampling (Appendix One, Box 2) are used for point and interval estimation and statistical inferences extend to other areas. B. The seven areas are viewed as constituting the spatial extent of the population. The formulas for stratified sampling (Appendix One, Box 2) are used for point and interval estimation,and statistical inferences extend solely to individuals in the seven areas.

$H_0{:}\bar{Y}=0$, regardless of whether investigators describe their statistical population correctly or incorrectly. One might argue that $\bar{Y}$ entails a particular population, and it certainly does. Nonetheless, to us it seems a little clearer to describe pseudoreplication as claiming that statistical inferences extend to a different population than the one to which they really apply. This view reflects our position, just described, that many data sets may legitimately be used to make inferences to more than one population (though the formulas used to estimate standard errors may differ). Clearly, it is incorrect to use the formula appropriate for one population but then claim that the statistical inferences apply to some other population. This, in our view, is the essence of pseudoreplication.

Discussions of pseudoreplication also often phrase the issue in terms of whether 'replicates are independent'. In some cases, however, statistical independence is a rather slippery concept. For example, with two-stage sampling and simple random selection of secondary units, the secondary units are a simple random sample with respect to the primary unit they are in but not with respect to the population. Furthermore, with stratified sampling valid inferences may be made even though the units in the sample are not a simple random sample from the population (selection is carried out one stratum at a time). Thus, we prefer to emphasize the more fundamental issue that a one-to-one relationship exists between any given population and sampling plan and the formulas used in the analysis. Thus, the analysis determines to what population the inferences apply and claiming that statistical inferences apply to other, usually larger, populations, is incorrect.

We have used the phrase 'from a statistical point of view' to emphasize that biologists have something to say about this issue too. Even if treating the data set above as a one-stage sample does not violate any statistical rules, one may still argue that it is inappropriate on the grounds that other investigators do not really care about the individuals in the study area at the time of the study. By the time the results are published, the individuals will probably be dead anyway. This issue, however, is quite complex. In a study of habitat preferences, in which densities are recorded by habitat, it might be true that little interest exists in the areas actually studied. They are useful only if we can extrapolate to other areas. In such a case, we may not be interested in a population that extends no further than the borders of the study area. If our interest is restricted to larger populations, then we may insist that the investigator adopt an analytical approach which lets him or her make inferences beyond the borders of the study area. If those inferences are inconclusive due to low power then we may not have much interest in the study.

The case above, however, is probably not very common. More often we are interested in processes not simply patterns (e.g., relative density across habitats), and we may well believe that a process occurring in one area probably occurs in other areas too. Under this point of view restricting the scope of inference to the particular study area may seem quite appropriate. As argued in Section 1.2, it is often useful to concentrate a study in a relatively small area because more can be learned about the process there.

Restricting the statistical population to the area studied amounts to doing exactly that, and the increased power appropriately reflects this.

As an admittedly artificial example of this point, suppose that three studies of the same issue were carried out in different areas, and each was able to make statistically significant inferences about a restricted population but not about a larger population. In this case, the generality of the conclusion begins to be established by repetition. But if each study attempted to make inferences about the broader population and fell short, then the only pattern would be that all studies of the issue had been inconclusive. One might note that all three obtained point estimates with the same sign; however, such an outcome could readily occur by chance. Furthermore, suppose that the signs differed. If one simply knew that all three were inconclusive, this would strengthen the view that no differences of biological importance existed, whereas if inferences were made to more restricted populations and all three were significant, then one might conclude that the effect was real but that its sign differed according to local conditions. This seems to be a real advance compared simply to concluding that all the studies were inconclusive. We do not claim that this is always the case. For example, one could make significant inferences about each of several individuals and then note that the effect in question varied between them for unknown reasons, and this line of reasoning really might not lead to any insights of value. The point, however, is that such issues should be decided on the basis of which populations are interesting to examine. Given a decision on that issue, the statistical analysis follows naturally, and assuming that the analysis is consistent with the sampling plan followed for the population, then no claim of pseudoreplication can be sustained.

We are not trying, in this section, to resolve the issue of power *versus* generality, and we doubt that a full resolution is possible. We are simply emphasizing: (1) that the issue is real; (2) that it must generally be resolved on the basis of biological, not statistical, reasoning; and (3) that it is quite complex so reasonable people may disagree. Given these points, particularly number (3), we believe that the best general advice is simply that investigators should clearly specify the population to which their statistical inferences apply. If that is done, and the analysis is consistent with this decision, then there should not be any question about whether pseudoreplication has occurred, although there certainly may be differences of opinion about whether investigators have made the wisest choice in the trade-off between power and generality. Basically, then, our view about how to avoid pseudoreplication is 'know thy population'. We now discuss three controversies involving pseudoreplication and attempt to explain how each might be viewed, and perhaps resolved, using the perspective developed here.

6.3 Fish, fish tanks, and fish trials

Dugatkin and Wilson (1992) captured 12 sunfish from Quaker Lake in Pennsylvania and tested them repeatedly in their laboratory. The fish were housed in two aquaria, six fish in each. Each fish was thus exposed to five other fish, but not to the other six (except during brief experiments). One of the tests involved exposing a focal fish to a familiar and an unfamiliar fish, and recording which one the focal fish spent more time with. The results were taken as a measure of preference for familiar – as opposed to unfamiliar – conspecifics. The study was carefully designed so that each fish was tested the same number of times, the tests (or trials) were randomized and so on. Below, we use the notation n = number of fish = 12, and m = number of trials per fish.

The authors viewed the trials as a one-stage sample and therefore used formulas for simple random sampling with the sample size being nm and degrees of freedom $nm - 1$. This approach was criticized by Lamprecht and Hofer (1994) who felt that the data set should be viewed as a two-stage sample (though they did not use this phrase) with fish as the primary units. With this approach sample size would be 12 and degrees of freedom 11. Subsequently, Lombardi and Hurlbert (1996) criticized both of these approaches, saying that the sample size really was 2, one for each tank, and the degrees of freedom was thus just 1. The original authors replied to both commentaries carrying out the analyses proposed, but not really accepting the views of either critic that only one analysis was valid. Their final response ended with the statement, 'our main point is that statistical questions such as this are no more black and white than the biological issues that they are intended to address' (Wilson and Dugatkin 1996).

Let us consider each of the suggested analyses, asking the sorts of questions we have used in previous sections to define sampling plans and appropriate analytical methods. Our basic questions are, 'is each analysis rigorous statistically?', and, if so, 'what population do the inferences apply to?'. The original approach amounts to defining population units as 'fish-trials' and to restricting the population to the 12 fish used as subjects. The population diagram would thus have the 12 fish as rows and the set of possible trials as columns. Its appearance would be the same as that of Fig. 6.1b (p. 179) except that rows would now be fish rather than 'areas', and columns would be trials rather than 'individuals'. Features of the study such as the lake from which all fish came, the date, the building in which the study was carried out, and the various tanks in which the fish were housed or tested are all part of the 'conditions of the study'. Population units in the sample, however, were not distributed randomly across this matrix because the trials were carefully balanced. The plan thus more closely resembles stratified sampling with fish as strata. The authors did not analyze their data this way, they treated it as a simple random sample, but if they had used the correct formula for stratified sampling (strata the same 'size' and all sample sizes within strata equal) the results would not have been much different from the results they obtained. Thus, other than this one relatively minor quibble, their analyses, properly viewed, are statistically rigorous in the sense that their point and interval estimates are unbiased, and their statistical tests (subject to the usual statistical assumptions) are valid. The inferences extend only to the 12 fish studied, not to any other fish in Quaker Lake, other aquaria, other buildings, etc.

The analysis proposed by Lamprecht and Hofer suggests that the 12 fish should be viewed as having been randomly selected from a larger population. While no formal selection plan was carried out, this is no different to hundreds of other studies in behavioral ecology which consider that the subjects may be viewed as a random sample. The population unit, in this case, is still a fish-trial, but the population includes fish that the investigators might have captured. The population diagram would still have fish as rows and trials as columns, but with this approach there would be many more than 12 rows, and the selected rows would be viewed as a random sample. The population diagram would have the same appearance as that in Fig. 6.1a except that rows would be 'fish' and columns would be 'trials'. Each fish (i.e., the set of measurements that might be collected from each fish) is a primary unit, and the analysis suggested is fully appropriate. Statistical inferences extend to the set of fish that might have been captured. As with the first approach, all results are conditioned on the particular environment, including the two tanks.

The analysis proposed by Lombardi and Hurlbert (1996) amounts to three-stage sampling with tanks as the first stage unit. A population unit thus becomes a 'tank-fish-trial'. The approach implicitly views tanks as having been randomly selected from – or at least as being representative of – a population of tanks. This is probably no more tenuous than viewing the fish as a random sample. A population diagram might have 'tanks' as rows – two of them selected at random – and 'fish' as columns, it being implicit that a cell represents the overall result for one fish. Trials would thus disappear (or could be viewed as a third axis). The analytical approach is statistically rigorous; inferences extend not just to other fish that might have been tested but to other tanks as well. The study, however, involves a sample size of two and thus has extremely low power and makes the greatest number of distributional assumptions (due to the low sample size). Nonetheless, many informed people would feel that the analysis provides useful information.

It thus seems to us that each of these analyses provides information of value. In the particular case, results were highly significant with fish viewed as primary units. Inferences could thus legitimately be made to a larger population than the 12 fish tested. To us, this seems a good reason for preferring this analysis. The analysis with tanks as primary units bothers us because power is so low. If we were to pick a single analytical approach before looking at the data, it would not be this one. We do not agree with the claim that tanks 'must' be viewed as part of the population unit. This amounts to saying that the population must always be defined in the broadest possible manner regardless of how this affects power. In this particular case, suppose all fish had been in one tank. In that event statistical inferences would have applied solely to that tank in the same way that they apply solely to fish in Quaker Lake. But if making statistical inferences that are restricted to one tank and one lake is acceptable, what is wrong with making the same sorts of inferences to two tanks and one lake? As we have stated previously (Section 1.4), generality is determined primarily by repeating the study with other species in other areas. Expanding the limits of the statistical population as much as possible in a single study, and causing power to decline almost to vanishing point, does not make much sense to us. Nonetheless, our major purpose in this section is to suggest how a survey sampling approach may help to clarify some of the issues related to pseudoreplication. We do not expect that all readers will agree with our particular preferences in this case.

In summary, statisticians rarely criticize biologists for defining their population in a particular way. The statistical issues are largely limited to whether the population and sampling plan are well defined and the analysis is consistent with these definitions. Thus, we agree with the view of Wilson and Dugatkin (1996) that some of these questions are inherently difficult. The real difficulty, however, is in deciding what populations are most interesting to make inferences about, not deciding how to make the inferences. Pseudoreplication, in our view, occurs when formulas appropriate for one population are used to make inferences about another, usually larger, population, not when an investigator makes statistically rigorous inferences about one population and a colleague suggests that it would be more interesting to make inferences about some other population – that is a nonstatistical issue, and one on which reasonable people will often disagree. Furthermore, the solution in many cases will be to carry out both analyses rather than trying to agree which of the analyses is most interesting.

6.4 The great playback debate

A series of publications in the field of behavioral ecology examined the issue of how to design playback experiments and analyze the resulting data without committing pseudoreplication. Kroodsma (1986) initiated the discussions in a paper describing what he viewed as common errors in this type of research. He described a hypothetical experiment to determine whether blue-winged warblers respond more strongly to local songs than to foreign songs. He supposed that two 'local' songs were obtained from Amherst, Mass., and two 'foreign' songs were obtained from Millbrook, New York. Four tapes were made, as shown below:

	Location	
Song type	Amherst	Millbrook
I	one tape	one tape
II	one tape	one tape

Trials were conducted in Amherst, used either type I or type II songs and involved playing the Amherst (local) and Millbrook (foreign) songs alternately from speakers located 20 m apart to male blue-winged warblers. The investigator estimated how close to each speaker each bird came. The experiment used a total of ten color-banded males, but 'trials' not 'birds' were used as the sample size.

The experiments were carried out first with the type I song. After 20 trials with 10 birds, the birds showed a statistically significant preference for local songs. The experiment was then repeated with type II songs. This time, the results were nonsignificant. The conclusion drawn from this hypothetical experiment was that blue-winged male warblers have a preference for local as opposed to foreign songs of type I but not type II.

Kroodsma then identified six weaknesses in his hypothetical study:

1. Distance estimates were subjective but were not done 'blindly'.
2. For a given type (I or II), tapes were obtained from only one location. Thus, there was no statistical basis for inferences about other areas.

3. Similarly, for a given type, only one song was recorded from the 'local' and 'foreign' categories so no statistical basis existed for generalizing to other songs in either group.
4. Birds, not trials, should have been the sample size.
5. Responsiveness may change seasonally and/or due to habituation but all type I songs were played before any type II songs were played. The design thus confounded time with the response variable.
6. Sample size should have been selected in other ways rather than continuing the first set of trials until a significant result was obtained.

Suggestions for correcting these and other problems were made.

Subsequently, some authors (e.g., Catchpole 1989; Searcy 1989) argued that designs such as this one, and real studies referred to by Kroodsma in a subsequent paper (Kroodsma 1989a), probably have substantial external validity despite these weaknesses. Kroodsma responded (1989b, 1990) with increasingly more detailed defenses of his original views. There was little if any debate, however, about the purely statistical issues. All, or at least most, participants seemed to recognize that statistical inferences applied solely to the sampled population, and if a single tape was used or work was done in a single locality then statistical inferences did not necessarily extend beyond that tape or location. This view was clearly expressed in a synthesis paper (McGregor *et al.* 1992) on which most of the participants in the earlier debate were authors. Thus, from the perspective developed in this Chapter, the 'playback debate' really centered on questions of external validity and were thus nonstatistical in nature, except for the caveat that authors should clearly identify the limits in all dimensions (e.g., tapes, locations, birds) of their statistical population. This being done, the subject-matter specialists could debate the issue of external (i.e., nonstatistical) validity without help from statisticians or recourse to statistical issues.

The debate, however, did involve one difficult statistical issue. Various suggestions were made about how many different tapes should be used, everyone agreeing that a new tape (i.e., song) should ideally be used in each trial. When only one tape was used, then clearly the statistical inferences apply only to that tape. When multiple tapes were used, then there was little recognition that investigators have the choices outlined in Section 4.10: (1) calculate means per male and use them in the standard formula for standard errors (i.e., Appendix One), in which case statistical inferences apply only to the set of tapes used; (2) calculate means per tape and use these in the standard formula for standard errors, in which case the statistical inferences apply only to the set of males used; (3) recognize the dependency in the data and use an analysis that incorporates the correlation which is present to make statistical inferences about the population of songs and males. This issue, however, as noted in Section 4.10, has gone almost totally unnoticed in behavioral ecology so it is not surprising that it was not mentioned in the papers discussed above.

6.5 Causal inferences with unreplicated treatments

In Hurlburt's original paper, he took a rather strong stand against the use of 'inferential statistics' when the purpose is to identify causal relationships and no replication or random assignment is present. He used, as an example, a suggestion by Green (1979) for how to study whether waste discharge affected downstream organisms. Green suggested that samples be taken upstream and downstream of the discharge site both before and after discharge began. Obviously, a significantly large change in the difference (or ratio) between locations following commencement of the discharge would suggest that the discharge caused the change. On the other hand, it is also obvious that the statistical analysis alone cannot prove that the discharge caused the changes because some other event might have occurred at the same time as the discharges began.

This issue, in our view, is basically one of not inferring cause from correlation (Section 1.4). Statistical analysis can properly be used to ask the questions: 'did the difference in population levels, between locations, change after discharge began? If so, in what direction and by how much?'. Deciding whether any documented change was caused by the discharge requires additional evidence and a nonstatistical (or at least additional) rationale. Nonetheless, the statistical analysis is useful because it may show that the observed change in relative population levels was not statistically significant, or if a significant change did occur, the analysis can be used to provide a confidence interval for the change, and hence provide us with information about the magnitude of the change. Thus, we agree that the statistical analysis does not permit identification of what caused any observed change, but feel that it is nonetheless useful as one part of the analysis.

6.6 Summary

Debates about how to avoid pseudoreplication (without stopping work altogether) continue to occur regularly. Many investigators are probably uncertain whether to agree with a comment made by Dugatkin and

Wilson (1994 pp. 1459) that in some studies one 'must accept a risk of pseudoreplication to do the study at all' or whether, on the other hand, to agree with Lombardi and Hurlbert (1996 p. 420) who disagreed, saying that, 'pseudoreplication is simply a type of incorrect statistical analysis. For no study design, however weak, is pseudoreplication an inevitable result'. We agree with Lombardi and Hurlburt that pseudoreplication is simply a mistake, and that it always can and should be avoided. On the other hand, we do not fully agree with what they seem to imply, i.e., that for every data set there is one and only one appropriate analysis. We would rephrase this by saying that for every data set *and population to which inferences are being made* there is one and only one analysis.

We suggest the following steps as a way to be certain that no legitimate charge of pseudoreplication can be made:

1. Define the conditions of the study, specifying where random sampling with more than one replicate occurred. Identify any nonrandom samples that will be treated as simple random samples and justify the assumption.
2. For a particular analysis, describe the limits of the statistical population in all dimensions. In the example with fish tanks above, we would describe the two tanks used as part of the conditions of the study, but the fish as having been randomly selected from a larger superpopulation in Quaker Lake at the time of the study.
3. Show that the formulas for point and interval estimates are appropriate, paying particular attention to the issue of independent selection within primary units. In general, if selection of secondary units was not independent in different primary units, then do not claim that statistical inferences extend beyond the particular secondary units measured. For example, in the playback experiments, one should not claim that statistical inferences extend to other songs; songs are part of the 'conditions of this study'. Imagining a 'population diagram' as illustrated in this Chapter and in Chapter Four may help clarify whether the population is correctly defined and all assumptions of the nominal sampling plan are met.
4. In discussing external validity, acknowledge that the inferences are not based on statistical analysis (except that such analyses justify conclusions about the statistical population).

If one follows all of these steps, others may still criticize a study for delimiting the statistical population too sharply, but this is a nonstatistical issue and, in our view, should not be described as a pseudoreplication error. Furthermore, in our opinion, those inclined to criticize the study should recognize the inherent difficulty of the issue and that reasonable people may disagree about it.

7
Sampling behavior

7.1 Introduction

In this Chapter we discuss methods for estimating the time spent in different behaviors. The Chapter is intended primarily for researchers who collect behavioral data, and we have therefore felt justified in discussing practical problems in some detail. We also assume a moderate familiarity with the goals of behavioral sampling and the practical difficulties often encountered. Lehner (1996) and Martin and Bateson (1993) discuss many other aspects of measuring and describing behavior as well as some of the points discussed here.

By 'sampling behavior' we mean that one or (usually) more types of behavior have been defined, and the objective is to estimate how often they occur. 'How often' may mean any of the following: proportion of time spent in the behavior, frequency of the behavior (e.g., number/hour), average duration of the behavior. All three estimates might be of interest for some behaviors (e.g., fights with intruders) while for other behaviors this might not be true. For example, it might be feasible to count the number of pecks/minute but difficult, due to the speed with which a peck occurs, to estimate either the average duration of a peck or the proportion of time spent pecking.

7.2 Defining behaviors and bouts

A first task in designing the sampling plan is to define the behaviors. The definitions should be specific enough that different observers would interpret them the same way and that they can be explained clearly. Defining positions of the body (head up/head down), type of movement (walking, running, flying), and different types of vocalizations generally meet these criteria better than broader terms such resting, feeding or scanning for predators. For example, when an animal stands in one place for a while but occasionally looks around or pecks at the substrate, different observers might classify the behavior as resting, feeding, or scanning for predators.

Once the data have been collected, it may be desirable to define 'bouts' (e.g. feeding bouts, singing bouts) so that the proportion of time spent in each general behavioral type may be estimated. A bout is a period during which some behavior is engaged in frequently, albeit not continuously. Three situations may be differentiated. First, clear breaks may occur between one behavior and the next. For example, a bird may sing from a perch for a period, and then leave to forage (without singing) for a period. In such cases, defining bouts presents no problems. Second, a single behavior may be recorded that tends to occur in groups separated by intervals in which the behavior does not occur. For example, the behavior may be pecking the substrate, and series of pecks in quick succession may be separated by periods with few or no pecks. In such cases, bouts may be defined by selecting a threshold time and stating that bouts begin when the behavior occurs and end when more time than the threshold elapse after the behavior occurs. Thus if the threshold is 10 s, and a bird pecks five times in quick succession but does not peck during the following 10 s, then the bout consists of the time during which it pecked five times. Various suggestions have been made for selecting the threshold time. Lehner (1996 pp. 117–23) provides a good introduction.

In the third situation more than one behavior is recorded and changes are not clear breaks. This case is probably the most common in behavioral ecology. Intermittent time-activity sampling is generally used in which the animal's behavior is classified periodically. For example, suppose we classify an animal as head up (i.e., alert), head down (i.e., foraging), preening, or resting. A feeding animal may spend much of its time with head up (e.g., scanning for predators) so 'alert' records will be interspersed with 'foraging' records. On the other hand, an alert bird may occasionally put its head down to feed. Defining the proportion of time spent in either behavior requires that a rule be devised to define foraging bouts and alert bouts. Additional analysis may then be carried out to describe the amount of time with head up during foraging bouts (e.g., in a study of mate guarding behavior). Defining bouts in this case is not a trivial effort especially with hundreds or thousands of observations. One approach is to declare that a particular type of bout occurs anytime the behavior was recorded at least k times in any n sequential observations. For example, if birds are recorded as feeding or not feeding every 60 s, the rule might be that anytime the animal was recorded feeding in at least three of any six sequential observations, then all six observations would be recorded as a feeding bout. To analyze the raw data and define bouts, a computer would count the number of feeding records in observations one to six. If there were three or more feeding records, then observations one to six would be recorded as part of a feeding bout. Then the computer would look at observations two to seven. If three or more were feeding records, then observations two to seven would be recorded as a feeding bout (of course two to six might already have been designated as a bout in which case these observations would not be changed). In this way, many observations when the animal was not feeding would be declared parts of a feeding bout. Additional procedures might be needed to ensure that every observation was assigned to one, and only one, bout type. Once the analysis was completed, quantities such as the proportion of time spent in each type of bout as well as the mean and standard deviation of the bout lengths could be calculated. A computer program is most helpful in carrying out the assignments because it is then easy to experiment with different values of k and n to determine what combination provides the most biologically meaningful description of the activities.

A simpler method, which requires less time in the field, is as follows. The approach works best when only two behaviors are being recorded. We use the example 'feeding' and 'not feeding' and assume that the behavior will be defined for each of several consecutive 1-min intervals. Within each interval the animal is watched for the entire 60 s or until it feeds (e.g., pecks) the first time. If it pecks once, the entire minute is assigned to the 'feeding bout' category and the observer does not have to continuously monitor the animal for the rest of the interval. This approach is easy in the field, especially if the behavior in question occurs frequently, and yields data with bouts already defined. Note, however, that the definition of a bout (any 60 s with at least one instance of feeding) cannot be altered in the analysis, for example to 'any 30 s with at least one instance of feeding'.

7.3 Allocation of effort

The terms scan sampling and focal individual sampling are widely used in the behavioral literature (Lehner 1996). The distinction has to do with whether one or a few individuals are followed (focal sampling) or, on the other hand, periodic scans of all or most individuals in a group are made (scan sampling). The distinction between the two is not hard and fast. For example, if there are several family groups in view and we watch one of them, keeping track of each member in the family, it may be a matter of personal choice whether the plan is called focal animal sampling (because only a few individuals of those in view are watched), or scan sampling (because the behavior of each member of the family is recorded at intervals).

Another distinction is between continuous and intermittent sampling. Continuous sampling is just what the name implies. The starting and ending times, or just the occurrence, of different behaviors are recorded throughout the observation interval. With intermittent sampling, the animal's behavior is recorded at regular intervals. Continuous sampling works best when only a few behaviors (e.g. sleeping, feeding, preening) are of interest, they are easy to recognize (e.g. fights, displays), and they either occur infrequently or last a long time. Continuous sampling is also easier when frequency, rather than duration, is being estimated because starting and ending times of the behavior then do not have to be recorded. Intermittent sampling is easier – and may be the only practical choice – for 'fine-scale' behaviors such as position of the head (up *versus* down) or measures of locomotion (stationary, walking, running, flying) that may alternate rapidly with each other.

A final distinction, or consideration, is whether individuals can be recognized or not. Typically, they are kept separate in focal animal sampling. The individuals may be recognizable because of location or by natural or artificial marks, but even if they are not recognizable, the observer may be able to keep track of one or two individuals for the duration of the observation interval. In scan sampling, individuals are not usually recognizable, though occasionally all members of the group are marked so that data for each can be maintained separately.

With these distinctions in mind, we turn to the issue of whether one should obtain lots of data on a few animals or look at lots of animals but obtain only limited data on each one. We first assume that animals are watched one at a time and can be returned to at will (e.g., because they hold known territories). We also assume that numerous animals are available so a decision must be made about how many animals to include in the sampling plan. The plan is multistage sampling (Section 4.5) with animals as primary units. We assume that the total amount of time for observations has been fixed and the problem now is to decide how many different animals to watch and, by implication, how long to spend with each animal.

We suggest that a two-stage process may be helpful for deciding how to allocate time in this sort of study. First, consider the purely statistical objective of estimating whatever parameters are of interest. Usually, the time required to switch attention from one animal to another can be estimated fairly easily. The time may vary from essentially zero when all subjects are visible from one blind, to very substantial amounts of time if the observer must change locations. Some idea of how much individuals vary is also of great value. For example, if feeding rates are being estimated, do some individuals consistently feed much less (perhaps because they are still in courtship) than others? The following guideline may be used to complete consideration of the purely statistical issues. Watch more animals (and spend less time on each one) to the extent that (1) it takes little time to select a new animal, and (2) animals are likely to vary with regard to the variable being measured. These qualitative suggestions can be converted into a quantitative analysis if prior information on variability within and between animals is available. We suggest Section 10.6 of Cochran (1977) for those interested in such an analysis. In practice one estimates many different quantities, and different allocations might be suggested by several of them. Thus, some compromise must be reached, and this limits the utility of a fully quantitative analysis.

A common misconception among biologists is that one somehow gets 'bad', presumably meaning biased, estimates of the frequency or proportion of time spent in rare behaviors if only a little time is spent watching each animal. The real issue is that the estimate for any given animal has high variance but it is unbiased as long as random selection of observation intervals is used, the observer does not influence the animal, and no measurement error occurs. If any of these requirements is not met, then the estimate may be biased regardless of how much time is spent with each animal.

The conviction above, expressed by many biologists, may actually be due to a feeling that the animal does respond to the observer and that an acceptance period is needed before reliable data can be obtained. One common response of animals is to become more wary, and such behavior may disappear after the observer has been present for some time. Another response is a reduced likelihood of engaging in certain behaviors such as aggressive interactions. A more subtle observer impact is that animals which tend to engage in long bouts of behavior may tend to change behavior type when mildly disturbed. Even if the animals being watched are unaffected by the observer the species' predators or prey may be affected by the observer's presence. Any of these responses may compromise the quality of observations, and may be a good reason for minimizing disturbance and waiting a substantial period before beginning observations (or at least analyzing the data to measure – and exclude – the acclimation period). If such effects occur, however, the corresponding periods should be excluded from the analysis completely rather than being used on the assumption that they will somehow be 'compensated for' by having long observation periods.

Once the purely statistical issues have been considered, we suggest putting the resulting conclusion aside and thinking about the biological trade-offs involved in the allocation issue. If the sole goal of a program was to estimate the proportion of time spent in some behavior, and if there was no hope of obtaining any other insights while doing the field work, then the statistical considerations described above would probably be sufficient to resolve the allocation issue. But in reality, of course, progress in understanding animals may occur anytime field work takes place, and one goal of a sampling plan should be to maximize the likelihood of making new discoveries. It is our experience that such discoveries are least likely to occur when biologists spend much of their time traveling from one sampling station to the next, watching each subject for only a short time. In contrast, when one comes to know a relatively few animals quite well, there is much more chance that new insights about questions or answers will occur. Thus, in many cases it seems best to watch fewer animals, but watch each one for longer, than might be suggested solely on the basis of minimizing standard errors of the resulting estimates. Furthermore, when animals can be identified from day to day, then repeated observations of a few animals, rather than selection of new ones each day, permits a much broader range of questions to be investigated because relationships between activities separated by many days (i.e., feeding activity as a function of time since nest failure) can be investigated.

A final decision, with intermittent sampling, is how often within observation periods to record behavior. Frequently one combines intermittent observations of some behaviors, continuous observation of others (e.g., fights, certain displays, copulations), and, if possible, general observations. Increasing the frequency of intermittent sampling provides better estimates of these variables but takes time and attention away from continuous observations and the ability to watch interesting incidental events. Thus this issue, like the question of how long to watch each animal, should be resolved on the basis of both statistical and nonstatistical considerations. The statistical considerations are that infrequent observations are sufficient for behaviors that last a long time (e.g., sleeping) whereas more frequent observations are valuable for behaviors that change rapidly (e.g., head up/down). For the latter type, the precision is likely to be similar to that in a simple random sample (even though the plan is actually systematic). The standard error of the estimate within an observation interval thus declines approximately as the square root of sample size so the same proportional increase in precision is accomplished by increasing the number of observations per interval from 16 to 25 as from 64 to 100. Note, however, that a great deal more time for other activities is lost by increasing the observations from 64 to 100 than is lost by increasing them from 16 to 25. In practice, the behaviors monitored by intermittent sampling usually include some that alternate rapidly but, even assuming this is true, it seems rare to need more than 60 observations per interval (or one per minute if intervals are longer than 1 h) or to have fewer than 20 per interval. With short intervals, this may dictate that observations should be made as often as every 10 s.

In summary, statistical considerations indicate that more animals should be watched (for shorter periods per animal) when little time is required to select new animals and when animals vary substantially. Investigators may wish to temper this advice by giving some priority to knowing a few animals well. It is common to have 20–60 observations per interval or a maximum of one per minute if interval lengths exceed 1 h.

7.4 Obtaining the data

We now discuss practical problems that arise in recording the data, emphasizing those that raise statistical issues. The most common problems stem from the fact that in most studies the animals one wants to observe are frequently out of sight. The individual one wants to watch may be completely out of sight so no data are recorded at the scheduled time (selection bias) or may be partially or completely out of sight during part of the observation interval (measurement bias). Selection bias, arising because individuals cannot always be found at will, is common and difficult to deal with. For example, waterfowl often rest and preen on banks in plain view but forage in thick vegetation where they cannot be seen. Obviously, a time budget to estimate what fraction of their time is spent feeding is not feasible in such cases (unless one knows how many individuals are present and can assume that any out-of-sight bird is feeding). Probably the best that can be said of such problems is, ‘don’t ignore them!’. Investigators are obliged to seriously consider the maximum likely extent of bias in such cases and decide whether the study is worthwhile.

Here is an example of how difficult the problem can be. In studying avian mate guarding behavior, data were collected on males using time budget methods to document the frequency of mate guarding, fights, copulations and so on. Males could be located easily if they were on their territories and gave aerial displays. Early in the season nesting was synchronized and most males were on their territories displaying. As nest failures and re-nesting occurred, however, males spent more and more time off their territories and by midway through the season observers usually needed 30–45 min to find their bird and frequently lost it repeatedly due to the male flying off the territory. In this case, time budget methods initially appeared feasible but eventually had to be abandoned completely due to the selection bias problems. Ultimately, the sample selection method was re-designed to focus on females, which usually did stay on territory. Examples such as this one, in which a method initially appears feasible but gradually is revealed to be unsuitable, are common in sampling behavior, and investigators must be prepared to abandon the methods even though doing so is difficult when substantial effort has already been investigated in data collection.

Here is an example in which measurement bias – rather than selection bias – became problematic. In studying territorial tundra swans from blinds, the birds could usually be located with enough work, but they were often out of sight. For example, when sleeping or preening they were usually in sight, but when feeding often only their heads could be seen as they scanned for predators. The initial list of behaviors included various types of feeding (head under, neck under, tip up, etc.) but these often could not be seen. The solution to this problem was designing a data collection form which indicated whether only the 'primary' behaviors (sleep, preen, feed, etc.) could all be seen, or whether all the specific feeding behaviors could be seen. If a feeding bird was in plain sight, then the specific feeding behavior was recorded, but if it was not, then the 'feed' category was used. Note that this scheme involves a hierarchy of behaviors with higher-level ones always recorded and lower-level ones recorded only when the birds were visible enough that all possible lower-level behaviors could be detected. This general approach may be useful in many situations involving animals that are difficult to keep in sight.

7.5 Analysis

Time-activity studies usually involve animals as primary sampling units. All or most individuals within the study area may be watched, or the 'first *n*' animals encountered (or encountered in favorable situations) may be monitored. As has been noted previously, investigators in such studies are nearly always interested in making inferences about larger, hypothetical groups or about the set of possible outcomes, so it is reasonable to view the data as a simple random sample from a larger population. In this case, animals comprise the first dimension, time the second, and there may be additional stages in the selection of the population units. In any case, animals (i.e., the set of times at which each animal might be watched) are the primary units. Results for each animal are nearly always equally weighted on the basis that each animal provides the same number of potential observations (primary units are the same size) or that each animal provides an equally good opportunity to study the behavior in question. Nonindependent selection of units in the second dimension, caused for example by watching all animals on the same days, is common and often unavoidable, but should be avoided whenever practical.

Here is a typical example in which equal weighting was appropriate. Clarke and Kramer (1994) studied scatter hoarding in eastern chipmunks. Several chipmunks were allowed to create scatter hoards and several variables describing the chipmunks' behavior were measured. Some animals were observed making more than one scatter hoard, and the authors computed means for each such animal and used the resulting statistics – rather than the original data – in their analyses. They thus implicitly defined animals as primary sampling units and weighted means per primary unit equally. In this case, the 'size' of each primary unit (i.e., the number of scatter hoards each animal might have been observed making) is not well defined. Furthermore, and of more fundamental importance, each animal provided an equally good opportunity to investigate hoarding behavior. Thus, weighting the results from each animal equally seemed appropriate.

Animals also may be selected in groups, in which case equal weighting may not be appropriate. For example, suppose that groups of animals are located and watched for 1-h periods to record the frequency with which individuals are involved in fights. For each interval, the number of animals and the number of fights is recorded. Fights involving two animals are counted as '2' (because two animals are involved), fights involving three animals are counted as '3', and so on. The number of animals in the group varies considerably (but for this example we assume that no animals enter or leave the group during the observation period). The population unit in this study is an 'animal-hour', the response variable is the number of fights the animal participated in (0, 1, 2,...), and the quantity being estimated is the population mean. The sampling plan depends on details not specified above (e.g., whether sites are revisited) but the means per observation period will probably be weighted by the number of animals present (Section 4.5).

Summarizing this section, equal weighting of the results for each primary unit is recommended when primary unit size is constant, poorly defined, or variable but unknown. When the size of primary units is well defined and variable, and is known for the units in the sample, then weighting by primary unit size is appropriate. Weighting by sample size (when sample size differs from primary unit size) is seldom appropriate (Section 4.5).

7.6 Summary

This Chapter describes issues that arise in estimating the duration or frequency of behavior. Behaviors should be defined objectively in ways that minimize measurement bias and maximize repeatability. Defining bouts is often useful. The rules for doing so depend on whether behavior changes sharply from one type to another and on whether more than one behavior is recorded. In gathering data, careful attention is needed to avoid selection bias. When this is a problem, defining a hierarchy of behaviors with more general categories that are easier to record may be helpful. Allocation of effort between the number of animals and the number of observations per animal calls for consideration of both statistical and nonstatistical issues. The statistical issues can be addressed using formulas, but doing so is often little better than careful thought about the issue. The basic nonstatistical issue is that valuable, if unanticipated, insights often result from prolonged observation on each of a few individuals. The sampling plan in such studies usually involves individuals as primary sampling units. Results per primary unit are usually weighted equally. When groups of individuals are selected, however, groups may be the primary units and weighting may then be by group size.

8
Monitoring abundance

8.1 Introduction

Monitoring abundance means estimating trends in abundance, usually through time though occasionally across space or with respect to some other variable. Estimating temporal trends in abundance of animal populations is a common objective in both applied and theoretical research. The most common design involves surveys in the same locations run once or more per year for several years. The data are often collected using 'index methods' in which the counts are not restricted to well-defined plots or if they are the animals present are not all detected and the fraction detected is not known. Results are usually summarized by calculating the mean number of animals detected per plot, route, or some other measure of effort, during each period. These means are then plotted against time (Fig. 8.1). When the counts come from complete surveys of well-defined plots, then the *Y*-axis is in density units. In the more common case of index data, the *Y*-axis shows the number recorded per survey route or some other measure of effort. We assume (or at least hope) that a 5% change in survey results indicates a 5% change in population size but we have no direct measure of absolute density.

The analysis of trend data raises several difficult issues. First, 'the trend' can be defined in different ways, and the choice among them may be difficult and subjective. Second, statistical difficulties arise in estimating precision because the same routes are surveyed each year so the annual means are not independent. Third, use of index methods, rather than complete counts on well-defined plots, means that change in survey efficiency may cause spurious trends in the data. These problems are often quite complex and consultation with a statistician is often required for rigorous analysis of

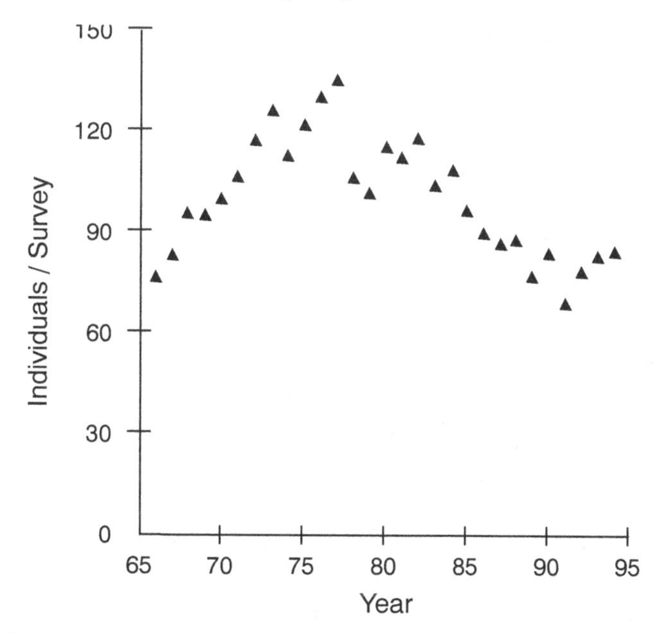

Fig. 8.1. Example of a scatterplot used to portray temporal trend in abundance.

trend data. Our goal in this Chapter is to describe some of the difficulties in enough detail that readers will be able to work effectively with a statistician. We assume in the following text that surveys are conducted once per year to estimate temporal trends, but the discussion generalizes easily to other situations.

8.2 Defining 'the trend'

Deciding how to describe the temporal trend in population size is surprisingly difficult, and with many real data sets it must be acknowledged that more than one reasonable definition exists. Regression of the means against time may be used, though appropriate calculation of standard errors is complex for reasons discussed in Section 8.3. Furthermore, several options exist for calculating the regression line and results may differ substantially depending on which option is used (e.g., James *et al.* 1996, Thomas 1996). An alternative approach, which simplifies error estimation but may yield different results from regression on the means, is 'route regression' (Geissler and Sauer 1990), developed at the Patuxent Wildlife Research Center for use with data from the Breeding Bird Survey. In this Section, we first describe methods based on standard regression and then describe route

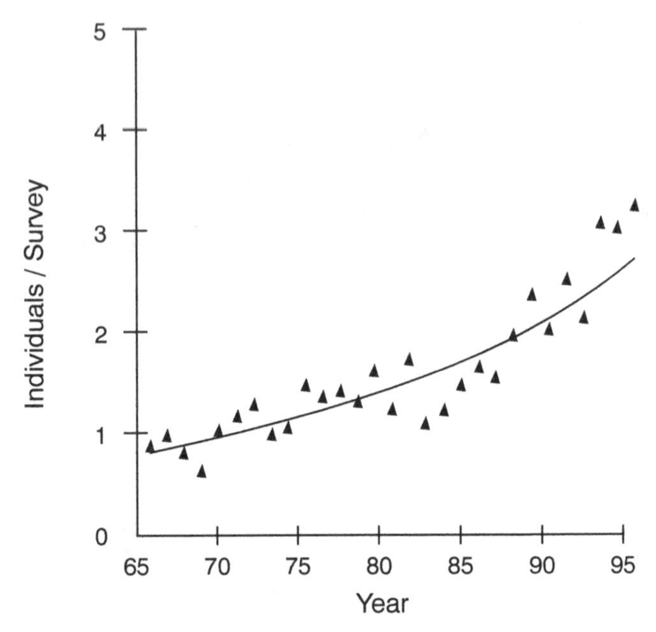

Fig. 8.2. A scatterplot in which an exponential curve fits the trend reasonably well.

regression, noting how it differs from regression on the means. Our main purpose is to emphasize that 'the trend' may be defined in numerous ways and that the choice among definitions should usually be based largely on nonstatistical bases. In this Section we assume that exact counts from well-defined plots are obtained. In reality, most surveys to estimate temporal trends use index data, and such methods introduce additional complexities. We address index methods in Section 8.5.

The usual approach for fitting a curve to a scatterplot is to use least-squares regression as described in Chapter Five. Trends in abundance, however, are often not linear, and even if they are approximately linear, an exponential curve may provide a more useful description of the trend. If this general approach (rather than route regression) is used, then we suggest first fitting an exponential curve to the data (Fig. 8.2), and using that curve unless good reasons exist to use some other curve. The reason for this suggestion is that the exponential curve has a simple interpretation which works well for index data and is meaningful to most biologists. The exponential curve has the feature of changing by the same proportional amount each year. Thus, if the curve increases 5% between the first and second year, then it increases by 5% between any other two sequential years. One may thus describe the trend by saying that the trend curve increased at an annual

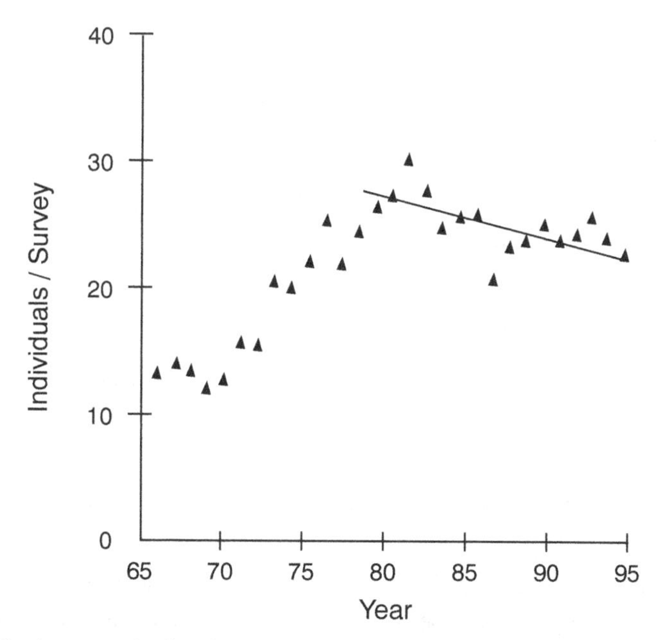

Fig. 8.3. A scatterplot in which the trend changed direction during the study but, during the most recent years, is described reasonably well by a linear curve.

rate of 5%. Notice, however, that the trend curve is just an approximation to describe population growth. Real populations almost never change in such a smooth fashion.

The form of the exponential curve is

$$Y_i = b_0 e^{b_1 X_i}, \tag{8.1}$$

where X_i is the year (e.g., 1992), b_0 and b_1 are values calculated using least squares methods (see later) and Y_i is the value of the trend curve in period *i*. The annual rate of change in the value of the trend curve is $\exp(b_1) - 1$. For example, if $\exp(b_1) = 1.05$, then the trend curve increases by 5% each year. If $\exp(b_1) = 0.98$, then the trend curve declines by 0.02 or 2% each year. Most statistical packages will fit exponential curves to data sets either by taking the ln of both sides of Eq. 8.1 and fitting a straight line to the transformed data or by using a nonlinear procedure. In some cases the rate of change itself may change through time, as when a population increases and then decreases. When the goal is deciding whether populations are increasing or decreasing *at present* – as is true in many monitoring studies – then one may fit an exponential or linear curve to the most recent years, choosing as a cut-off date the time when the exponential curve first fails to fit the data (Fig. 8.3).

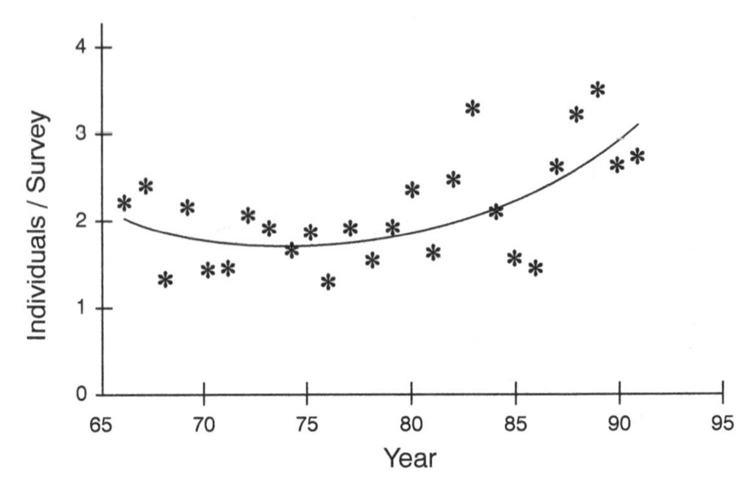

Fig. 8.4. A trend described using a second-degree log-linear curve.

When the exponential curve does not fit the data in the period of interest well, then other curves may be considered. One approach is to add higher-degree terms to the exponential curve. The exponential curve is a polynomial curve of degree one (because time enters the equation raised only to the first power). Using a subscript to indicate degree, the second- and third-degree polynomials are

$$Y_{2i} = b_{20}e^{b_{21}X_i + b_{22}X_i^2} \tag{8.2}$$

$$Y_{3i} = b_{30}e^{b_{31}X_i + b_{32}X_i^2 + b_{33}X_i^3}.$$

The first-degree exponential curve only goes upward or downward, it cannot reverse direction. The second-degree curve (Fig. 8.4) can reverse direction once (though it does not necessarily do so), and the third-degree curve (Fig. 8.5) can reverse direction twice. The second-degree, and especially the third-degree, curves tend to produce sharp changes in direction at the start and end of the study period. This tendency can be illustrated with real data by truncating the analysis a few years before the period actually ended. In many cases, the curve based on truncated data will show an upward or downward trend that is not followed when all the data are included (Fig. 8.6). When part of the purpose of the analysis is to infer how the population may change in the future, these tendencies should be recognized, and the second- and third-degree curves should not be used if influential changes in the trend curve occur only a few years before the most recent survey. These curves are usually drawn by taking logs of the annual

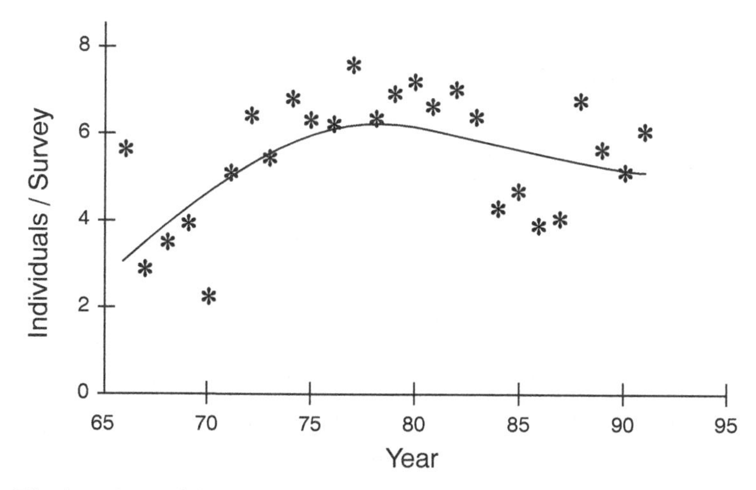

Fig. 8.5. A trend described using a third-degree log-linear polynomial curve.

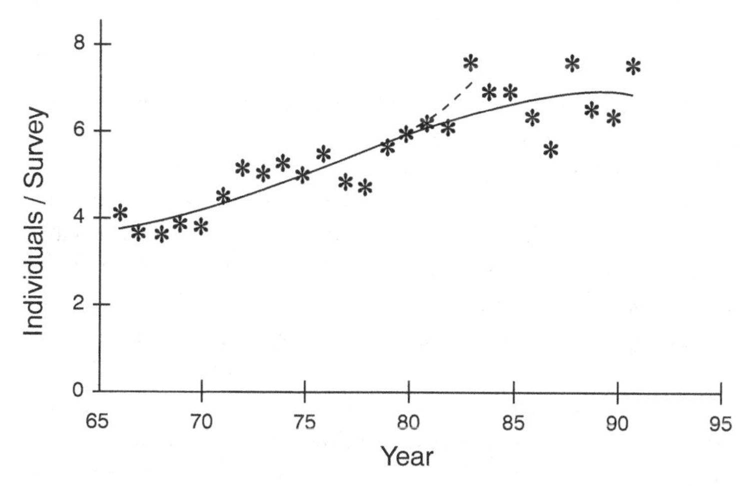

Fig. 8.6. A third degree log-linear trend fitted to years 1 to 17 inclusively (dashed line) and to the entire data set (solid line). Note the upturn in the shorter curve, suggesting a sharp change in population trend that is not confirmed by the larger data set.

means and then using linear regression to estimate the coefficients (which are then back-transformed). The curves are therefore often referred to as first-, second-, and third-degree log–linear curves.

A fourth curve that is sometimes helpful is the linear curve (Fig. 8.7)

$$Y_{Li} = b_{L0} + b_{L1}X_i, \tag{8.3}$$

where L indicates 'linear'. When either an exponential or a linear curve fits the data well (as often happens) then we prefer the exponential curve

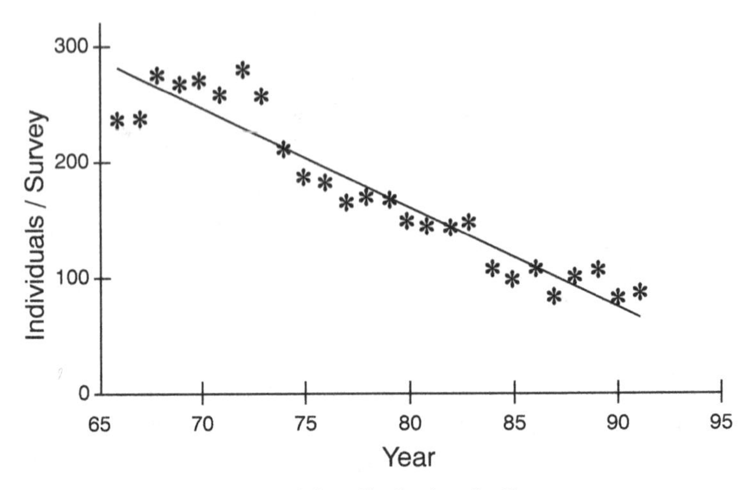

Fig. 8.7. A trend described using the linear curve.

because the trend can be summarized easily and in terms (i.e., annual proportional change) that biologists find more meaningful. A linear curve does not change the same proportional amount each year. It does change by the same absolute amount but change in absolute terms is usually not very satisfactory as a description, especially with index data. Thus, saying that the number of birds recorded per route declined by one each year leaves the reader wanting to know what the percentage change was. That is, readers want the *proportional* change.

As already mentioned and described in more detail in the next Section, calculating standard errors for estimates of a trend calculated using the methods here is difficult for statistical reasons. The 'route regression' method developed at the Patuxent Wildlife Research Center provides an alternative approach which avoids this problem. In this approach, an exponential curve is fitted to the counts from each route. The average of the slope parameters, $\exp(b_1)$, which measure annual proportional trend for each route, is then used as the overall measure of trend and the usual formulas for standard errors of means can also be used. This approach has been generalized for use with stratified sampling and possibly different sampling intensity between strata (as occurs in the Breeding Bird Survey). Differences in skill between observers has also been incorporated into the analysis by assigning an indicator variable for each different observer on a given route (Sauer *et al.* 1994). The approach thus handles complexities that the regression methods already given here do not handle well. Route regression is widely used to analyze Breeding Bird

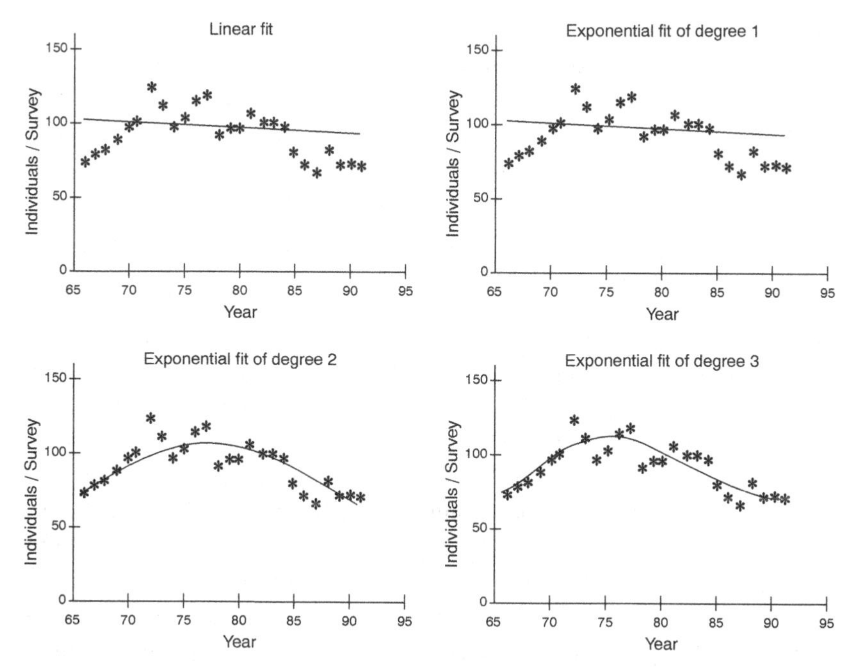

Fig. 8.8. First-, second-, and third-degree log-linear curves and a linear curve fit to the same data. Note how different 'the trend' is with these different curves.

Survey data but is not generally available in standard statistical software packages.

It is important to realize that 'the trend' for a given population may be quite different according to which method for describing the trend is used (Fig. 8.8), and no one of them will necessarily be clearly superior to the others. This situation is similar to other cases in which summary statistics are used to describe complex quantities. For example, in describing central tendency (i.e., a 'typical' value) in a given case some investigators may prefer the mean and others the median; in describing variability some investigators may prefer the standard deviation and others the mean absolute deviation. Guidelines may be given for when each measure is most appropriate but the choice is still partly one of personal preference. The same is true in describing 'the trend'.

This point may seem difficult because repeated reference has been made here to use of regression methods and these methods, as described in Chapter Five, often assume that the population means all fall on a given line. If this assumption were true, then 'the trend' would be a precisely defined quantity rather than a subjective quantity which different people might reasonably define in different ways. The problem with applying standard regression logic is that animal populations seldom change in a smooth manner. Population sizes are affected by numerous year effects which move them away from the long-term trend. They thus 'bounce around', and even if we had actual population sizes in each year, we would not find that the means were all on any simple line such as those used in the methods above. Thus, a summary statistic, such as the proportional change in an exponential curve fitted to the annual means, is a descriptive statistic of a complex, and poorly defined, quantity we call 'the trend' rather than being a precisely defined quantity such as the variance of a mean or the correlation coefficient for two variables. As a result, people may disagree on the best way to summarize the trend, and in many cases it is difficult to give statistical reasons for preferring one approach over another. One analyst may prefer fitting an exponential curve to the annual means, another may feel that a second-degree polynomial describes the curve better, even though the description is more complex, and a third may prefer route regression because it facilitates calculation of precision even though the resulting curve may not fit the annual means as well as the other two approaches. The problem is particularly difficult when an exponential curve yields one trend for the population at the end of the study period (e.g., 'decreasing') and a second-degree polynomial curve yields a different description (e.g., 'stable'). Users of the data are likely to want a simple answer to the question, 'is the population decreasing or stable?' and may find it difficult to understand that the answer may really depend on what curve is used to describe the trend.

Given these complexities, a few guidelines may be helpful in choosing which model to use. We begin by clarifying the difference between route regression and regression on the means. Although route regression fits an exponential trend line to each route, and then uses the average of these values as the overall trend, it does not, in general, produce the same results as fitting an exponential trend to the annual means. Route regression thus estimates a different parameter or feature of the process compared with regression on the means. The parameter, in the case of route regression, is complex and may best be thought of simply as the result that would be obtained with a very large sample. Alternatively, route regression may be thought of as estimating the exponential trend in the population means, but with some degree of bias.

When trends are clear and consistent through time, and sample sizes are large, then route regression and exponential regression on the means tend to give similar results. However, with smaller sample sizes, or very different trends on different routes, this may not be true, and it is helpful to understand that the two methods really are estimating different quantities. It should thus not be surprising that they sometimes lead to different conclusions (e.g., one may indicate a significant change when the other method does not).

We suspect that when describing trends biologists are usually interested in a curve that fits the annual means. Thus, if the trend departs strongly from exponential (e.g., Fig. 8.5) then we suggest either using only the most recent years during which the trend was exponential or using a higher-degree trend. If the means being analyzed do show an exponential trend, then one might compare the exponential and route regression results. If the two differ substantially, then we suspect that biologists will generally prefer fitting a regression line to the means. On the other hand, if the two methods produce similar results, the greater ease of variance estimation using route regression may make this approach more attractive.

8.3 Estimating standard errors

When standard regression methods are used to fit curves to annual means, the fact that routes are surveyed repeatedly each year must be acknowledged in the formulas for standard errors and quantities derived from them such as the estimated change in the trend line between two years. This can be done by including covariance terms in the formula for the variance of the regression coefficients. The sum of these terms must be estimated. Two approaches are discussed in this Section though neither has been widely used in behavioral ecology studies, many of which still inappropriately use standard regression formulas. One is a 'nonparametric' approach based on regression techniques developed in survey sampling; the other is a 'parametric' approach based on traditional regression methods. We recommend consultation with a statistician for assistance in implementing either approach.

In survey sampling, methods have been developed for estimating means and totals using regression methods without assuming that the mean response for a given X value falls on the regression line (Cochran 1977; Thompson 1992). These methods can be extended to estimate the standard errors of the regression coefficients without assuming that the annual means are independent. The formulas, however, are complex and to our knowledge are not available in any existing software package at the time of writing, so users may have to write their own programs to carry out the analyses.

The parametric approach involves assuming that a relatively simple model comes close enough to fitting the data that any errors caused by lack of fit can be ignored. Thus, one assumes that the mean of all possible counts for each year falls on the regression line, even though this is not strictly true. The sum of the covariance terms may be estimated by assuming that these terms also follow some simple pattern. For example, one may assume that they are all constant or that they decline as a smooth function of the time between years. One major advantage of this approach is that some statistical packages (i.e., SAS, SAS Institute, Cary, NC, USA) permit estimation using this approach.

Analysis with the route regression approach is simpler. Each route produces one regression coefficient; the mean of these coefficients is an unbiased estimate of the population mean (i.e., the mean of the coefficients for 'all possible' survey routes), and the sample standard error it provides is an appropriate estimate of precision (assuming sample sizes per route are the same). No problem is caused by surveying the same routes each year because each route produces only one outcome (the regression coefficient). This approach can be incorporated into existing software such as SAS fairly easily. One has only to obtain the regression coefficients from each route and then treat these as a simple random sample.

8.4 Outliers and missing data

Outliers and missing data occur frequently in animal surveys. The mean for one year may be far from the trend line, and may have a small standard error indicating that the disparity is not due simply to sampling error. When this is true, performing the analysis with and without the outlier may be useful to indicate how much effect this single mean has on the trend (Section 5.3, 'Examining assumptions'). It is also useful to determine how consistent the results were across routes in the outlier year. This sometimes identifies recording errors (e.g., 5 may have been recorded as 50 during data entry) and may help one decide what course to follow. Biological information is often helpful here. For example, the study area might be on the edge of a species' range but a vagrant flock of the species might have been observed on a single route in one year causing the mean per route for the year to be abnormally high. An investigator might note the existence of this record and that it had been excluded in the analysis.

In estimating precision, the effect of including outlier years differs according to which analytical method is being used. Regression on the means, using the nonparametric approach for variance estimation, and route regression make no assumption that the true means for each year fall on the regression line. Thus, the existence of strong year effects in some years does not indicate a failure of assumptions or any other purely statistical problem. In the parametric regression approach, one assumes that all the annual means do fall on the regression line, and the calculation of standard errors relies on this assumption.

Missing data are also often encountered in these data sets. The main point to consider is how evenly the missing data are distributed over the sample of possible route-years. Ideally, missing data are few and are not concentrated in any portion of the data such as one part of the study area early in the interval, or on routes with low numbers of animals actually present. With surveys conducted by volunteers, one problem is that routes with few animals present may be covered less frequently than routes with many animals present. The effect of such nonrandom distribution of the missing data varies according to which method is being used, and careful thought may be needed to decide how serious a problem is caused by the missing data.

Approaches for dealing with missing data include calculation of expected results for missing route-years, exclusion of the missing route-years from the analysis, and forming 'super-routes' which combine results from several routes. The parametric approach for using standard regression methods is complex with missing data in the sample because the route-years matrix must be inverted. Problems also exist over how to reduce the degrees of freedom appropriately in this case. The nonparametric regression and route regression approach can both accept missing data. Missing data for a series of consecutive years may have large effects on route regression, especially when the trend along the route departs significantly from an exponential curve. Thus, if the true trend is up and then down, but the 'up portion' of the data is missing, then the calculated regression coefficient for that route will be far from the value that would have been obtained if all the data were present. Super-routes are formed by combining results from routes with missing data and calculating the mean/year, for each year, using whatever routes were surveyed. Super-routes may be formed by combining routes in the same part of the study area. The resulting data may then be used with any of the three analytical approaches described above.

8.5 Index methods

Index methods, in which counts do not involve thoroughly searched, well-defined plots, are widely used to monitor population trends of animals. Index methods raise a special set of problems which we discuss in this

Section. We assume in the following text that the same routes are surveyed each year (or at some other interval). This case – annual counts of animals along the same routes – is by far the most common case in behavioral ecology, but most of the ideas discussed in this Section also apply to variations of this situation, and some of these are briefly discussed at the end of the Section.

Imagine a survey in which the animals are counted on several routes, each with several stations, for several years. The animals might be detected by visual or auditory cues, traps (e.g., snap traps, mist nets), or by signs (e.g., tracks) that they leave. The number of detections per unit of effort varies substantially during the study interval, and we believe this variation reflects variation in population size. On the other hand, we do not detect every animal present during the intervals; we may not even detect a large fraction of them, thus variation in the numbers detected might be caused by variation in how efficiently we sampled. High numbers caught might reflect high efficiency rather than high numbers present, and so on. The example illustrates both the essential feature of index methods (an *unknown* fraction of the animals present is detected in surveys) and the primary concern about them (how well does variation in the index reflect variation in population size?).

The notion of how many animals are 'present' in examples like the one just given may be confusing since no well-defined plots are established. We can imagine, however, delineating a plot with boundaries sufficiently far from the survey stations that no animal outside the boundaries will be detected at the station. The number of animals 'present' at a station is then the number of animals within this hypothetical plot. This hypothetical plot is just a useful mental construct; its only point is to show that we can attach a specific meaning to the notion of how many animals are 'present' at each survey station. This, in turn, lets us talk meaningfully about the detection rates. Our concern is not with absolute detection rates, which are arbitrary since they depend on how large the hypothetical plots are, but rather with how much the rates vary through time. Do they vary 10% or 500% through time, and does a consistent trend exist for the rates to increase or decrease through time? The answers to these questions would be the same regardless of how large we imagined the hypothetical plots to be.

The critical issues in deciding how reliable a given index is involves the ratio, for each year, (count)/(number actually present). We refer to this quantity as the 'index ratio'; it depends on the average detection rate for animals. We emphasize average because the detection rates for individual animals or groups of animals nearly always vary greatly in these surveys but this variation alone may not cause much variation from one year to another in the index ratio. Biologists and others thinking about index methods frequently confuse these issues. In particular, they often assume that high variation in detectability, from one animal to another, automatically means that index methods are unreliable. This view, however, is incorrect. If the index ratio is constant then the index yields results that are the same as complete counts on well-defined plots. To some people, this seems intuitively obvious. To others, however, it does not seem obvious so we present the following algebraic argument.

Let $\bar{v}_i$ = the mean number of animals recorded per survey in year i, $\bar{y}_i$ = the mean number of animals that would be recorded if a complete count was made at sampled stations in year i, and define $p = \bar{v}_i/\bar{y}_i$ as the index ratio; p is not subscripted because we are assuming here that it is constant. First consider estimating the ratio of counts in any two years i and j. With the index survey, this ratio would be estimated by $\bar{v}_i/\bar{v}_j$. Since $\bar{v}_i = p\bar{y}_i$ and $v_j = p\bar{y}_j$, we have $\bar{v}_i/\bar{v}_j = p\bar{y}_i/P\bar{y}_j = \bar{y}_i/\bar{y}_j$ which is the same estimate that we would have obtained with complete counts. A more relevant case concerns the ratio of values for the trend line fit to the data. Consider fitting a straight line or an exponential curve to the data. Using the fact that $\bar{v}_i = p\bar{y}_i$, it is easy to show that the ratio of the fitted curves at two times is the same whether we used the index values, $\bar{v}_i$, or the complete counts, $\bar{y}_i$, to estimate the fitted curves.

These arguments show that index surveys have the same reliability as complete counts if *no* variation exists in the index ratio. In reality, however, the index ratio always varies from year to year to some degree, and this raises the question of how, and how much, such variation affects the validity of estimates of population changes based on indices. If the correlation between the index ratio and time is 0.0, then estimates of the population change based on indices are valid estimates of the true change in the population. However, if detection rates are low and variable, then precision of any estimates made from the index survey data will be correspondingly low although this will be fully reflected in the analysis of the data.

On the other hand, any *temporal trend* in the index ratio reduces the reliability of the analysis. This point is not complicated: if detection rates steadily rise during a 10-year survey from, say 10% to 20%, then survey results will tend to increase even if population size is stable during the 10-year period. Clearly, in this case we would be misled by the index if we accepted it at 'face value'. Strictly speaking, the statistical analysis is still valid, but it does not answer the question we are really interested in. The trend being estimated by the index numbers is not the same as the trend in the population if detection rates change consistently through time. Thus, the real problem, in this case, arises if we *interpret* change in the index as being equal to change in population size.

Summarizing these points: (1) the critical issues in evaluating index results involve the 'index ratio', average count in a given year/average number present that year; (2) in real surveys the index ratio almost always varies from year to year, though the amount of variation may be small if many routes are surveyed under standardized conditions; (3) if no temporal trend exists in the index ratio [i.e., $Cov(P_i,X_i)=0$] then the statistical analyses described above provide essentially unbiased estimates of trend and precision; (4) any temporal trend in the index ratio does compromise the statistical analysis of index results.

Having established the general principles to consider in evaluating an index, we now briefly discuss factors that may cause the index ratio to change consistently with respect to time or other factors of interest. As an example, we use singing bird surveys, in which most detections are made by hearing the songs of birds. The quantity of interest is nearly always the size of the population at the time of the survey, usually the beginning of the reproductive season. Many birds, including females of most species, non-breeding males, and males whose females are incubating in many species, sing little if at all and thus have low probabilities of being detected. If the fraction of the population comprising these individuals varies substantially between years, then the index ratio will also vary substantially. Furthermore, surveyors vary greatly in their ability on many singing bird surveys; if their average ability varies between years then this will also contribute to variation in the index ratio. Other factors such as change in habitat, and corresponding change in detection rates, change in extraneous noise, especially from traffic, and different trends along roads, where the surveys are usually conducted, and throughout the region of interest may also compromise the index. These problems often combine to cause substantial variation from year to year in the index (Ralph and Scott 1981; Lancia *et al.* 1994), but in many cases they even out across years so that the correlation between the index ratio and year is zero or negligible. Factors that may change progressively include correlation between density of birds and detection rate, progressive change in average observer skill, and different temporal trends in habitat – and thus bird abundance – along roads and throughout the region. Many if these issues have been investigated, and results indicate that small biases are often (probably always) present but that major trends are often nonetheless detectable. For example, Bart and Schoultz (1984) concluded that variation in detection rates with density might cause trends to be underestimated by up to 25%. Thus, if the true rate of annual change was 4%, the expected value of the estimate would be 3% or more, a level of bias which, however regrettable, would probably not be too serious in most applications. Opinions on the reliability of index methods vary, however, with some researchers expressing considerable doubt about index methods in general (e.g., Lancia *et al.* 1994). Everyone can agree that careful scrutiny of index data is needed before assuming that trends in the index reflect trends in the population of interest.

A few additional points may be made about index surveys. The index ratio does not have to be less than 1.0; even if some fraction of the animals present are counted more than once, this does not compromise the index as long as the index ratio has zero correlation with time. Furthermore, the ratio does not need to be a true count of animals or their sign, it can be any quantity as long as the index ratio does not have a temporal trend in magnitude. For example, wildlife agencies often use the number of complaints from the public about deer damage as a very rough indication of change in deer herd size. The index ratio, for a given year, in this case is number of complaints/number of deer present, and clearly provides only a very rough indication of population trends. Under some situations, however, the correlation between the index ratio and time may be sufficiently small, and the trend in deer population size sufficiently large, that the index does provide useful information.

If a temporal trend in average observer skill is thought to be present, then regression methods including indicator variables for different observers can be used on each route to estimate the trend correcting for different observer abilities (Sauer *et al.* 1994). This method was developed for the route regression approach. Analyses of Breeding Bird Survey data using this approach generally include coefficients for all observers. An alternative approach (Fancy 1997) is to include coefficients only if the regression with observer coefficients included provides a significantly better fit than the regression obtained without observer coefficients included. If observer coefficients are included but regression on the means is to be employed, then the counts on each route should be adjusted using the coefficients for each observer prior to calculating the annual means.

Emphasis so far has been placed on estimating temporal trends because that is the most common use of index data. The counts, however, may be plotted against any other variable such as geographical location, a measure of productivity the previous year, deviation of last year's index from the trend line, etc. The critical question is still whether the index ratio, count/actual number present, has zero correlation with whatever value is plotted on the X-axis. For example, if geographical location (e.g., distance from the center of the range) is plotted on the X axis, then the issue in evaluating the index is whether the index ratio has zero correlation with the location measure. A nonzero correlation could well occur. For example, habitat may affect detection rate and may vary consistently across the range or observer skill may tend to be better in one portion of the range than in others.

The examination of outliers is often particularly difficult with index data. The reason for this is that many factors other than actual density may affect the index ratio. For example, if surveys for breeding birds are conducted at the same calendar time each year, it may happen in one year that the season is far ahead of or behind the average and this may have a major effect on survey results. Thus, many birds sing much more prior to incubation than later. Suppose that the survey is timed so that in most years most birds are in incubation, and thus have low detection rates, but in one year the season is late so many birds have not begun incubation. The index results for that year may be inflated solely due to the high detection rate. In many real surveys, it is difficult or impossible to rule out such variation in detection rates for a single year, so it may not be possible to decide whether an outlier is caused by a true difference in population size or just a difference in the index ratio.

Emphasis has been placed in the preceding text on estimating long-term temporal trends in abundance because this is the most common use of index data. In designed experiments, however, biologists often use index data gathered in only one or two periods. For example, a series of plots may be surveyed before and after treatment with a pesticide. In such cases, careful thought about how the index ratio may have varied is necessary. As an obvious example, if one technician surveyed plots before treatment and another surveyed them after treatment then the treatment is confounded with the observer effect.

8.6 Pseudoreplication

Analyses ignoring the fact that the same plots or routes are surveyed each year produce biased estimates of standard errors, and the bias is nearly always negative (i.e., the standard error tends to be underestimated so precision is overestimated). This error thus resembles pseudoreplication in the sense that inappropriate formulas are used to estimate standard errors and precision is generally overestimated. The problem, however, is not one of failing to recognize primary sampling units, but rather failing to recognize the lack of independent sampling in primary units. Thus the error is not really pseudoreplication.

True pseudoreplication is seldom a problem. If routes are not distributed randomly across the study area, but instead are selected within primary units which are themselves a sample from the area, then the two-stage nature of the design needs to be recognized and failing to do so would amount to pseudoreplication. We have seldom seen this error in behavioral ecology however.

8.7 Summary

Estimating trends in abundance raises several difficult conceptual and statistical issues. The true densities in each period seldom fall on a simple line. Instead, 'year effects' are pronounced. As a result, no single definition of 'the trend' exists and the choice between alternative definitions, and the analytical approaches they lead to, may be largely subjective. Unfortunately, however, different analytical approaches often give widely divergent results, not because some are wrong but simply because they estimate different parameters. Among standard regression methods, we recommend first fitting an exponential curve to the data because this curve changes by a constant proportional amount which makes possible a simple summary of the trend. If the exponential curve does not fit the data well, then we suggest adding a second-degree, and possibly a third-degree, term. A straight line also sometimes provides a good description of the trend though it lacks the simple interpretation of the exponential trend. Error estimation is difficult with all of these approaches because the same routes are surveyed in different years. An alternative is route regression in which a single measure of trend is obtained from each route. The overall trend is taken as the average of the trends on each route, and variance estimation is straightforward. Route regression estimates a different parameter than exponential regression on the means but results are often quite similar. Most data on temporal trends in abundance are collected using index methods in which counts are not complete and restricted to well-defined plots. The critical question with index methods is whether any temporal trend occurs in the index ratio, defined as (index value)/(actual number present). If no such trend occurs, then point and interval estimates of trend (however it is defined) are both essentially unbiased. This is true even if the index ratio varies substantially from year to year. Gaining confidence that no temporal trend exists in the index ratio is usually difficult, calling for intensive studies from which generalizations are difficult to make. Thus, index results should generally be regarded as suggestive rather than conclusive unless the estimated trend is substantially greater than the maximum likely temporal trend in the index ratio.

9
Capture–recapture methods

9.1 Introduction

The phrase 'capture–recapture methods' refers to studies in which a sample of animals is marked and then some, but usually not all, of them are recovered on one or more subsequent occasions. Goals of capture–recapture studies include estimating population size, survival, and certain other variables discussed in this Chapter and studying associations between these and other variables such as how survival rates vary with age or across years. In this Chapter we review the basic concepts that underlie capture–recapture studies, identify major branches of the field, and describe recent developments. Although the methods were originally used primarily to estimate population size and survival rates, contemporary methods provide a rigorous approach for studying a wide variety of other issues of interest in behavior, ecology, and evolution (Nichols 1992). The methods, however, are complex and new refinements appear constantly. We therefore suggest consulting a specialist in capture–recapture methods before undertaking work in this field, and we do not attempt to provide a comprehensive guide to the methods.

9.2 Rationale

The basic rationale in capture–recapture methods is to estimate what fraction of the animals marked and present in the study area was counted during each sampling period. This fraction is then used to estimate quantities of interest. Two general approaches might be distinguished for estimating the fraction. First, suppose we can assume that all animals (marked and unmarked) present at the start of the study are still present on each recapture occasion. We will also assume that the marked and unmarked animals, in a given cohort, have the same recapture probabilities. The fraction of a sample bearing marks is thus an unbiased estimate of the fraction of the population bearing marks, and this lets us estimate the fraction of the population that we counted in the sample. For example, if we mark 100 animals, then capture a subsequent sample of 200 animals and find that 50 of them are marked, then $50/200=25\%$ is an unbiased estimate of the fraction of the population that we marked. Accordingly, $100/0.25=400$ is a reasonable estimate of population size.

The second approach permits change in population size due to births, deaths and movements. Assume, initially, that animals leaving the area do not return, and animals entering the area do not leave again, during the course of the study. Suppose we mark 100 animals on the first capture occasion and recapture some of them on two subsequent occasions. We denote the number of marked animals captured on occasions two and three as m_2 and m_3. In general, some of those with marks captured on the third occasion will not have been recaptured on the second occasion. The proportion of the m_3 that were also captured on occasion two is an unbiased estimate of what fraction of the marked animals were captured on occasion two. Thus, the fraction of marked animals captured may be estimated by examining how many marked animals were captured on a subsequent occasion but not captured on the occasion of interest. Given an estimate of this fraction, the total number of marked animals alive and in the study area may be estimated, and from this survival rates can be estimated. In this case, 'mortality' actually means either dying or surviving but leaving the study area. Mortality thus has the meaning 'loss from the population' through either death or emigration.

While these comments are intended to show the conceptual basis of capture recapture estimation, the actual calculations of point and interval estimates are quite complex. It is thus important for behavioral ecologists to become familiar with the methods and use one of the several comprehensive software programs (see Sections 9.5 and 9.6) rather than attempting to develop their own methods.

9.3 Capture histories and models

The results of a capture–recapture study are often summarized by a series of capture histories. The summary is an array with rows representing individuals and columns representing capture and recapture occasions. In the simplest case, the entries are either 0, meaning that the animal was not captured or recaptured, or 1, meaning that the animal was captured or recaptured. Thus, with four occasions for marking and recovery, the history

1001 refers to animals that were captured on occasions 1 and 4 and not on 2 or 3. The results from a capture–recapture study may be summarized using a separate line for each animal, or the number of individuals having each history may simply be noted. For example, 1001 23 might mean that 23 individuals had the history 1001.

Corresponding to the capture history is a statistical model specifying the probability of obtaining each row in the array. The statistical model depends on what events are assumed to be possible. For example, suppose all animals alive and in the study population at the start of the study remain in the population throughout the study. Thus, there are no changes in population size due to births, deaths, or movements. Suppose, too, that all animals have the same probability, p, of being captured on each capture occasion. The probability of obtaining capture history 1001 is then $p(1-p)(1-p)p$. The probability of obtaining the history 1100 is $pp(1-p)(1-p)$. Given this type of description of each possible capture history, statistical methods (based on maximum likelihood theory) can be developed to estimate the parameters given the number of individuals having each capture history. This is the basic analytical approach in capture–recapture methods.

The simple capture histories described so far can be expanded to describe many other situations and to estimate many other parameters. For example, suppose that some individuals die or leave the population. Let the probability of surviving and remaining in the study area be ϕ, and assume that it is the same for all animals during all intervals (i.e., between consecutive capture occasions). The probability of obtaining capture history 1011 is then $p\phi(1-p)\phi p\phi p$. This expression, in words, means the probability of being captured on occasion one (p), surviving until occasion two (ϕ), not being captured on occasion two ($1-p$) and so on (where 'surviving' means surviving and remaining in the study area). If the survival or recapture probabilities vary through time, then subscripts can be used to indicate that the values are, or may be, different, e.g., $p_1\phi_1(1-p_2)\phi_2 p_3\phi_3 p_4$. In this case we would estimate seven parameters rather than the two when rates were assumed not to vary through time. Alternatively, if the rates are thought to vary in a constant, for example linear, manner, then we may replace p_t, where the subscript t indicates time (1 to 4 inclusively in the example), with an expression such as $p_t = \beta_0 + \beta_1 t$ in which recovery is viewed as a linear function of time. The statistical methods would then estimate the two parameters that define the recovery rate for any time rather than the single rate or the time-specific rates.

Probabilities of movement can also be incorporated into this approach. For example, with two recapture sites the capture histories could be written using A for 'site one' and B for 'site two'. Thus, the capture history A0BA would refer to animals caught on occasion one in site A, not caught on occasion two, caught at site B on occasion three and caught at site A on occasion four. The statistical model for this case is more complex because so many different movements are possible. The approach, however, is similar in principle (e.g., Brownie *et al.* 1993; Nichols *et al.* 1993).

9.4 Model selection

Much of the new work in capture–recapture methods during the past few decades has been devoted to developing comprehensive models that can be used to analyze and compare many different submodels. In the example above, we described models in which capture and recovery rates were time specific or constant. A model with both rates varying through time would be considered a general model because it includes, as special cases, models in which one or both rates were constant and models in which one or both rates were the same for some but not all occasions. These models are thus nested in the sense that the models with more assumptions (e.g., all recovery rates the same) have the same structure as the general model but with some constraints (e.g., all p_t = a constant value, p). Nested models can be compared using procedures similar to the ones used in multiple regression. For example, a model with only a few parameters may be preferred to one with more parameters unless the one with more parameters achieves a significantly better fit to the data (generally evaluated use likelihood ratio tests and a quantity known as Akaike's Information Criterion or AIC). An overall goodness-of-fit test is also often carried out. These procedures are described in detail in reviews of capture–recapture methods such as Burnham *et al.* (1987) and Pollock *et al.* (1990). We now briefly describe some of the general categories of models and the factors they are designed to study.

9.5 Closed population models

Closed population models assume that no births, deaths, or movements into or out of the population occur during the study. Population size is thus constant throughout the study. The parameter of interest in these models is usually population size, though differences in capture probabilities are also estimated by some of the models and are occasionally of interest. Models for this case were developed by Otis *et al.* (1978) and White *et al.* (1983) and incorporated into the program CAPTURE. More recent refinements are described by Rexstad and Burnham (1991), Nichols (1992) and White (1996).

In closed population models, much emphasis is placed on variation in recapture probabilities. Variation in capture probabilities can easily cause serious bias in the estimates of population size unless models recognizing the variation are used. For example, if some animals consistently have a high probability of being captured, then the estimate of what fraction of the population is captured could be seriously in error, leading to a seriously biased estimate of population size. The same problem could occur if capture probabilities change through time.

The program CAPTURE permits examination of three types of variation in capture probabilities: variation between individuals, variation through time, and variation in response to being trapped. Models are provided for each type of variation and each pair of types (e.g., variation through time and in response to being trapped). A model with no variation is also provided, and a model with all three types of variation is discussed but the parameters cannot be estimated.

9.6 Open population models

In open population models, births, deaths, and movements into or out of the population are permitted. Population size thus may change during the study. Early models developed in the mid-sixties by Cormack (1964), Jolly (1965) and Seber (1965), often referred to as the Cormack–Jolly–Seber model, assumed that arrivals to, and departures from, the population were permanent. Thus 'survival' in these models should be interpreted as meaning surviving and remaining in the study area and should not be interpreted solely as survival unless rates of emigration (and survival) are very low.

SURGE

Several programs have been developed specifically for open population models. One of the most widely used is SURGE (Lebreton *et al.* 1992) which is reasonably easy to use and is extremely flexible. The user's manual for SURGE is rather difficult for beginners to understand but Cooch *et al.* (1996) provide a detailed practical guide to using SURGE. Parameters are specified in SURGE by describing arrays for each time period and group of animals (classified by time of first capture). Separate arrays are declared for cohorts defined by age, sex, location and so on. For example, suppose we captured males and females during each of 7 years, some as hatching year birds, and others as adults of unknown age. We are interested in age-specific rates for up to three age classes: hatching year (HY), second year (SY), and after second year (ASY). Birds can be aged only in their hatching year. We assume that all birds first captured after their hatching year are ASY (though in reality some of them will be SY birds). Survival and recovery rates are sex, age, and year specific. The parameter diagram for this model, in SURGE format, is shown in Table 9.1. Parameters are numbered within sex and age cohorts. Thus, parameters 1–6 refer to males first captured during their hatching year. Parameters 7–11 refer to these birds during the year after their first capture. Many of the parameters appear multiple times in the arrays. For example, parameter 15, the survival rate of ASY males in year 1993, appears 10 times.

This notation, while somewhat complex, provides the basis for evaluating a great many models by specifying that certain parameters are equal to each other. For example, to compare the model in Table 9.1 to one in which survival rates for second year males are the same as those for adults, we would include the following constraints: $8=12$, $9=13$, $10=14$, $11=15$. SURGE prints the AIC for each model and, if we used this criterion for model selection, we would simply adopt the model with the lowest AIC. A little thought will suggest numerous other comparisons. For example, we could set the survival rates of second year males equal to that of HY males, thereby obtaining a third model which could be compared to the other two. In a real evaluation of the general model above, applied to capture – recapture data for Kirtland's warblers, more than 40 submodels were evaluated. The final model selected had just four parameters: survival for first year birds, survival for all older birds, recovery rates for males (all ages), recovery rates for females (all ages). SURGE provides various shortcuts and additional information and runs on a personal computer. It does not, however, carry out general goodness-of-fit tests and is thus often used in combination with RELEASE (Burnham *et al.* 1987) which does perform these tests.

Other parameters

Many other special cases have been investigated by specialists in capture–recapture methods. For example, a 'robust design', combining closed and open population models was introduced by Pollock (1982). White (1983) developed a general program, SURVIV, that is more difficult to use but permits estimation of parameters with numerous models including ones that model survival rates in terms of external covariates. Models

Table 9.1. *Model of survival and recovery rates with separate rates for each sex, age (HY, SY, ASY[a]), and calendar year. Row headings indicate year of first capture; column headings indicate years of capture and recapture. Entries indicate which parameters are assumed to have the same values.*

1. Rates for males first captured during their hatching year

	Survival rates						Recovery rates					
	'88	'89	'90	'91	'92	'93	'88	'89	'90	'91	'92	'93
'87	1	7	12	13	14	15	35	41	46	47	48	49
'88		2	8	13	14	15		36	42	47	48	49
'89			3	9	14	15			37	43	48	49
'90				4	10	15				38	44	49
'91					5	11					39	45
'92						6						40

2. Rates for females first captured during their hatching year

	Survival rates						Recovery rates					
	'88	'89	'90	'91	'92	'93	'88	'89	'90	'91	'92	'93
'87	16	22	27	28	29	30	50	56	61	62	63	64
'88		17	23	28	29	30		51	57	62	63	64
'89			18	24	29	30			52	58	63	64
'90				19	25	30				53	59	64
'91					20	26					54	60
'92						21						55

3. Rates for males first captured after their hatching year

	Survival rates						Recovery rates					
	'88	'89	'90	'91	'92	'93	'88	'89	'90	'91	'92	'93
'87	31	32	12	13	14	15	65	66	46	47	48	49
'88		32	12	13	14	15		66	42	47	48	49
'89			12	13	14	15			37	43	48	49
'90				13	14	15				38	44	49
'91					14	15					39	45
'92						15						40

4. Rates for females first captured after their hatching year

	Survival rates						Recovery rates					
	'88	'89	'90	'91	'92	'93	'88	'89	'90	'91	'92	'93
'87	33	34	27	28	29	30	67	68	61	62	63	64
'88		34	27	28	29	30		68	57	62	63	64
'89			27	28	29	30			52	58	63	64
'90				28	29	30				53	59	64
'91					29	30					54	60
'92						30						55

Note:
[a] HY = hatching year, SY = second year, ASY = after second year.

for harvested birds were developed by Brownie *et al.* (1985). Methods for accommodating temporary emigration have been studied by several authors (e.g., Kendall and Nichols 1995, Kendal *et al.* 1997). Skalski and Robson (1992) developed methods for testing hypotheses about differences in population size with closed populations. Methods for separating immigration from births were studied by Nichols and Pollock (1990). Other developments prior to 1991 are described by Nichols (1992). During the past several years, specialists in capture–recapture statistics have continued to adapt these methods for studying such issues such as recruitment (Pradel *et al.* 1997), individual variation in demographic parameters (Skalski *et al.* 1993), age-specific breeding rates (Clobert *et al.* 1994), detection of senescence (Burnham and Rexstad 1993, Nichols et al. 1997), and estimating population size in plants (Alexander *et al.* 1997). In general, however, theoretical studies of behavior, ecology and evolution have made little use of modern capture–recapture methods (Lebreton *et al.* 1993), a situation that may change rapidly during the coming several years as researchers working on theoretical topics realize the great advances that are possible using these methods.

9.7 Summary

Capture–recapture methods provide ways for converting index data pertaining to population size, survivorship and other parameters into rigorous, often essentially unbiased, estimates of the parameters. The basic principle in capture recapture is that the fraction of the animals present on a sampling occasion, but not detected on surveys, can be estimated from the capture–recapture data and used to obtain parameter estimates. Capture–recapture analysis involves summarizing the capture data in capture histories and writing down expressions for all possible capture histories in terms of the unknown parameters. Given this model, maximum likelihood methods may be used to obtain point and interval estimates. In recent years, a great deal of work has been carried out to develop powerful, efficient, and flexible computer programs that provide maximum likelihood estimates of population size and survival rates. These methods are beginning to be used to estimate other demographic parameters such as movement rates and measures of productivity. Furthermore, the methods are suitable for studying how these variables are associated with other factors such as weather and dominance status or for estimating derived quantities such as the cost of reproduction or lifetime reproductive success. These methods thus may be of great value to behavioral ecologists when counts are incomplete and the fraction of individuals missed on surveys may vary and thus compromise the data set. They are now routinely used in fisheries and wildlife but have been much less used by behavioral ecologists working in theoretical research. We urge readers who must deal with incomplete counts and the biases that may result to become acquainted with the literature in this exciting and rapidly developing field, learn the basic approaches, and then contact one or more of the specialists working on capture–recapture methods for further assistance.

10
Estimating survivorship

10.1 Introduction

Survivorship, the proportion of individuals surviving throughout a given period, may be estimated simply as $p=x/n$ where $n=$the number alive at the start of the period and x equals the number alive at the end of the period, or it may be estimated using capture–recapture methods as discussed in Chapter Nine. Two additional issues, however, often arise in behavioral ecology. One is that in many studies a measure of overall survival across several periods, each having a separate survival estimate, may be desired. The second issue is that in studies of nesting birds or other animals, information is often incomplete because many nests are not discovered until well after they have been initiated and may not be followed to completion. In this Chapter we discuss methods developed to handle both of these cases. We focus on telemetry studies, which raise the first issue, and studies of nesting success which raise both issues.

10.2 Telemetry studies

In telemetry studies, transmitters are attached to animals which are then monitored for various purposes, including the estimation of survival rates. The simplest case arises when all transmitters are attached at about the same time and animals are checked periodically at about the same times until they die or the study ends. Cohorts based on age, sex, or other factors may be defined.

Data of this type are basically binomial (White and Garrott 1990; Samuel and Fuller 1994). On any sampling occasion, t, the proportion of animals still alive is the appropriate estimator of survivorship to that time. The standard formulas (e.g., Appendix One) for proportions apply to the calculation of standard errors and confidence intervals and to the carrying

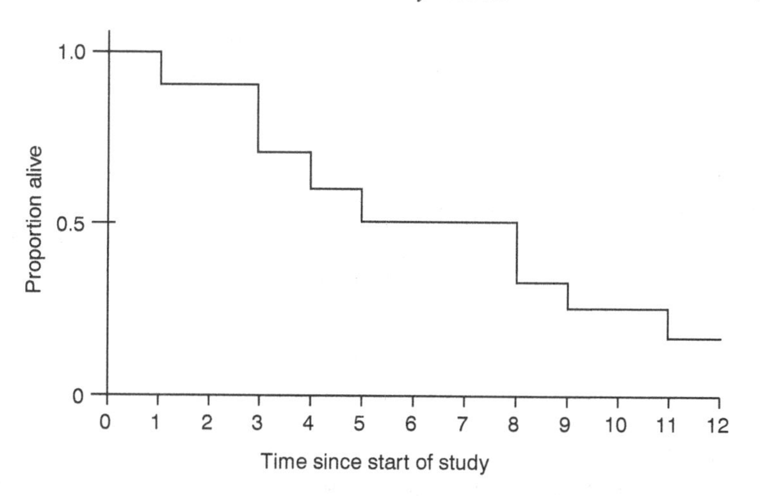

Fig. 10.1 Temporal decline in population size as portrayed using Kaplan–Meier estimates. (Stair-step diagram.)

out of statistical tests. Thus, if time starts at 0, ends at T, and the number still alive at time $t \leq T$ is n_t, then survivorship to time t may be estimated as $s_t = n_t/n_0$ with standard error $[s_t(1-s_t)/n_0]^{0.5}$.

Suppose we calculate the survivorship each time animals are checked say at times $1,\ldots,t$. As long as no animals have died, the estimated survivorship will be 1.0. When the first animal dies, the estimate will drop to $(n_0-1)/n_0$. When the second animal dies, the estimate will drop again to $(n_0-2)/n_0$ and so on. The plot of estimated survivorship against time thus has a stair-step appearance (Fig. 10.1), and provides a convenient graphical summary of the data set. This approach is a simple case of the Kaplan–Meier method (Kaplan and Meier 1958) which allows one to incorporate into the analysis more complicated features such as animals which leave the study area.

A practical problem with using $s_t = n_t/n_0$ as the estimator of survivorship to time t is that in most studies animals do not all enter the study at the same time. Furthermore, in many cases animals drop out of the sample due to causes other than death such as transmitter failure and emigration. Notice, however, that we may rewrite n_t/n_0 as

$$s_t = \frac{n_t}{n_0} = \frac{n_t}{n_0}\left(\frac{n_1 n_2 \ldots n_{t-1}}{n_1 n_2 \ldots n_{t-1}}\right) = \frac{n_1 n_2 \ldots n_{t-1} n_t}{n_0 n_1 \ldots n_{t-1}} \tag{10.1}$$

$$= \prod_{i=1}^{t} \frac{n_i}{n_{i-1}} = \prod_{i=1}^{t} s_i^*$$

in which the Π sign indicates multiplication. Thus, with no additional assumptions, we may express the overall survivorship from the start of the study to any time $t \leq T$ as the product of the proportions surviving the separate intervals. This form of the basic binomial estimator, however, accommodates changes in sample size during the study. We simply use whatever animals are observed in each interval to calculate the estimate s_t^*. With changes in sample size, we need a slightly different notation. Let $s_i^* = 1 - d_i/n_i$ where d_i is the number of deaths during interval i and n_i is the number alive at the start of the interval. Then we may rewrite Eq. 10.1 as

$$s_t = \prod_{i=1}^{t} s_i^* = \prod_{i=1}^{t} \left(1 - \frac{d_i}{n_i}\right) = \prod_{i=1}^{t} \frac{n_i - d_i}{n_i}. \tag{10.2}$$

The Kaplan–Meier formula for survivorship to time $t \leq T$, with changes in sample size, is generally presented in this form (e.g., White and Garrott 1990 p. 234, Samuel and Fuller 1994 p. 407, Bunck *et al.* 1995 p. 791). The variance of s_t may be estimated as

$$v(s_t) = \frac{s_t^2 (1 - s_t)}{n_t}. \tag{10.3}$$

When animals enter or leave the sample (due to causes other than death), then an adjustment to the variance formula is needed. Pollock *et al.* (1989a,b) referred to this as the 'staggered entry' case and pointed out that a formula in Cox and Oakes (1984 p. 51) is appropriate for this case. If no changes in sample size occur, then $s_t = n_t/n_0$ and this formula reduces to the familiar equation for a binomial proportion, $v(s_t) = s_t(1 - s_t)/n_0$. Many software programs (e.g., SAS, BMDP) provide graphs and variance estimates for each period using the Kaplan–Meier approach, but often under the assumption that all animals enter at the start of the study. White and Garrott (1990) provide a SAS program to compute estimates for the staggered entry design. Tests for differences in survivorship between cohorts have also been developed, primarily in the medical field. White and Garrott (1990 pp. 232–50) and Samuel and Fuller (1994 pp. 403–9) describe these methods and their utility in behavioral ecology. Bunck *et al.* (1995) provide modifications for the case in which some tagged animals are missed during surveys. The variance estimates described here assume that all fates are independent. Flint *et al.* (1995) discuss application of the Kaplan–Meier approach for estimating brood survival, and note that the fates of individuals in a brood are usually not independent whereas the traditional Kaplan–Meier approach assumes that all fates are independent. They suggest treating broods as unequal-sized primary units, calculating separate survival estimates for each brood, and then using the formula for multistage (i.e., in this case, cluster) sampling to obtain standard errors for interval-specific survivorship. Standard errors for overall survivorship are obtained using a bootstrapping approach (Efron and Tibshirani 1993 pp. 43–56).

10.3 Nesting success

Studies of nesting success by birds and other animals are similar to telemetry studies in that animals are visited periodically and the goal is to estimate survivorship through one or more periods of interest. When all nests are found upon initiation and followed until failure or successful completion, then the simple binomial methods, including the Kaplan–Meier approach, already described provide the best method of analyzing results. The exact method depends on what measure of success is desired. If the average number of young produced per nesting attempt is desired, then the data set may be viewed as a simple random sample of nesting attempts with the variable defined as 'number of young produced'. The standard formulas for simple random sampling are then used to obtain point and interval estimates. If the objective is estimating what fraction of eggs survive to produce young leaving the nest, then nesting attempts are best viewed as unequal-sized primary units with primary unit size being the number of eggs and the variable being the proportion of the eggs that produce young. The formula for multistage sampling with primary units weighted unequally (Appendix One, Box 3) may then be used to obtain unbiased point and interval estimates. Investigators sometimes estimate the proportion of nests that produce at least one young leaving the nest (i.e., 'successful nests'). In this case, the variable is dichotomous ($0=$ none produced/$1=1^{+}$produced) and the formulas for proportions with simple random sampling (Appendix One, Box 2) are used.

Finding all attempts, including those that fail rapidly, is often difficult or impossible. However, in some of these studies successful attempts can be counted because, for example, food-bearing adults can be detected. The number of successful attempts per unit area can then be estimated. If the number of pairs in the study area is known and the parents in successful attempts can be identified, then pairs can often be defined as the population unit (i.e., invoking the superpopulation concept, Section 1.2) and used to estimate success/pair and its standard error. If successful attempts cannot be assigned to pairs, the point estimate for success/pair or success/area can still be calculated. If multiple sites are included in the study, and can be viewed as a random sample of sites, then sites may be defined as primary units and used to calculate the standard errors. Sometimes nests can be found after the season has ended and examined to determine how many young left each nest successfully. This approach is widely used in studying waterfowl, the estimates being based on counting membranes left in the nest after the young have departed. If all nests in a study area are found, then this approach yields success/attempt and success/area for the study area and period. If the nests are not all found, but those that are found may be viewed as a random sample from all nests, then the approach yields an unbiased estimate of success/attempt. Thus, careful thought about possible problems during data collection and about parameters of interest often leads to the identification of parameters that can be estimated using binomial formulas or other formulas from survey sampling theory.

In many studies, however, investigators wish to estimate success per nesting attempt and cannot do so using any of the approaches described here because attempts are not monitored from start to finish. Instead, many nests are not found until well after initiation and some nests are not found at all because they fail before they would have been found. As a result, the quantity (number of eggs or young alive when observation ended)/(number of eggs or young found) would underestimate mortality and overestimate survivorship, often by a large but unknown amount. This summary statistic would not even be a good index to survivorship in most cases because other studies to which results might be compared might involve either more or less intensive searching and thus more or less missed mortality.

Mayfield (1961, 1975) suggested a solution to the problem of incomplete coverage, and Johnson (1979) provided a formula for calculating standard errors of Mayfield estimates. Mayfield organized the analysis in a way that facilitates addressing questions that are often of interest to behavioral ecologists. The cost of this flexibility, however, is that the method can appear rather complex on first acquaintance. Consequently, we first present the method in a simpler form than Mayfield did.

Let us consider the incubation period and assume that survivorship is thought to be approximately constant, that is, that the same fraction of eggs is lost on each day. We will call this fraction the daily mortality rate (DMR); 1–DMR is the proportion that survive each day which we will call the daily survival rate, DSR. The probability of surviving the entire period, under this set of assumptions, is DSR^t where t is the length of the period. Thus, if we could estimate the DMR we could estimate the proportion of eggs surviving the entire period as $DSR^t = (1 - DMR)^t$. Mayfield proposed a very simple estimate of the daily mortality rate

$$DMR = \frac{\text{deaths}}{\text{days}}, \tag{10.4}$$

Table 10.1. *Example of the Mayfield method for estimating survivorship during a period in which survival rates are assumed to be constant*

Nest	Date	No. of eggs	Interval length	No. of deaths	No of exposure days	
					Survivors	Fatalities
1	May 1	4	—	—	—	—
	8	4	7	0	28	0
	12	4	4	0	16	0
2	May 5	3	—	—	—	—
	9	0	4	3	0	6
3	May 7	4	—	—	—	—
	11	3	4	1	12	2
	17	1	6	2	6	6
4	May 4	5	—	—	—	—
	11	5	7	0	35	0
Total				6	97	14

Notes:
Daily mortality rate: $6/111 = 0.054054$.
Daily survival rate: 0.945946.
Estimated survival for 15-day period: $0.945946^{15} = 0.435$.

where days means the number of days that eggs were alive and under observation. Thus, Mayfield proposed counting all deaths that were observed, dividing this by the number of days for which the eggs were under observation, which he termed ‘exposure days’, and using the ratio of these numbers as the estimate of the DMR. The calculation of days is somewhat complex, however, because nests are usually not visited daily. If they were visited daily, then obviously we would stop visiting them, and thus stop counting days, once mortality had occurred. When nests are visited only periodically, so we do not know the exact time of death, Mayfield suggested using half the number of days in the last observation interval, assuming that, on average, this would give a good estimate of the number of days that would have been recorded if daily visits had been made. Calculations for the incubation period with a small hypothetical sample of nests using this approach are shown in Table 10.1.

When the attempt includes two or more periods, for example incubation and nestling periods, then the overall rate of success is estimated as the product of the period-specific rates. For example, if the DSRs for incubation and nestling periods are s_I and s_N, and the average durations of these periods are T_I and T_N then overall survivorship (s), the proportion of eggs at the start of incubation that survive until nest departure, is

$$s = s_I^{T_I} s_N^{T_N},$$

Table 10.2. *Data from Table 10.1 presented with whole and partial mortality distinguished*

				Nests		Individuals	
Nest	Date	Number of eggs	Interval length	Deaths	Days	Deaths	Days
1	May 1	4	—	—	—	—	—
	8	4	7	0	7	0	28
	12	4	4	0	4	0	16
2	May 5	3	—	—	—	—	—
	9	0	4	1	2	—	—
3	May 7	4	—	—	—	—	—
	11	3	4	0	4	1	14
	17	1	6	0	6	2	12
4	May 4	5	—	—	—	—	—
	11	5	7	0	7	0	35
Totals				1	30	3	105

Notes:
Daily mortality rate: nests 1/30=0.033333.
Daily mortality rate: individuals 3/105=0.028571.
Daily survival rate: individuals 1−0.033333=0.966667.
Daily survival rate: nests 1−0.02857=0.971429.
Overall daily survival rate: (0.966667)(0.971429)=0.939048.
Estimated survival for 15-day period: $0.939048^{15}=0.389$.

and the average number produced per nesting attempt is $\bar{c}s$, where $\bar{c}$ is the average clutch size.

We now turn to the main additional feature that Mayfield suggested. Biologists are often interested in whether nests failed completely or only some of the eggs or young were lost. For example, complete failure may indicate predation and partial failure may indicate starvation. Mayfield therefore suggested dividing deaths and days into cases in which the entire attempt failed and those in which only some of the eggs or young died during the interval. To avoid double counting, deaths and days for partial losses were only recorded when the nest (i.e., at least one egg or young) survived. Calculations for this approach are illustrated in Table 10.2. The point estimates may vary slightly depending on whether mortality is divided into nest and individual categories or defined as a single category.

Various additional complexities may occur in applying the Mayfield method. Investigators often want to estimate the success per egg laid, rather than success per egg at the start of incubation. Visits during laying, however, tend to cause abandonment in many species. If this visitor-induced mortality is not excluded the estimated nest success may be seriously biased. If visits during laying are feasible, then the approach for counting days has to be adjusted because the number of eggs present increases during the observation interval. Also, single egg loss during laying is often difficult to detect because the egg may be laid and lost during a single interval. Problems may occur at hatch too, caused, for example, by uncertainty over to which period intervals spanning hatch should be assigned and because hatching failure may cause mortality rates during this time to be high (which calls into question the assumption of constant mortality). Finally, visits near the end of the nestling period may cause premature nest departure which has undesirable effects on the nestlings and means that mortality which would have occurred prior to nest departure is missed. Careful consideration of problems such as these is thus needed in designing a study of nesting success.

Various refinements and extensions to the basic Mayfield method have been proposed. For example, Trent and Rongstad (1974) applied essentially the Mayfield approach to telemetry data. Bart and Robson (1982) developed maximum likelihood estimators under the assumption of constant survivorship, evaluated sensitivity to these assumptions and presented models for testing whether visitor impact occurred in collecting the data. Heisey and Fuller (1985) extended the Mayfield approach to include more than two (mutually exclusive) classes of mortality and developed a program, MICROMORT, for obtaining maximum likelihood point and interval estimates of nesting success. They also discuss related statistical issues such as bias, transformations, and the relationship of precision to visitation frequency.

Estimating standard errors of Mayfield estimates has been problematic because the method assumes that fates of eggs or young in nests are independent. This is appropriate under the assumption of constant daily survivorship (fates on each day are then analogous to flipping a coin), but this assumption is not realistic, and while departures from the assumption generally cause little if any error in point estimates, variance estimates are generally underestimated. The error amounts to pseudoreplication. Nests, not days, are selected randomly but days are used in variance calculation. A more appropriate approach could be developed by viewing the population units as individual-days and the variable as 'survived/did not survive'. The population diagram (Chapter Four) would be an array with eggs down the side, separated into nests, and days across the top. Rows would be of unequal length because not all individuals survive until nest departure. If the cell entries are 0 = died and 1 = survived, then it is easy to show that the mean of the cells in the statistical population equals the DSR. If nests may be viewed as a random sample from the population and observation intervals a random sample from the possible intervals then the estimate deaths/days is an unbiased estimate of the population mean. The sampling plan is two stage with unequal-sized primary units (nests, or more precisely the set of individual-days for each nest), and the formula for multistage sampling would provide unbiased point and interval estimates. Essentially this suggestion was made recently by Flint *et al.* (1995). The major problem with this approach is whether the observation days for a given nest are really an unbiased estimate of the mean of the variables for that nest. For example, if the start of the observation interval is selected randomly but nests are then followed until completion, then obviously days early in the attempt have a much smaller probability of selection than later days. Thus, more work is needed in this area, but the general approach of treating nests as clusters of days seems promising.

Although the Mayfield approach is a great improvement over simply reporting the success rate in the sample (Miller and Johnson 1978; Johnson and Shaffer 1990), careful thought should be given to whether this approach is necessary (White and Garrott 1990). Several examples have been given in this Chapter in which simpler methods may be used to estimate quantities of interest. Furthermore, success/attempt does not give success/season if re-laying occurs, yet in many investigations (e.g., demographic analyses, forecasting population size) success/season is much more useful than success/attempt. Thus, while the Mayfield method is certainly preferable to simply ignoring the incomplete count of deaths, other methods may be more feasible, and in some cases may yield more relevant results.

10.4 Summary

Estimates of survivorship are often simple proportions (Chapter Four) or are based on capture–recapture methods (Chapter Nine). This Chapter discusses two other methods that are widely used in behavioral ecology to estimate survivorship. The Kaplan–Meier method is useful when individuals are checked at about the same time and when time (e.g., date) is known. The method permits graphic display of estimated survivorship through time, even if new animals enter the sample after the beginning of the study and other animals leave the study through causes other than death. Confidence intervals and tests for a difference in survivorship between cohorts are also available. The Mayfield method for estimating nesting success is useful when many individuals are not found at the beginning of the period of interest and time (e.g., age of the nest) is not always known. It may also be useful in telemetry studies when individuals are visited at different times. Both the Kaplan–Meier and Mayfield methods have traditionally required the assumption that fates of individuals are independent, but methods have been suggested recently for removing this often unrealistic assumption. The Kaplan–Meier method is nonparametric, making no assumption about temporal trends in actual survivorship. In the Mayfield method, the overall period of interest is divided into intervals during which survivorship is assumed to be constant. This introduces an element of subjectivity into the method since different investigators may delineate different intervals and this may affect the overall survivorship estimate. Furthermore, if survivorship varies during any of these periods, and monitoring effort is not distributed evenly, then the point estimate will be biased, sometimes severely. The Mayfield method thus requires stronger assumptions than the Kaplan–Meier method, a fact which has caused investigators to prefer the Kaplan–Meier method when they have a choice between the two methods. While the Mayfield method is a great improvement over ignoring incomplete coverage of nest attempts, practical difficulties often arise such as how to estimate survivorship during laying. Furthermore, in some cases other parameters, such as the number of young produced per pair, can be estimated with less difficulty and may provide results that are nearly as useful – or even more useful – than success per nesting attempt. We therefore recommend that careful thought be given to goals and practical issues before adopting the Mayfield approach.

11
Resource selection

11.1 Introduction

We use the phrase resource selection to mean the process that results in animals using some areas, or consuming some food items, and not consuming others. In some studies, resources are defined using mutually exclusive categories such as 'wooded/nonwooded' in a habitat study or 'invertebrate/plant/bird/mammal' in a diet study. In other studies, resources are defined using variables that are not mutually exclusive and that define different aspects of the resource such as elevation, aspect, distance to water, and cover type. In some studies, only use is measured. In many others, availability of the resources is also measured and analyses are conducted to determine whether the resources are used in proportion to their availability or whether some are used more – and some less – than would be expected if resources were selected independently of the categories defined by the investigator. In this Chapter, we concentrate on methods in which resources are assigned to mutually exclusive categories because this approach has been by far the most common in behavioral ecology. However, studies in which resources are defined using multivariate approaches are discussed briefly, and we urge readers to learn more about this approach since it can be formulated to include the simpler approach but offers considerably more flexibility.

The investigator has great flexibility in defining use. In the case of habitat studies use may mean that an area is used at least once during the study or (less often in practice) used more than some threshold number of times, or used for a specific activity (e.g., nesting). When the population units are possible prey items, used items might be those attacked, consumed, partly consumed, etc. In nearly all studies of resource use, population units are classified as used or not used. Methods can be developed in which use is viewed as a quantitative variable (e.g., Manly *et al.* 1993, Chapter Ten), but relatively few studies have adopted this approach so we do not discuss it in this Chapter.

11.2 Population units and parameters

In this Section we define population units and parameters for the case in which only one animal is being monitored or individual animals are not distinguished. For example, data may be collected by making several surveys and recording the type of habitat in which each animal is found, but individual animals are not identified. In Section 11.3 we extend the discussion to consider the case in which separate data sets are obtained for each animal and the goal is to combine these data sets in a single analysis.

In some diet studies, prey are individuals which are either consumed or not consumed during the study, and all prey are present at the start of the study and do not leave during the study period (except as a result of predation). In these cases, population units are easily defined as the individual, potential prey.

When prey enter and leave the study area, in diet studies of herbivores, and in habitat use studies a slightly more complex approach is needed in defining population units. We consider the case of habitat studies, but the rationale developed applies easily to diet studies with immigration and emigration.

Defining the population diagram may be helpful in thinking about the parameters 'use' and 'availability'. We suggest viewing population units as area-times, with the areas being so small that they can hold only one individual at a time. Similarly, the intervals may be viewed as being too short for an animal to enter or leave an area during an interval. This is equivalent to viewing the population units as dimensionless points in space and time. In defining use, the variable is (in use)/(not in use) by an individual. In defining availability, the variable is (type of habitat). If the resource type in each population unit does not change throughout the study, then the temporal dimension is not needed in defining availability. On the other hand, allowing the possibility of change provides a more general approach, and permits the same notation for defining use and availibility. We therefore refer to the population as the set of all 'area-times' in defining both use and availibility.

Let $i=1,\ldots,I$ denote the resource types and

U = number of used population units

U_i = number of used population units of type i ($\sum U_i = U$)

A = number of units available (population size)

A_i = number of units available of type i ($\sum A_i = A$).

Many different parameters have been estimated in resource selection studies. Some of the most common parameters of interest are discussed in the following subsections.

Proportion of type i *units that are used,* U_i/A_i

In some diet studies the number of prey of each type is estimated or known at the start and end of the study. For example, Bantock *et al.* (1976) released known numbers of snails of different types (A_i) into an area in which they were preyed upon by song thrushes. The number of prey of each type was counted again at the end of the study period and used to obtain U_i.

The number of units of each type used and available provides the best information on resource use. All of the quantities discussed in the following text can be expressed in terms of these numbers. In habitat studies, however, estimating U_i is usually not feasible. In general, when we obtain a count, u_i say, and wish to estimate a population total, U_i , we use u_i/f, where f=fraction of the population in the sample (with complex sampling designs a different approach may be needed). But in habitat studies, as already noted, the population units are simply defined as being 'very small' in space and time so f is not well defined. The U_i could be estimated by monitoring the study animals during randomly selected periods, recording their habitat use continuously. The fraction of the study period contained in the observation intervals would then provide a basis for estimating U_i (i.e., if the observations comprised 1% of the study period we would multiply sample results by 100). However, this is seldom practical. Furthermore, the methods discussed here permit estimating most of the parameters in which the behavioral ecologists are interested.

Proportion of units used, $P_{Ui}=U_i/U$

This quantity can obviously be estimated if estimates of U_i are available since ($\Sigma u_i=u$) as in the example of predation on snails by song thrushes. More commonly, a random sample of used units is selected, for instance by conducting a survey of animals and recording the resource type that each is using. Any of the sampling plans discussed in Chapter Four might, in principle, be used to obtain the point and interval estimates for P_{Ui}.

Proportion of the population comprising each type, $P_{Ai}=A_i/A$

In habitat studies, the A_i are often measured exactly from a cover map using a planimeter or Geographic Information Systems methods. In other cases, a sample of areas (or area-times if the type changes during the study period) is used to estimate the A_i. Any of the sampling plans discussed in

Chapter Four may be used, and point and interval estimates are calculated accordingly.

Proportion used/proportion available, $W_i = P_{Ui}/P_{Ai}$

The W_i provide a way to account for availability when comparing resources. If animals selected resources without regard to the types that the investigator has defined, then we would expect all W_i to be 1.0 except for the effects of sampling error. For example, if a particular habitat type covers 20% of the study area and animals distribute themselves randomly with respect to the types, then we would expect them to be found in that habitat type about 20% of the time. The same rationale holds for subsets of the types. Thus, if we have five types and the study species does not distinguish between types two and five, then W_2 would be equal to W_5 except for the effects of sampling error. Type *i* is sometimes said to be 'preferred' if W_i is shown to be above 1.0 and to be 'avoided' if W_i is shown to be below 1.0. These terms, of course, only denote departure from the null model 'selection is independent of type'. They do not imply that one habitat is more important than another. Under many general models of habitat selection, one would expect marked changes in W_i when A_i changes, so generalizations about which habitats are preferred or avoided usually require replication of the study in several locations and empirical demonstration that the W_i tend to be about the same despite changes in availability of the type. Even within the area and period studied, it may be difficult to derive conclusions about which habitats are most and least important to the animal simply from the W_i. Thus, an animal may be dependent on a particular habitat (i.e., for escape from predators), but may not spend a great deal of time in the habitat, thus causing W_i to be small. Thus, the results from use-availability analyses may be difficult to interpret in the absence of additional biological information.

The point and interval estimates of W_i are obtained from the point and interval estimates of P_{Ui} and P_{Ai}. When simple random sampling is used to estimate P_{Ui} and P_{Ai} is a known constant, then a chi-square goodness-of-fit test is generally used for a comprehensive test of the null hypothesis 'all W_i are equal'. When the P_{Ai} are also estimated, using simple random sampling, then a chi-square test of independence is generally used for the comprehensive test. Manly *et al.* (1993, Chapter Four) describe the procedure for these two cases in detail. In either case, pairwise tests and tests involving groups of resource types, may be carried out with further chi-square tests. Alternatively, Bonferroni adjustments may be made to the significance level, in which case there is no need to carry out the comprehensive test. As noted in Section 3.7, this point is important because when more complex sampling designs are used to estimate P_{Ui} or P_{Ui} and P_{Ai}, then designing the comprehensive test may be quite difficult. Use of the Bonferroni test avoids this difficulty.

In many studies, designs other than simple random sampling are employed to estimate use. For example, if surveys are made on each of several days, and habitats are recorded for all animals sighted, then we have multiple stage sampling with days as primary units. The analysis of such data depends on the details of the sampling plan and the parameter being estimated. Tests concerning a single parameter (e.g., the null hypothesis, H_0: $W_i = 1.0$) involve $V(w_i)$, where w_i is the estimate of W_i. When availability, P_{Ai}, is known, then, as explained in Chapter Two, $V(w_i) = (1/P^2_{Ai})V(p_{Ui})$, where p_{Ui} is the estimate of P_{Ui}. If P_{Ai} is estimated, then procedures for ratios apply. Tests involving two parameters (e.g., H_0: $W_i = W_j$ or H_0: $W_i/W_j = 1.0$) are more complex because covariance terms must be evaluated. The formulas can be derived from the material in Chapters Two to Four, but consultation with a statistician is recommended.

Other parameters

The W_i may also be 'standardized' by dividing each term by the sum of the terms, e.g., $W_i/\sum W_j$. The results, which sum to 1.0, are often referred to as α_i (Chesson's Index, Chesson 1978) or B_i (Manly's standardized selection index, Manly *et al.* 1972). These indices are sometimes viewed as 'the probability that a category *i* resource unit would be the next one selected if somehow it was possible to make each of the types of resource unit equally available' (Manly *et al.* 1993 p. 41). This interpretation, however, requires the assumption that the quantities U_i/A_i would not change if we somehow made all the types equally likely. Many other indices have been proposed using the basic variables U_i, U, A_i and A. Manly *et al.* (1993 p. 10) provide a comprehensive review. They note that many of the indices do not have a clear biological interpretation and recommend use of W_i or quantities closely associated with them (or the methods discussed in the remainder of this Chapter).

Defining availability

The definition of availability, which determines P_{Ai}, is often difficult and subjective but, unfortunately, may have great effect on the parameters. Including or excluding an area covered largely by type *i* habitat has an obvious effect on P_{Ai} and on P_{Ui} if this area is either heavily or lightly used relative to other type *i* areas. In many cases it is difficult to decide whether whole types should be included or not. For example, if one habitat or one prey type is barely used, then investigators may have difficulty deciding whether they want to consider it as 'available'. The decision, however, affects all of the W_i. For example, if an abundant type is not used, or only rarely used, then all the other W_i may be above 1.0 when the type is included but one or more of them may be below 1.0 when it is excluded. Thus, whether a type is defined as 'preferred' or 'avoided' may depend on which other types are included in the analysis. When deciding whether to include one or more types is difficult and subjective, but has a large influence on conclusions about which types are preferred or avoided, then it may be best to recognize that use in relation to availability cannot be investigated in a satisfactory manner and to concentrate instead on measures of use (P_{Ui} and U_i).

11.3 Several animals

In many studies, several animals are monitored and separate results are obtained for each one. If animals are viewed as having been randomly selected, with subsequent sampling being confined to the selected individuals, then the population unit is an 'animal-time' rather than an 'area-time' as described in Section 11.3. The population diagram should be visualized as having animals, rather than areas, down the side, and times across the top. The variable is 'resource type'. Treating animals as primary units is equivalent to imagining that we randomly select rows and then monitor the animals to which they correspond.

Such data are usually analyzed by treating each animal separately and reporting simple summary statistics to characterize the group. Thus, the investigator may report that one type was used more (or significantly more) often than another by eight of ten animals. Use, in this case, might be defined either as P_{Ui} or as W_i. The mean w_i may also be reported, along with its standard error. In some studies availabilities, P_{Ai}, are calculated or estimated for each animal separately; in others a single common value is used. If animals actually have different A_i values but we use the same set of values for all animals then the biological interest of the analysis will obviously be compromised. The parameter A_i is still well defined, however, and may be measured or estimated using reliable methods. The point to remember is that if animals are treated as primary units, then the sample size is the number of animals and statistical tests and other calculations are based on the results per animal (i.e., these results become the 'y_i' of Appendix One, Box 2).

Conducting a comprehensive statistical test, when animals are primary sampling units, is often difficult. When simple random sampling is used to select times at which animals will be observed, then chi-square methods described by Manly *et al.* (1993, Chapter Four) may be used to conduct certain tests. The Bonferroni approach may provide adequate power and avoids the comprehensive test. Occasionally, the pairwise tests of interest are independent in which case the approach in Section 3.7, based on the binomial distribution, may be sufficient for the comprehensive test. An example is now provided.

Suppose that eight cohorts of animals have been defined based on two sex classes and four age classes. We have estimated P_{Ui} or W_i for each cohort using a complex design and unequal sample sizes. Variances appear to differ between cohorts. Our objective is to compare use of habitat 'i' by males and females within age classes. We will thus carry out four independent pairwise tests, one for each age class, with $\alpha = 0.05$. The complex sampling design, unequal sample sizes, and unequal variances all make conducting a comprehensive test difficult. Bonerroni adjustments of the α-level is one possibility but the tests may then have very low power, especially if we have only a few degrees of freedom. Since the tests are independent, however, we may determine the probability of achieving two or more significant t-values, if all null hypotheses are true. From expression 3.13, the probability of achieving two or more significant results with four tests, is only 0.014 if all null hypotheses are true. We may thus reject the comprehensive null hypothesis and use the results of the pairwise tests just as we would if we had actually carried out an ANOVA and obtained a significant result. Section 3.7 provides additional explanation of this rationale.

11.4 Multivariate definition of resources

A different approach to studying resource selection is to define a model specifying the degree of use or probability of being used as a function of the variables measured on each population unit. In habitat use studies these variables might be amounts of different habitats, distance to water, patch size, and elevation. In diet studies these variables might be species, size, color, and age of the prey. If use is a quantitative variable, then a multiple regression model might be used; if use is a dichotomous variable, then logistic multiple regression might be used. In either case, the regression model (i.e., the 'true') model identifies which variables are related to resource selection and describes how they affect the degree of use. The regression coefficients (including 0.0 for variables with no effect) thus constitute the parameters using this approach. When sampling extends over a considerable period, time can be included in the model as an explanatory variable, and its effect on the selection process can be investigated. In some diet or foraging studies that use this general approach, one or more counts of prey still alive are made. The dependent variable is 'number still alive' or 'probability of still being alive' and is thus a measure of which resource units are not used, rather than which ones are used. Manly *et al.* (1993, Chapters Five to Ten) discuss many specific cases of using multivariate methods in this manner.

Model-based parameters are usually estimated with standard regression, logistic regression or other multivariate methods. The usual cautions and procedures applicable to regression apply to the case of resource selection (see Section 5.5). One caveat is particularly important. As noted in Chapter Five, most regression analyses assume that the data comprise a simple random sample. When the data are collected using multistage sampling (e.g., selection of animals; selection of used items), substantial difficulties may arise with using the model-based approaches (i.e., especially in variance estimation), so a statistician should be consulted for advice before embarking on this approach if multistage sampling is to be used to gather the data.

Several other multivariate approaches have been used in resource selection studies. Discriminant analysis may be used to classify population units as used or unused. The methods, which are somewhat complex, are described in texts on multivariate statistics (e.g., Johnson and Wichern 1982). Application of this method to resource selection studies is described by Manly *et al.* (1993, Section 8.2). The discriminant function may be used either to classify population units as used or unused or as a measure of resource quality.

A related technique which employs only population units known to have been used is described by Knick and Dyer (1997). These authors studied jackrabbit habitat in Idaho. Surveys were used to identify used areas and to identify habitat variables that appeared to be important in determining habitat suitability. The means of each of these variables, among population units known to have been used by rabbits, $\bar{x}_1,\ldots,\bar{x}_k$ say, were then used to obtain a single measure ($\bar{x} = \Sigma\bar{x}_i/k$) of habitat quality. The variables were measured (using Geographic Information Systems methods) for each population unit in the study area, and the mean of the variables was calculated ($\overline{x}_h = \sum \overline{x}_{hi}/k$ for population unit h). The difference, $\overline{x}_h - \overline{x}$, was then used as a measure of habitat quality. Population units with a small difference were given high ranks (i.e., habitat was assumed to be of high quality). To evaluate this measure of habitat quality, the authors conducted a new survey of rabbits in the study area. The results showed a strong tendency for rabbits to be found in high-ranked cells. For example, when the cells were ordered by rank, the 30% with the highest ranks contained 75% of the rabbit sightings. Thus, this procedure identified a small proportion of cells (30%) that contained a high proportion (75%) of the sightings.

Arthur *et al.* (1996) and Elston *et al.* (1996) discuss additional extensions and improvements in resource selection methodology.

11.5 Summary

Studies of resource selection usually involve monitoring habitat use or foraging behavior, recording which resource types, as defined by the investigator, are used. In many studies, resources are classified using a single series of mutually exclusive categories such as 'crop land/pasture/scrubgrowth/woodland'. In other studies many different variables may be defined and categories may not be mutually exclusive (e.g., elevation, cover type, patch size, distance to water). Population units may be defined by imagining small plots and periods of time, such that each plot at each time can hold at most one animal and animals do not change areas during an interval. The variable for defining use is 'used/not used'. The variable for defining resource availability is 'resource type'. Studies in which resources are defined with a single series of categories generally involve estimating the use and availability of each resource type. For example, for resource type 'i' P_{Ui} is the proportion of the used population units that are classified as resource type 'i', and P_{Ai} is the proportion of the study area classified as type 'i'. The P_{Ui}, and a measure of use in relation to availability, $W_i = P_{Ui}/P_{Ai}$, may be compared using chi-square tests if simple sampling designs are used, or methods from survey sampling theory (Chapter Four) if more complex designs are used. The statistical methods for these cases have been covered in Chapters Two to Five. Deriving the correct formulas for specific cases, however, may be complex because some of the parameters involve ratios of correlated variables, so consultation with a statistician may be advisable. Studies in which numerous variables are recorded generally use regression methods, including logistic regression, to identify combinations of variables that have high power to separate used from unused population units. We recommend Manly *et al.* (1993) for a detailed discussion of the statistical issues. These approaches can be formulated to include the case with mutually exclusive categories but provide more flexiblity and thus may warrant attention from behavioral ecologists interested in resource selection.

12
Other statistical methods

12.1 Introduction

Several statistical methods not previously discussed are briefly described in this Chapter. Our goal is to introduce the methods and indicate where more information on them may be found rather than to present in-depth discussions of the techniques. The first three Sections discuss relatively new methods that have not been widely used in behavioral ecology, or at least some of its subdisciplines, but that may be useful in a variety of applications. The subsequent three Sections discuss other branches of statistics with well-developed methods that we have not had space in this book to cover in detail.

12.2 Adaptive sampling

In conventional survey sampling plans, such as those discussed in Chapter Four, the sampling plan and sample size are determined before collecting the data and, in theory, all of the population units to be included in the sample could be identified before data collection begins. This approach, however, is sometimes unsatisfactory when the population units are uncommon and clumped in space and/or time. For example, suppose we are estimating the proportion of trees in an orchard damaged by rodents, and the damage occurs in widely scattered patches. We select rows of trees to inspect. Most trees are undamaged but occasionally we encounter an area in which most trees are damaged. Under conventional sampling plans we can only include trees in the selected rows in our sample. Yet we may be able to see that the damage extends to nearby rows and feel that some way should exist to include those trees in the sample as well. As a second example, suppose we are observing animals for fixed observation intervals to record some aspect of behavior such the outcome of fights. The behavior is uncommon but, when it occurs, may occur several times in fairly short succession. With fixed observation intervals, a bout of the behaviors may begin shortly before the end of the scheduled period. We thus may miss several observations that would have added valuable information to the sample if they could be included.

Adaptive sampling methods permit modification of the sample design so that additional observations of the kind described above may be included without causing bias in the point and interval estimates. The methods are relatively new and have not been widely applied in behavioral ecology. Thompson (1992, Chapters 23–26) provides an introduction to the subject with many examples from behavioral ecology, particularly fisheries and wildlife.

12.3 Line transect sampling

In line transect, or distance, sampling, an observer proceeds along a series of randomly selected lines recording the number of items of interest detected and the perpendicular distance to each item. The data are used to estimate the density of the items. The method is generally used as an alternative to searching well-defined plots thoroughly. It applies when observers can detect each item directly on the transect line, but cannot detect all items away from the line. Thus, the proportion of items detected must be estimated and used to correct the count. This is done by deriving a function that describes the decline in numbers detected as a function of perpendicular distance from the transect line. Under the assumption that all items are detected on the transect line, the overall proportion of items detected out to any given perpendicular distance may be calculated and used to correct the count. For example, suppose that all detections are within 200 meters of the transect line and the function describing the number detected is a straight line declining to 0.0 at a distance of 200 meters from the line. In this case, we would estimate that half the items actually present within 200 meters of the transect line were observed and the count would be multiplied by 2.0 to obtain an estimate of the number actually present within 200 meters of the transect line. A special case of this approach, in which counts are made from fixed points rather than by traversing transects, is known as the variable circular plot method. Line transect and variable circular plot methods are described by Thompson (1992, Chapter 17), Lancia *et al.* (1994 pp. 230–4), and Buckland *et al.* (1993). Anderson and Southwell (1995) and Quang and Becker (1996) provide examples of recent developments in this area.

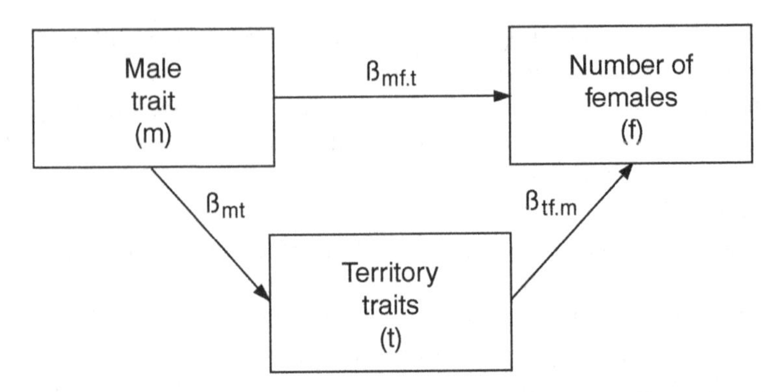

Fig. 12.1. A path analysis model of the importance of male quality and territory quality in determining male pairing success, defined as the number of females paired with males.

12.4 Path analysis

Path analysis is a method for evaluating how well a data set fits a specified causal model and estimating the strength of the causal relationships in the model. The model specifies variables that are linked by causal (as opposed to purely correlative) relationships and the 'directions' of the relationships. The method was developed by Sewall Wright (1921) for use in genetics research. It has been widely used in some fields (e.g., business, sociology) but relatively little in behavioral ecology. Some authors (e.g., Sokal and Rohlf 1981) feel that it might be applied more widely in behavioral ecology and a few applications using this approach have appeared recently (e.g., Thery and Vehrencamp 1995, Thomson et al. 1996).

The general approach in path analysis is to specify variables that are involved in a causal network of relationships. As a simple example, consider the issue of whether male pairing success, in a polygynous species, is determined by a given male trait, a given territory trait, or a combination of both. We assume that the male trait does not change during mate selection, that the species is migratory and territorial, and that males arrive prior to females. Thus, males arrive at the breeding ground and select territories, perhaps in competition with other males, and females then arrive and choose where to settle. Females may base their choices on the territory trait (independent of the males present), on the male trait (independent of the territory trait), or on a combination of both territory and male traits. The causal model for this situation (Fig. 12.1) has arrows connecting male trait, territory trait, and the number of females with which the male pairs. The purpose of the analysis is to estimate the magnitude of each relationship. The key contribution of path analysis is estimating the magnitude of the indirect effects, in this model the extent to which males obtain females by obtaining territories preferentially selected by females.

Causal models are evaluated using regression techniques. In the example just given, a multiple regression analysis would be carried out with the number of females as the dependent variable and territory trait and male trait as the independent variables. The resulting equation provides estimates of the amount by which the number of females increases when the value of the territory trait increases but the male trait remains constant ($\beta_{tf.m}$) and the amount by which the number of females increases when the value of the male trait increases but the territory trait is held constant ($\beta_{mf.t}$). The effect of the male trait on the territory trait (β_{mt}) is estimated by regressing territory trait on male trait. The resulting equation provides an estimate of β_{mt}. Variables are usually standardized (i.e., transformed by subtracting the mean and dividing by the standard deviation) to facilitate comparisons. On the transformed scale 1.0 means 1.0 standard deviation. Standard regression methods are used to determine whether relationships and coefficients are statistically significant.

The indirect effect of male trait on the number of females via territory is the product of the two 'pathways', 'male trait ⇛ territory trait' and 'territory trait ⇛ number of females' and is estimated as $b_{mt} b_{tf.m}$ where b_{mt} is the estimate of β_{mt} and $b_{tf.m}$ is the estimate of $\beta_{tf.m}$. This effect may be compared to the direct effect of male trait on number of females, $b_{mf.t}$, as a measure of whether males obtain females primarily by obtaining favored territories, primarily by possessing traits that females select preferentially, or by both mechanisms. More complex models of the pairing process may also be evaluated with path diagrams and compared to the rather simple model described in Fig. 12.1. Cohen and Cohen (1983) and Li (1975) provide extended discussions of path analysis and its application in behavioral research.

12.5 Sequential analysis

The basic goal in sequential analysis is to reduce uncertainty about the value of an observation by using information about past observations. The word 'past' often refers to time, but observations can be ordered in any other way; for example, by position in space. Here is a simple example, from our own research, in which sequential analysis was useful. Tundra swans on their breeding grounds in northern Alaska usually rotate basic behaviors in a predictable way. They forage for a few hours, then preen for an hour or more, then sleep for a few hours, and then return to foraging. This pattern seems to persist throughout the 24-h cycle (of continuous daylight). Suppose we divide all behavior into these three types and let the proportions of time spent in foraging, preening, and sleeping be p_f, p_p, and p_s respectively ($p_f + p_p + p_s = 1.0$). If an instant in time is selected randomly, then the probability that a given swan is foraging is p_f. But due to the pattern of behavior described above, if we knew the previous behavior of the swan, then if that behavior was sleeping, we would be fairly sure that the current behavior was foraging, and if the previous behavior was preening, we would be fairly sure the current behavior was not foraging. Sequential analysis provides a series of methods for investigating whether patterns of the sort described above are present in a data set and for describing the strength of the associations.

Sequential analysis may also be viewed as a means of understanding biological patterns in time, space, or with respect to some other dimension. Thus, these analyses help us determine whether changes are predictable and how other factors affect predictability. For example, change might be quite predictable early in the season, but not later, or in one habitat but not another. Alternatively, the nature of the pattern might vary between cohorts being compared, such as territorial and nonterritorial individuals. Sequential analysis is somewhat similar to regression in that an external variable (e.g., time) is used to help understand or predict values of a variable. Sequential analysis differs from regression in that no overall trend is estimated or defined in the analysis. Instead, time (or whatever dimension is involved) is used to identify population units occurring immediately prior to a focal unit and the values of these units are used in the analysis.

Sequential analyses may be carried out in many different ways. We use, as an example, recording a series of behaviors through time. The example generalizes easily to other observations and dimensions (e.g., space). Suppose the behaviors are coded as A, B, C, ... Data are usually recorded periodically, generating a data stream such as BBBACCBCAAA. The data may then be reduced to show only changes in behavior, BACBCA in this example. Most analyses involve comparing the unconditional frequency of a given type with the proportion of the records that are the given type when a stated prior event or set of events occurs. For example one might compare the proportion of records that are type B with the proportion of the records immediately following a type A record that are type B (i.e., $n_{B,1}/n_A$ where n_A

is the number of type A records and $n_{B,1}$ is the number of records immediately following a type A record that are type B). This pattern is referred to as a first-order Markov chain, the phrase 'first-order' meaning that events are modelled as a function of the most recent events. A second-order Markov chain includes the two most recent events, and so on.

Statistical issues not previously considered arise for several reasons in sequential analysis. Perhaps the major reason is that each record is used more than once in calculating probabilities of different events. For example, in a first-order Markov chain analysis each record is used twice, once as the 'current' event and once as the 'immediately preceding' event. Thus, the proportions used to estimate probabilities are not independent. Other statistical issues include deciding whether to combine data from different individuals, whether the pattern changes through time (i.e., whether it is 'stationary'), how to analyze data sets in which more than one behavior (or other variable) is recorded, and how to compare different groups to identify differences in pattern or the strength of relationships.

A large amount of literature exists on sequential analysis, particularly in ethology and sociology. Bakeman and Gottman (1986) and a sequel by Gottman and Roy (1990) provide comprehensive discussions of the statistical and nonstatistical aspects of sequential analysis. Lehner (1996 pp. 444–64) provides a brief review which also discusses the relationship of information theory to sequential analysis. Haccou and Meelis (1992) provide a much more detailed treatment of the statistical issues in sequential analysis.

12.6 Community analysis

Several multivariate statistical methods are commonly used by community ecologists to identify environmental variables responsible for the distribution and abundance of plant and animal species and to study relationships between species. Gauch (1982) distinguishes three approaches, generally used in combination: direct gradient analysis, ordination, and classification or cluster analysis.

Direct gradient analysis is essentially a regression technique for studying the relationship between the abundance of a taxon in relation to a (possibly composite) environmental variable. When the samples include only a small portion of the range of the environmental variable, then the relationship may be monotonic (constantly increasing or decreasing) or even linear. More commonly, measurements are made across a wider range of the environmental variable, and the relationship is more complex, abundance of the taxon increasing to a maximum and then decreasing. Statistical analysis of such data thus requires fitting a model that permits such variation.

Ordination involves arranging samples along a small number of axes, typically one to three, in such a way that samples with similar species and species' abundances are close together and samples with different species are farther apart. The analysis is usually based solely on the occurrence and abundance of species in the samples, not on environmental data, though the results of the ordination are commonly interpreted using knowledge of environmental features at the sites. Ordination techniques use a variety of multivariate statistical methods including multidimensional scaling, component analysis, factor analysis and latent-structure analysis (ter Braak 1995b).

Classification or cluster analysis results in species and/or locations being assigned to classes, as when vegetation types are defined and a cover map using these types is constructed. Like ordination, classification is generally carried out using only the data on species occurrence and abundance. Numerous methods have been proposed (Gauch 1982, ter Braak 1995b) but most of them are nonstatistical in the sense that they do not involve explicit consideration of random variables, estimation, and statistical inferences.

12.7 Summary

This Chapter briefly considers several statistical methods not discussed elsewhere in this book. Adaptive sampling is a relatively recent development in survey sampling theory that permits investigators to modify the sampling plan part way through data collection. Distance sampling provides a way to estimate density when data are collected by either traversing transects or at fixed points and when individuals away from the observer are difficult to detect. Path analysis is a method based on regression techniques for evaluating causal models and estimating the strength of causal (as opposed to purely correlative) relationships. Sequential analysis provides methods for determining whether the value of a variable depends on past events; for example, whether the vocalizations by one individual tend to occur in response to vocalizations from another individual. Community analysis involves numerous multivariate techniques. Direct gradient analysis is basically a regression technique for studying the relationship between occurrence and abundance of a taxon and one or more environmental variables. Ordination, used to identify similar species or locations, uses a variety of multivariate techniques not discussed in this book, including multidimensional scaling, component analysis, factor analysis and latent-structure analysis. Classification, or cluster analysis, is used to assign species or locations to classes. Numerous methods have been used in this process but they generally make little use of statistical analysis.

Appendix One

Frequently used statistical methods

This Appendix contains a summary of the statistical methods described in Chapters One to Five. Table A1.1 contains a brief key to the methods. Box 1 contains a more detailed guide, arranged in the form of a dichotomous key that may be helpful in deciding which analytical methods to use. Detailed instructions for carrying out the analyses are contained in subsequent Boxes and other Sections of the Book.

Table A1.1. *Quick guide to statistical methods in this book.*

Method	Location
A. Estimating means and standard errors	
1. Simple random sampling	Box 2
2. Other survey sampling methods	Box 3
B. Tests and confidence intervals	
1. For one parameter	
a. Proportions estimated using the formulas for simple random sampling	Box 5
b. Otherwise	Box 4
2. For the difference between two parameters	
a. Proportions estimated using the formulas for simple random sampling	Box 7
b. Other quantities	
(1) Parametric methods	
(a) Paired data	Sec. 3.5
(b) Unpaired data	Box 6
(2) Nonparametric methods	Box 8
3. For the ratio of two parameters	Box 9
4. Multiple pairwise comparisons	Sec. 3.7
C. Sample size and power calculations	Box 10

Box 1. A dichotomous key to statistical methods with emphasis on those described in this Appendix.

A1.	Calculations to decide how much data (or more data) should be collected (e.g., sample size and power calculations)	Box 10
A2.	Calculations to estimate quantities of interest or compare estimates	B
B1.	Purpose is to estimate a quantity (e.g., mean, density, proportion, correlation coefficient)	C
B2.	Purpose is to construct a statistical test or confidence interval	F
C1.	Estimating means, proportions, survival rates, totals	D
C2.	Estimating measures of relationship (e.g., correlation/regression coefficients	Chapter 5
D1.	Survey sampling methods used (e.g., simple random sampling, stratified sampling)	E
D2.	Other methods used (e.g., capture–recapture, line transect)	Individual chapters
E1.	Simple random sampling	Box 2
E2.	Other survey sampling methods	Box 3
F1.	For one estimate	G
F2.	To compare two estimates	H
F3.	Multiple pairwise comparisons	Section 3.7
G1.	Proportions estimated with simple random sampling	Box 5
G2.	Other estimates	Box 4
H1.	Difference between two estimates	I
H2.	Ratio of two estimates	Box 9
I1.	Proportions estimated with simple random sampling	Box 7
I2.	Other quantities	J
J1.	Parametric methods	K
J2.	Nonparametric methods	Box 8
K1.	Paired data	Section 3.5
K2.	Unpaired data	Box 6

Box 2. Formulas for means and proportions, and their standard errors, with simple random sampling or multistage sampling with equally weighted primary units (Chapter Four)

A. Definitions

For one-stage sampling, let

n = number of units measured (sample size)

y_i = measurement from (or value of) the ith unit

For multistage sampling, let

n = number of primary units measured (sample size)

y_i = estimate from the ith unit

B. The estimated mean and its standard error are

$$\bar{y} = \sum y_i / n \tag{A1.1}$$

$$se(\bar{y}) = sd(y_i)/\sqrt{n} \tag{A1.2}$$

where

$$sd(y_i) = \sqrt{\sum(y_i - \bar{y})^2/(n-1)} \tag{A1.3}$$

$$= \sqrt{(\sum y_i^2 - n\bar{y}^2)/(n-1)} \tag{A1.4}$$

and all sums are from $i = 1$ to n.

C. The degrees of freedom (df), used in tests and to constrict confidence intervals, are usually calculated as

$$df = n - 1 \tag{A1.5}$$

Strictly, this formula is valid only if the population has a normal distribution (see pp. 75–78).

D. Notes

1. The formula for se($\bar{y}$) assumes that the population is 'large' (Section 4.3). If this is not so, then multiply $se(\bar{y})$ as calculated in Eq. A1.2 by the square root of $(1 - n/N)$, where N = number of units (primary units for multi-stage sampling) in the population. This case seldom occurs in behavioral ecology.

2. In estimating proportions with simple random sampling, the y_i are usually coded as 1 if the ith unit is in the category of interest and 0 otherwise. With this convention $\sum y_i/n = \bar{y}$ = the estimated proportion, usually denoted by p. Furthermore, $\sum y_i^2 = \sum y_i = n\bar{y}$ so that Eq. A1.4 above may be written

$$sd(y_i) = \sqrt{n\bar{y}(1-\bar{y})/(n-1)} \tag{A1.6}$$

Box 2 ***(cont.)***

and Eq. A1.2 may be written

$$se(\bar{y}) = \sqrt{\frac{\bar{y}(1-\bar{y})}{n-1}} \tag{A1.7}$$

This expression is commonly written as

$$se(p) = \sqrt{\frac{p(1-p)}{n-1}} \tag{A1.8}$$

3. In hypothesis testing, with proportions we calculate $SE(p)$ under the assumption that the proportion equals the value specified by the null hypothesis. In this case the $SE(p)$ is known and is

$$SE(p) = \sqrt{\frac{P(1-P)}{n}} \tag{A1.9}$$

Box 3. Formulas for multistage sampling with unequally weighted primary units (Section 4.5) and stratified sampling (Section 4.6).

A. Notation

N = number of groups (strata or primary units) in the population (sometimes indefinitely large)

n = number of groups sampled

w_i = number of population units, or proportion of the population, in the ith group

$\overline{w}$ = average of the w_i in the sample (except occasionally in C2 below)

y_i = estimate from the ith group

Note: If primary units were selected with unequal probabilities (Section 4.4), see Part D below.

B. The estimated mean is

$$\bar{y} = \frac{1}{n\overline{w}} \Sigma w_i y_i \qquad \text{(A1.10)}$$

Note: if w_i = proportion of the population in the ith group then $n\overline{w} = 1$ and can be omitted.

C. The standard error and degrees of freedom (df) depend on whether stratified or multistage sampling was used.

1. Stratified sampling, $n = N$ (modified from Cochran 1977: p. 95).

$$se(\bar{y}) = \sqrt{\frac{1}{(n\overline{w})^2} \Sigma w_i^2 [se(y_i)]^2} \qquad \text{(A1.11)}$$

$$df = \frac{(\Sigma g_i)^2}{\Sigma[g_i^2/(n_i - 1)]} \qquad \text{(A1.12)}$$

where $g_i = w_i^2[se(y_i)]^2$, n_i = sample size within the ith stratum, and $se(y_i)$ is calculated as follows:

a. If one-stage sampling (or multistage sampling with equally weighted primary units) is used within strata (the most common case), then Eq. A1.2 in Box 2 can be used to obtain the $se(y_i)$

b. If multistage sampling with unequally weighted primary units is used within strata, then the $se(y_i)$ may be obtained with the formulas in Section C2 below.

c. The formulas above assume that stratum sizes are known. If they are estimated from a sample of size n^* (see Sec. 4.6), then

$$se(\bar{y}) = \sqrt{\frac{1}{(n\overline{w})^2} \Sigma w_i^2 [se(y_i)]^2 + \left[\frac{1}{n^*} - \frac{1}{N^*}\right] \Sigma w_i^2 (y_i - \bar{y})^2} \qquad \text{(A1.13)}$$

where N^* = the number of population units in the entire population. N^* is usually indefinitely large and is then ignored in the formula above. If the stratum sizes are known (i.e., $n^* = N^*$),

Box 3 (*cont.*)

then the right-hand term in the formula drops out (see Section 4.6 for additional explanation).

2. Multistage sampling, $n<N$ (modified from Cochran 1977, Eq. 11.30).
   a. If $n/N<0.1$, then use

$$se(\bar{y})=\sqrt{\frac{1}{(n\bar{w})^2}\Sigma w_i^2(y_i-\bar{y})^2/(n-1)} \tag{A1.14}$$

$$df=n-1 \tag{A1.15}$$

Note: if the average size of the primary units in the entire population is known, then one may use this value for $\bar{w}$ in calculating $\bar{y}$ (A1.10). For $se(\bar{y})$, let $y_i^*=w_iy_i/\bar{w}$ ($\bar{w}$=population average) and $\bar{y}^*$ = the simple mean of the y_i^*. The formulas for simple random sampling are then used with these y_i^* to calculate the standard error (Cochran 1977, Section 11.7). That is

$$se(\bar{y})=\sqrt{\frac{\Sigma(y_i^*-\bar{y}^*)^2}{n(n-1)}} \tag{A1.16}$$

$$=sd(y_i^*)/\sqrt{n} \tag{A1.17}$$

The number of degrees of freedom is $n-1$. In most studies this $se(\bar{y})$ is larger than that in Eq. A1.11 (see Section 4.5).

   b. If $n/N\geq 0.10$ then use

$$se(\bar{y})=\sqrt{(n/N)se_1^2+(1-n/N)se_2^2} \tag{A1.18}$$

where se_1 is the $se(\bar{y})$ calculated using Eq. A1.14 and se_2 is the $se(\bar{y})$ calculated using Eq. A1.11.

D. Multistage sampling when primary units are selected with unequal probabilities (and subsequent selection is by equal probability methods)

1. Additional notation.
   $\bar{w}$ = average w_i in the *population*
   z_i = probability of selecting the ith primary unit
   x_i = estimate from the ith primary unit
2. Let

$$y_i=\frac{w_ix_i}{N\bar{w}z_i} \tag{A1.19}$$

3. The estimated population mean and standard error are

$$\bar{y}=\frac{1}{n}\Sigma y_i \tag{A1.20}$$

Box 3 (*cont.*)

$$se(\bar{y}) = sd(y_i)/\sqrt{n} \qquad \text{(A1.21)}$$

$$df = n - 1 \qquad \text{(A1.22)}$$

where, as usual,

$$sd(y_i) = \sqrt{\Sigma(y_i - \overline{Y})^2/(n - 1} \qquad \text{(A1.23)}$$

Box 4. Statistical tests and confidence intervals for a single parameter.

A. Definitions (for paired data, see Part D below)

$\bar{y}$ = the parameter estimate (typically a mean).

$t_{df}(\alpha/2)$ = value in Table A2.1, Appendix Two, with level of significance $= \alpha$ (use $2Q = \alpha$) and degrees of freedom $= df$ (see Boxes 2 or 3).

$\bar{Y}$ = the true value under the null hypothesis

B. Reject the null hypothesis if and only if

$$\frac{|\bar{y} - \bar{Y}|}{se(\bar{y})} \geq t_{df}(\alpha/2) \qquad \text{(A1.24)}$$

where $se(\bar{y})$ is calculated following the guidelines presented in Boxes 2 and 3.

C. The $1 - \alpha$ confidence interval for $\bar{y}$ is

$$\bar{y} \pm t_{df}(\alpha/2)[se(\bar{y})] \qquad \text{(A1.25)}$$

D. Paired data.

If pairs of population units have been randomly selected calculate $\bar{y}$ as the mean of the *differences* between the measurements on the two units of each pair and then follow the procedures above with $\bar{Y}$ = the value of the difference in population means under the null hypothesis (often zero). For example, with a simple random sample of size n, and measurements x_{1i}, x_{2i}, $i = 1,\ldots,n$, $\bar{y}$ would be

$$\bar{y} = \frac{1}{n}\sum_{i=1}^{n}(x_{1i} - x_{2i})$$

For partially paired data, see Section 3.5.

E. The formulas above are exact if the population has a normal distribution. Effects of non-normality are discussed on pp. 75–78

Box 5. Statistical tests and confidence intervals for proportions estimated with one-stage sampling (Chapter 4).

A. Definitions

p = the estimated proportion
q = $1-p$
n = the sample size
P = the true proportion under the null hypothesis (often 0.5)
$t_{df}(\alpha/2)$ = the value in Table A2.1, Appendix Two, with level of significance
α (use $2Q=\alpha$) and degrees of freedom = df (see below)

B. Guidelines for when to use the normal approximation (modified from Cochran (1977 p. 58).

p or $1-p$, whichever is smaller	Minimum n for normal approximation to be used
0.5	30
0.4	50
0.3	80
0.2	200
0.1	600
0.05	1400

C. Analysis using the normal approximation

1. p is significantly different from P if and only if

$$\frac{|p-P|}{\sqrt{PQ/n}}-\frac{c}{\sqrt{nPQ}}\geq t_{\infty}(\alpha/2) \qquad \text{(A1.26)}$$

where c, the 'correction for continuity' (Snedecor and Cochran 1980, Section 7.6, see note in Appendix Three), is defined as follows

a. For two-tailed tests, c depends on f = the fractional part of the quantity, $n|p-P|$. If $f \leq 0.5$, then $c=f$; otherwise, $c=f-0.5$. Examples: If $n(p-P)=7.3$ or -7.3, then $f=0.3$ so $c=0.3$ If $n(p-P)=7.9$ or -7.9 then $f=0.9$ so $c=0.4$.

b. For one-tailed tests (rarely used – see Section 3.2), $c=0.5$.

2. The $1-\alpha$ confidence interval for p (Cochran 1977, Eq. 3.19) is

$$p \pm t_{n-1(\alpha/2)}\sqrt{\frac{pq}{n-1}+\frac{0.5}{n}} \qquad \text{(A1.27)}$$

(See Appendix Three)

3. If the finite population correction is appropriate (which is rare – see Section 4.3) then in Eq. A1.26 replace 'PQ/n' with '$(1-n/N)$

Box 5 (*cont.*)

PQ/n' and in Eq. A1.27 replace 'pq' with '$(1-n/N)pq$' where N = population size.

4. Note that the significance test uses PQ/n for $se(p)$ because under the null hypothesis, the standard deviation is known. When it is unknown, as in constructing the confidence interval, then (A1.27) is recommended (Cochran 1977 pp. 52, 57) however the expression $p \pm t_{\alpha,\infty}\sqrt{pq/n}$ is widely used.

D. Analysis when the normal approximation is not appropriate.

1. Hypothesis tests require tables of the binomial distribution (see Hollander and Wolfe 1973). An alternative (Steel and Torrie 1980 p. 484) is to compute confidence intervals (see below) and reject the null hypothesis if the confidence interval does not include P.
2. Confidence intervals.
   a. Approximate limits may be determined using Fig. A2.1, Appendix Two, or by interpolation of values of binomial confidence limits (e.g., Steel and Torrie 1980, Table A.14).
   b. Exact limits (modified from Hollander and Wolfe 1973 p. 24) may be calculated using a table of values from the F distribution (Table A2.2, Appendix Two). The F value depends on the level of significance and 'numerator' and 'denominator' degrees of freedom, ν_1, and ν_2, respectively.

$$\text{lower endpoint} = [1 + F_{\alpha/2|\nu 1,\nu 2}(q + 1/n)/p]^{-1} \quad \text{(A1.28)}$$

$$\text{where } \nu_1 = 2(nq+1)$$

$$\nu_2 = 2np$$

$$\text{upper endpoint} = \left[1 + \frac{q}{(1/n + p)F_{\alpha/2|\nu_1,\nu_2}}\right]^{-1} \quad \text{(A1.29)}$$

$$\text{where } \nu_1 = 2(np+1)$$

$$\nu_2 = 2nq$$

(See note in Appendix Three.)

Box 6. Parametric methods (t-tests) for comparing two population parameters based on independent estimates. For paired data, see section 3.5.

A. Notation

$\bar{y}_1, \bar{y}_2$ = the estimates from populations 1 and 2

$\bar{Y}_1, \bar{Y}_2$ = the true but unknown values (i.e., the parameters)

se_1, se_2 = the standard errors of the estimates (obtained using the formulas in Boxes 2 or 3)

n_1, n_2 = sample sizes for the two estimates (see Boxes 2 and 3)

df_1, df_2 = the number of degrees of freedom for the estimates $\bar{y}_1$ and $\bar{y}_2$ (see Boxes 2 and 3)

$t_{df}(\alpha/2)$ = the value in Table A2.1, Appendix Two, with level of significance α and degrees of freedom = df

Y_d = value of $\bar{Y}_1 - \bar{Y}_2$ under the null hypothesis (usually 0.0)

B. General formulas for tests and confidence intervals

1. The null hypothesis that $\bar{y}_1 - \bar{y}_2$ equals Y_d is rejected if and only if

$$\frac{|\bar{y}_1 - \bar{y}_2| - Y_d}{se(\bar{y}_1 - \bar{y}_2)} \geq t_{df}(\alpha/2) \quad \text{(A1.30)}$$

$$df = df_1 + df_2. \quad \text{(A1.31)}$$

2. The $1-\alpha$ confidence interval on $\bar{y}_1 - \bar{y}_2$ is

$$\bar{y}_1 - \bar{y}_2 \pm t_{df}(\alpha/2)[se(\bar{y}_1 - \bar{y}_2)] \quad \text{(A1.32)}$$

$$df = \frac{(se_1^2 + se_2^2)^2}{se_1^4/df_1 + se_2^4/df_2} \quad \text{(A1.33)}$$

C. Formulas for $se(\bar{y}_1 - \bar{y}_2)$

1. Hypothesis testing when the data are from one-stage samples (or multistage samples with equally weighted primary units – see Section 4.5) and samples sizes, n_1 and n_2 are unequal.

$$se(\bar{y}_1 - \bar{y}_2) = \sqrt{\left(\frac{n_1 + n_2}{n_1 n_2}\right)\left(\frac{n_1(n_1 - 1)[se(\bar{y}_1)]^2 + n_2(n_2 - 1)[se(\bar{y}_2)]^2}{n_1 + n_2 - 2}\right)} \quad \text{(A1.34)}$$

Note: Eq. A1.35 may be used for simplicity but usually yields a slightly larger value.

2. All other cases.

$$se(\bar{y}_1 - \bar{y}_2) = \sqrt{[se(\bar{y}_1)]^2 + [se(\bar{y}_2)]^2} \quad \text{(A1.35)}$$

D. The formulas above are exact if the population has a normal distribution. Effects of non-normality are discussed on pp. 75–78.

Box 7. Comparison of two proportions estimated using one-stage sampling (from Fleiss 1981).

A. Definitions

p_1, p_2 = the estimated proportions

n_1, n_2 = the sample sizes ($n_1 = n_2 = n$ for paired data)

$t_{\infty(\alpha/2)}$ = value in Table A2.1, Appendix Two with level of significance α and degrees of freedom ∞ (infinity)

c = the 'correction for continuity' (see Snedecor and Cochran 1980 p. 117). Let $c = 1/n$ if $n_1 = n_2$ or $0.5(1/n_1 + 1/n_2)$ if $n_1 \neq n_2$

B. Testing whether the proportions are equal.

1. The general formula for tests with large samples is to reject the null hypothesis that the proportions are equal if and only if

$$\frac{|p_1 - p_2| - c}{se(p_1 - p_2)} \geq t_{\infty(\alpha/2)} \quad \text{(A1.36)}$$

Specific guidelines for use of this formula are given below.

2. Independent estimates.
   a. Use Fisher's Exact Test (rather than the formula above) whenever possible. Table A2.3, Appendix Two, explains the test and gives critical values for samples sizes of $n_1, n_2 \leq 15$. Many statistical packages provide the exact p-values for larger sample sizes.
   b. If sample sizes are too large to use Fisher's Test, then use Eq. A1.36 above with

$$se(p_1 - p_2) = \sqrt{pq(1/n_1 + 1/n_2)} \quad \text{(A1.37)}$$

where $p = (p_1 + p_2)/n$ if $n_1 = n_2$. If $n_1 \neq n_2$, then let

$$p = \frac{n_1 p_1 + n_2 p_2}{n_1 + n_2} \quad \text{(A1.38)}$$

In either case, $q = 1 - p$.

3. Paired estimates. The two outcomes or measurements on each unit in the sample are classified as 'success' (or 'in the category of interest') or 'failure' (or 'not in the category of interest'). Let n_{sf} = the number of units in which the two outcomes were 'success, failure' and let n_{fs} = the number of units in which the two outcomes were 'failure, success'. Calculate the test statistic with Eq. A1.36 letting

$$se(p_1 - p_2) = \frac{1}{n}\sqrt{n_{sf} + n_{fs}} \quad \text{(A1.39)}$$

C. Confidence intervals for the difference between proportions.

1. Confidence intervals are difficult to calculate for small samples. The following large-sample approximation is appropriate if the smallest of $n_1 p_1$, $n_1 q_1$, $n_2 p_2$, and $n_2 q_2$ is ≥ 5. The $1 - \alpha$ confidence interval on

Box 7 (*cont.*)

$p_1 - p_2$ is

$$p_1 - p_2 \pm t_{\infty(\alpha/2)}[se(p_1 - p_2)] + c \qquad \text{(A1.40)}$$

where $se(p_1 - p_2)$ is calculated as shown in 2 or 3 below.

2. For independent estimates

$$se(p_1 - p_2) = \sqrt{\frac{p_1 q_1}{n_1 - 1} + \frac{p_2 q_2}{n_2 - 1}} \qquad \text{(A1.41)}$$

where $q_1 = 1 - p_1$ and $q_2 = 1 - p_2$.

3. For paired estimates,

$$se(p_1 - p_2) = \frac{1}{n}\sqrt{n_{sf} + n_{fs} - \frac{(n_{sf} - n_{fs})^2}{\sqrt{n}}} \qquad \text{(A1.42)}$$

where n_{sf} and n_{fs} are defined as in Part B3 above.

Box 8. Nonparametric methods for comparing two estimates

A. Paired data (Wilcoxon signed rank test).

1. Rank the nonzero differences, ignoring sign, giving the smallest difference a rank of 1. If two or more of the differences are equal, assign to each of them the average of the ranks they would have received if they had differed slightly (e.g., if the four smallest differences were 3, 6, 6, 7, their ranks would be 1, 2.5, 2.5, 4 respectively).
2. Compute the sum of the ranks assigned to the positive differences and the sum of the ranks assigned to the negative differences.
3. The test statistic, T_+, is the smaller sum, and is used with the number of nonzero differences (n) in Table A2.4, Appendix Two, to determine whether the median of the differences is different from zero.
4. If $n > 20$, a large sample approximation to the distribution of T_+ is used. The median of the differences is significantly different from zero at level α if and only if

$$\frac{[n(n+1)/4] - T_+ - 0.5}{\sqrt{n(n+1)(2n+1)/24}} \geq t_{\infty(\alpha/2)} \qquad \text{(A1.43)}$$

where $t_{\infty(\alpha/2)}$ is the value in Table A2.1, Appendix Two, with $2Q = \alpha$ and degrees of freedom ∞ (infinity).

B. Unpaired data (Mann–Whitney test).

1. Rank the observations giving the smallest observation a rank of 1. If two or more of the observations are equal, assign to each of them the average of the ranks they would have received if they had differed slightly (e.g., if the five smallest observations were 5.1, 6.0, 6.0, 6.0, 6.1, their ranks would be 1, 3, 3, 3, 5 respectively).
2. Calculate the test statistic, T_1 as follows
   a. $n_1 = n_2$. Calculate the sum of the ranks received by the observations in each sample. $T_1 =$ the smaller sum.
   b. $n_1 \neq n_2$. Calculate T, the sum of ranks for the sample with fewer observations, say n_1. Next calculate $T^1 = n_1(n_1 + n_2 + 1) - T$. $T_1 =$ the smaller of T and T^1.
3. T_1 is used with n_1 and n_2 in Table A2.5, Appendix Two, to test the null hypothesis that the populations are identical against the alternative that one population is shifted to the right of the other.
4. For values of n_1 and n_2 outside the limits of Table A2.5, Appendix Two, a large-sample approximation to the distribution of T_1 is used. The populations are significantly different at level α if and only if

Box 8 (*cont.*)

$$\frac{|n_1(n_1+n_2+1)/2 - T'_\alpha| - 0.5}{\sqrt{n_1 n_2(n_1+n_2+1)/12}} \geq t_\infty(\alpha/2) \quad \text{(A1.44)}$$

where $t_\infty(\alpha/2)$ is the value in Table A2.1, Appendix Two, with $2Q = \alpha$ and degrees of freedom ∞ (infinity).

Box 9. Confidence intervals for the ratio of two parameters.

A. Definitions

y, x = the two estimates (e.g., means, regression coefficients, survival rates)

$\frac{y}{x}$ = the estimated ratio

$se(x)$, $se(y)$ = the estimated standard errors of x and y

B. Continuous estimates (all cases except proportions estimated from one-stage samples – Cochran 1977:156).

1. If n > 30 and both $se(y)/y$ and $se(x)/x$ are <0.1, then

   a. Calculate se(y/x) as

$$se\left(\frac{y}{x}\right) = \sqrt{\left(\frac{y}{x}\right)^2\left[\left(\frac{se(y)}{y}\right)^2 + \left(\frac{se(x)}{x}\right)^2 - \frac{2cov(x,y)}{xy}\right]} \quad \text{(A1.45)}$$

   The cov term depends on the relationship between the two estimates, x and y, as follows:

   (1) If x and y are independent estimates then cov(x,y) = 0 so the term, 2cov(x,y)/xy, drops out.

   (2) If x and y are the means of two measurements from each unit in a sample of size n, then

$$cov(x,y) = \frac{\sum_i^n (x_i y_i - xy)}{n(n-1)} \quad \text{(A1.46)}$$

$$= cov(x_i, y_i)/n$$

   where x_i and y_i are the measurements from the ith unit (or the means from the ith primary unit if multistage sampling was used). Many calculators and computer printouts provide cov(x_i,y_i); be sure to divide this quantity by n to obtain cov(x,y).

   (3) For other cases see Chapter Two.

   b. Calculate the confidence interval as

$$\frac{y}{x} \pm t_\infty(\alpha/2)se(y/x) \quad \text{(A1.47)}$$

   where α = the level of significance (usually 0.05).

2. If the conditions in point1 above are not met, then calculate the limits of the confidence interval as

$$\left(\frac{y}{x}\right)\left[\frac{(1 - t_\infty(\alpha/2)^2 c_{xy}) \pm t_\infty(\alpha/2)\sqrt{(c_{yy} + c_{xx} - 2c_{xy}) - t_\infty(\alpha/2)^2(c_{yy}c_{xx} - c_{xy}^2)}}{1 - t_\infty(\alpha/2)^2 c_{xx}}\right]$$

(A1.48)

where c_{xx} = [se(x)/x]^2, c_{yy} = [$se(y)/y$]^2, c_{xy} = $cov(x,y)/xy$ and $cov(x,y)$ is as defined in Eq. A1.46 above.

Box 10. Sample size and power calculations (modified from Snedecor and Cochran 1980 pp. 102–4, 129–30 – see Section 3.4).

A The guidelines in this Box can be used for tests and confidence intervals for evaluating a single estimate or for comparing two estimates. They assume that simple random sampling (or multistage sampling with equally weighted primary units – Section 4.5) has been used to estimate means or proportions. See Part F2 if a more complex plan (e.g., stratified sampling) has been used, or a different quantity (e.g., slope in a linear regression analysis) is being estimated.

B. Definitions

g_1, g_2 = the estimates (for a single estimate, g_2 is the true value = under the null hypothesis)

δ = the assumed true difference (= 0 if a single estimate is being evaluated). In many studies, δ is the smallest difference felt to be of biological importance.

α = the level of significance

P' = power, the probability of obtaining a significant difference between g_1 and g_2 if the true difference is δ

n_1, n_2 = the sample sizes. If a single estimate is being evaluated, then n_2 is not defined. For power calculations, n_1 and n_2 are the samples sizes you *expect to obtain* by the end of the study. For sample size calculations, n_1 and n_2 are the sample sizes *required* by the end of the study if power is to equal P′

s_1^2, s_2^2 = estimates of the population variances (see Part B below). If a single estimate is being evaluated, then s_2^2 is not defined

s^2 = a measure of how variable the populations are (see Part C)

$se(g_1 - g_2)$ = the estimated standard error of the difference, $g_1 - g_2$, assuming that the final samples sizes are n_1 and n_2 (or just n_1 if only one estimate is being evaluated)

C. Estimation of s^2 (required for either sample size or power calculations)

1. Two independent estimates

a. Let

$$s^2 = \frac{1}{2}\left(s_1^2 + s_2^2\right) \qquad \text{(A1.49)}$$

where s_1^2 and s_2^2 are calculated as explained below.

b. For means (or proportions estimated using multistage sampling) s_1^2 and s_2^2 are usually estimated with data from a pilot study or the first part of an ongoing study. Let

$$s_1^2 = \frac{\sum_i^{n_{p1}} (y_{1i} - \bar{y}_1)^2}{n_{p1} - 1} \qquad \text{(A1.50)}$$

Box 10 (*cont.*)

and

$$s_2^2 = \frac{\sum_i^{n_{p2}} (y_{2i} - \bar{y}_2)^2}{n_{p2} - 1} \quad \text{(A1.51)}$$

where y_{1i} and y_{2i} are the measurements on the ith unit in samples one and two respectively (or the means of the measurements from the ith primary units if multistage sampling was used); n_{p1} and n_{p2} are the sample sizes from the pilot study. If no preliminary data from the study area are available, data from other published studies may be used. If the ranges of the y_{1i} and y_{2i} can be predicted, an alternative method is to assume that s_1^2 equals 0.25 times the range in population one and s_2^2 equals 0.25 times the range in population two (Snedecor and Cochran 1980 p. 442).

c. For proportions estimated using one-stage sampling, let

$$s_1^2 = p_1 q_1 \quad \text{(A1.52)}$$

and

$$s_2^2 = p_2 q_2 \quad \text{(A1.53)}$$

where p_1 and p_2 are estimates of the true proportions in the populations, and $q_1 = 1 - p_1$, $q_2 = 1 - p_2$. If p_1 (or p_2) is between 0.30 and 0.70, then $p_1 q_1$ (or $p_2 q_2$) is between 0.21 and 0.25. Thus, even imprecise estimates of p_1 and p_2 often suffice to estimate s_1^2 and s_2^2. If data from a pilot or preliminary study are available, then the resulting values for p_1 and p_2 may be used in Eqs. A1.52 and A1.53.

2. One estimate

Let $s^2 = s_1^2$, where s_1^2 is calculated as in Part C1b or C1c above.

3. Paired estimates

a. Methods to use if data are available from a pilot study:

For means, let

$$s^2 = \frac{\sum_i^{n_p} (d_i - \bar{d})^2}{(n_p - 1)} \quad \text{(A1.54)}$$

where d_i is the difference between measurements on the ith unit and n_p is the sample size (number of pairs) in the pilot study. If multistage sampling is employed, then d_i is the average difference for the ith primary unit. In either case, $\bar{d}$ is the mean of the d_i.

For proportions (estimated using one-stage sampling), let

$$s^2 = \sqrt{n_{pd} / n_p} \quad \text{(A1.55)}$$

Box 10 (*cont.*)

where n_p is the sample size in the pilot study and n_{pd} is the number of the n_p units in which the two measurements (or outcomes) were different.

b. If no data from a pilot study are available, one possibility is to use the methods in Part C1 above for independent estimates. This will usually result in overestimating s^2 (because the benefits of pairing are ignored) which causes sample size requirements to be overestimated and power to be underestimated.

D. Sample size calculations [for estimating the sample size(s) *required* by the end of the study]

1. One estimate or two estimates (paired or independent) with $n_1 = n_2$ (n is then the required sample size for *each* estimate).

a. Determine $K_{\alpha,P'}$, a constant used below. If $\alpha = 0.05$, and the test is two-tailed, then $K_{\alpha,P'}$ may be determined from the following table:

Desired power	$K_{\alpha,P'}$	Z_β
0.50	3.84	0.000
0.60	4.90	0.253
0.70	6.18	0.525
0.80	7.87	0.845
0.90	10.50	1.282
0.95	13.00	1.645

For other cases define $K_{\alpha,P'}$ as $(Z_\alpha + Z_\beta)^2$, where Z_β (whose value depends only on power) is obtained from the Table above, and Z_α is obtained from the table below.

	Level of significance (α)		
Test	0.10	0.05	0.01
Two-tailed	1.645	1.960	2.576
One-tailed	1.282	1.645	2.326

Snedecor and Cochran (1980 p. 102) describe how to obtain other values of Z_α and Z_β.

b. The sample size required to achieve power P' if the true difference is δ, is

$$n = \frac{K_{\alpha,P'} s^2}{\delta^2} + 2 \qquad \text{(A1.56)}$$

The '+2' in (A1.56) compensates for the fact that standard deviations will be estimated, not known. More detailed

Box 10 (*cont.*)

procedures for making this compensation are contained in Snedecor and Cochran (1980 p. 104).

2. Two estimates, $n_1 \neq n_2$
   Sample size requirements are best calculated by trial and error using the power formula in Part E below. Select two sample sizes, calculate power, adjust n_1 and n_2 up or down, and then calculate power again. A few iterations will identify combinations of n_1 and n_2 that achieve the desired level of power.
3. Sample size to achieve a specified confidence interval
   The confidence interval is usually calculated as: (estimate) $\pm (t_{df}(\alpha/2))$ [se(estimate)]. Let w = a desired half-width for the confidence interval.

$$w = t_{df}(\alpha/2)\, se(\text{estimate}). \quad (A1.57)$$

The general approach is to insert the formula for the standard error in Eq. A1.57, and then solve the equation for n. The standard error is

$$se(g_1 - g_2) = \sqrt{\frac{s^2}{n}}, \quad (A1.58)$$

and therefore

$$n = \frac{[t_{df}(\alpha/2)]^2 s^2}{w^2}. \quad (A1.59)$$

E. Power calculations [based on the sample size(s) *expected* by the end of the study]

1. Calculate $se(g_1 - g_2)$
   The formula for a single estimate, paired estimates, or independent estimates with $n_1 = n_2$, is

$$se(g_1 - g_2) = \sqrt{s^2/n}. \quad (A1.60)$$

The formula for independent estimates with $n_1 \neq n_2$ is

$$se(g_1 - g_2) = \sqrt{\frac{s_1^2}{n_1} + \frac{s_2^2}{n_2}}. \quad (A1.61)$$

2. Calculate $\delta/se(g_1 - g_2)$, where δ is the assumed, or minimum important, difference between parameters
3. Calculate the degrees of freedom for the test ($n_1 + n_2 - 2$ for independent estimates or $n - 1$ for paired estimates – see Boxes 6 or 7)
4. Power, for a two-tailed test with level of significance $(\alpha) = 0.05$, is given in the following Table

Box 10 (*cont.*)

$\delta\, se(g_1 - g_2)$	Degrees of freedom						
	5	7	10	15	30	60	∞
1.00	0.06	0.09	0.11	0.13	0.15	0.16	0.17
1.25	0.09	0.13	0.16	0.19	0.21	0.23	0.24
1.50	0.14	0.19	0.23	0.26	0.29	0.31	0.32
1.75	0.21	0.27	0.32	0.35	0.39	0.40	0.42
2.00	0.28	0.36	0.41	0.45	0.48	0.50	0.52
2.25	0.37	0.46	0.51	0.55	0.58	0.60	0.61
2.50	0.47	0.56	0.61	0.64	0.68	0.69	0.71
2.75	0.57	0.65	0.70	0.73	0.76	0.77	0.79
3.00	0.67	0.74	0.78	0.81	0.83	0.84	0.85
3.25	0.75	0.81	0.85	0.87	0.89	0.89	0.90
3.50	0.82	0.87	0.90	0.91	0.93	0.93	0.94
3.75	0.88	0.92	0.94	0.95	0.96	0.96	0.96
4.00	0.92	0.95	0.96	0.97	0.98	0.98	0.98

5. Power may be calculated more exactly as

$$power = P\left[Z > t_{\infty}(\alpha/2) - \frac{\delta}{se(g_1 - g_2)}\right] \quad \text{(A1.62)}$$

where $t_{\alpha,df}$ is the critical value in Table A2.1 in Appendix Two (e.g., for a two-tailed test at $\alpha = 0.05$, use $t_{\infty}(\alpha/2) = 1.96$). Calculate the quantity to the right of the >, and then use a table giving areas under the normal curve to determine the probability that a standard normal variable exceeds this value. The result equals power. For additional explanation, see Snedecor and Cochran, 1980 p. 68).

F. Notes

1. The formulas above assume that *t*-tests and confidence intervals based on the data having a normal distribution will be used. With proportions and small sample sizes, exact methods are preferred (Boxes 5 and 7), and may have slightly different power. The formulas above give adequate approximations for practical purposes.
2. For other sampling plans or estimates, the following procedures may be used. Calculate sample size requirements iteratively by calculating power as suggested in Part D2 above for $n_1 \neq n_2$. To calculate power, estimate all quantities in the formula for the standard error (e.g., for stratified sampling, stratum sizes must be estimated), estimate the standard error and δ, and then follow the guidelines in Part E4 or E5 above.

Appendix Two
Statistical tables

Fig. A2.1: Confidence limits for p in binomial sampling *page* 280
Table A2.1: Percentage points of the t-distribution 284
Table A2.2: Points of the F-distribution 286
Table A2.3: Percentage points for Fisher's exact test 290
Table A2.4: Percentage points for the signed rank statistic T 302
Table A2.5: Percentage points for the rank sum statistic T 304
Table A2.6: The binomial distribution function 309

Figure A2.1 *Chart providing confidence limits for p in binomial sampling, given a sample fraction c/n. Confidence coefficient, $1 - 2\alpha = 0{\cdot}95$.*

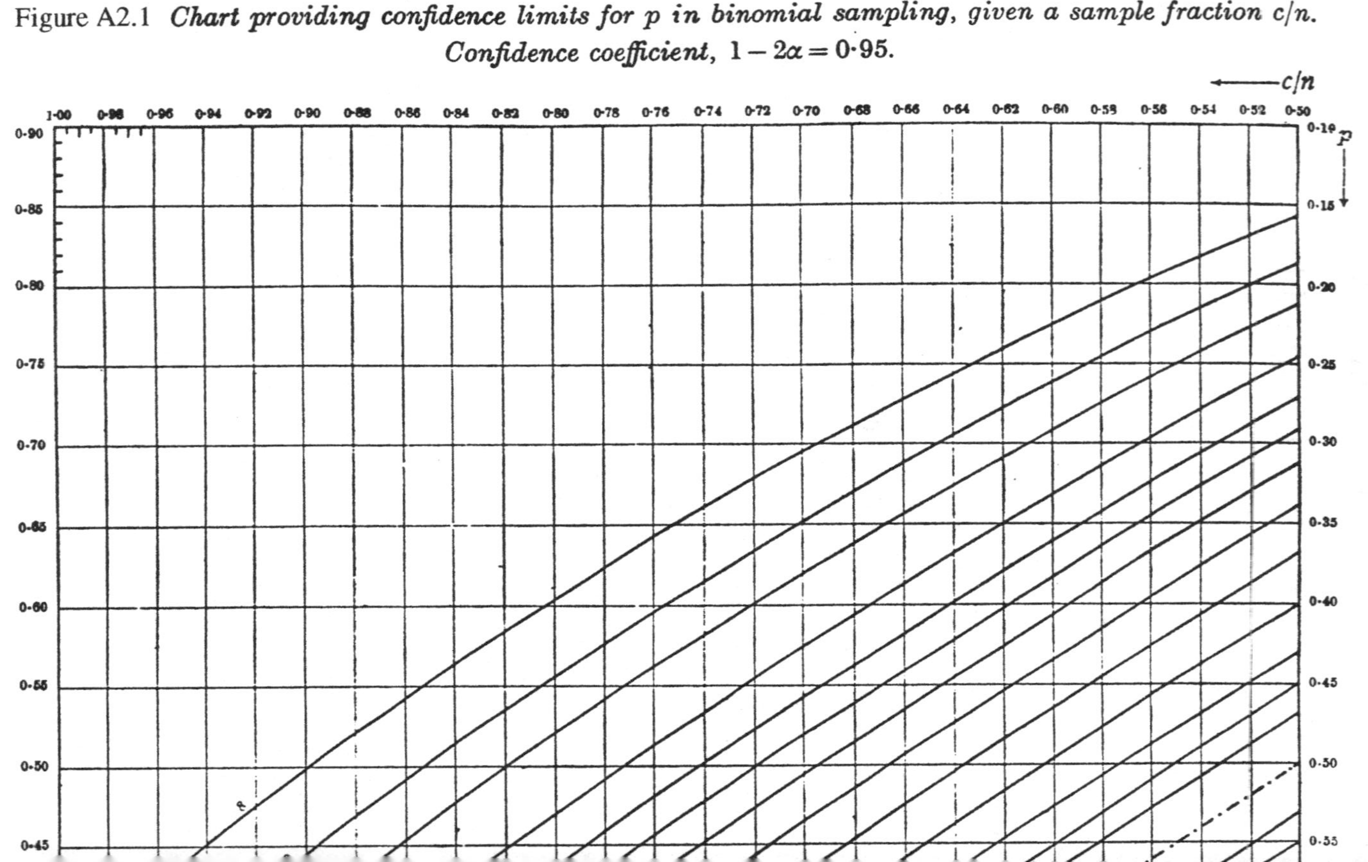

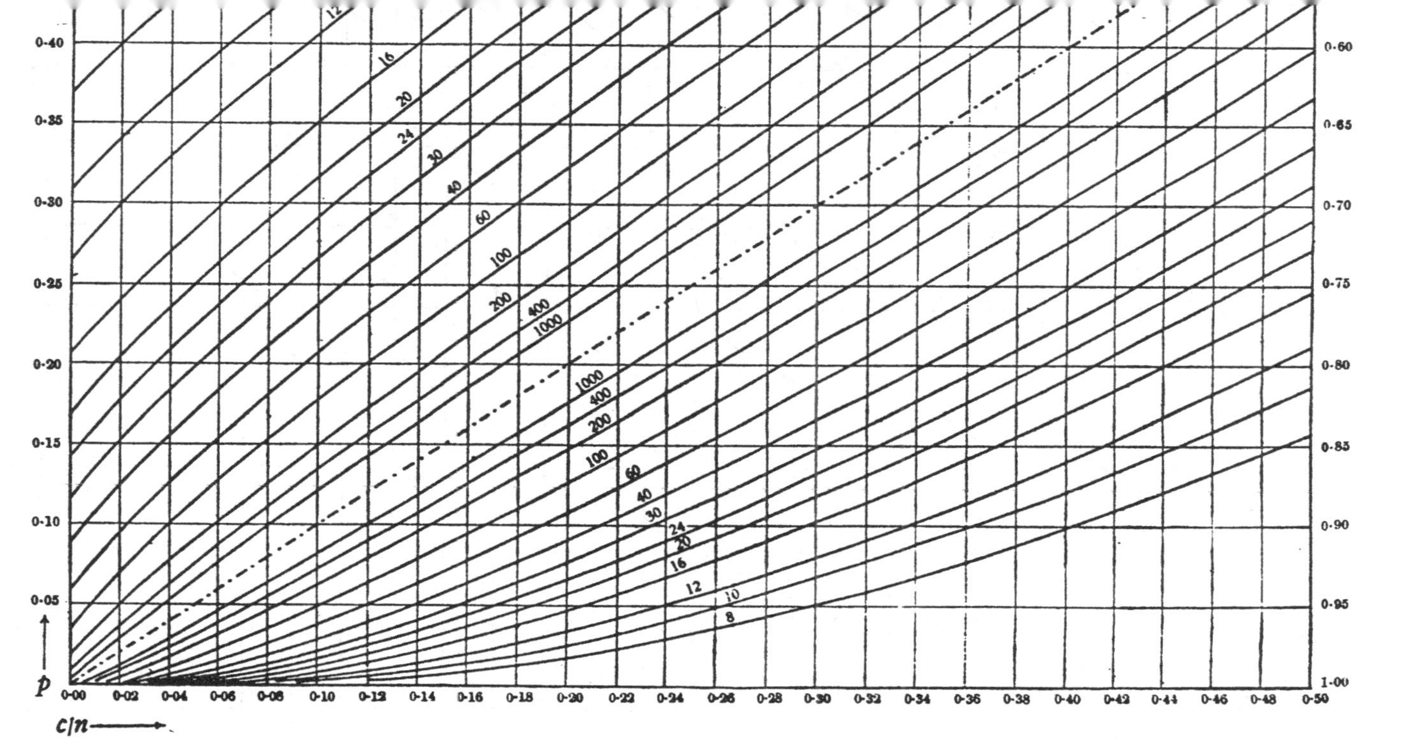

The numbers printed along the curves indicate the sample size n. If for a given value of the abscissa c/n, p_A and p_B are the ordinates read from (or interpolated between) the appropriate lower and upper curves, then

$$\Pr\{p_A \leqslant p \leqslant p_B\} \leqslant 1 - 2\alpha.$$

Figure A2.1 (*continued*). *Confidence coefficient*, $1-2\alpha=0{\cdot}99$.

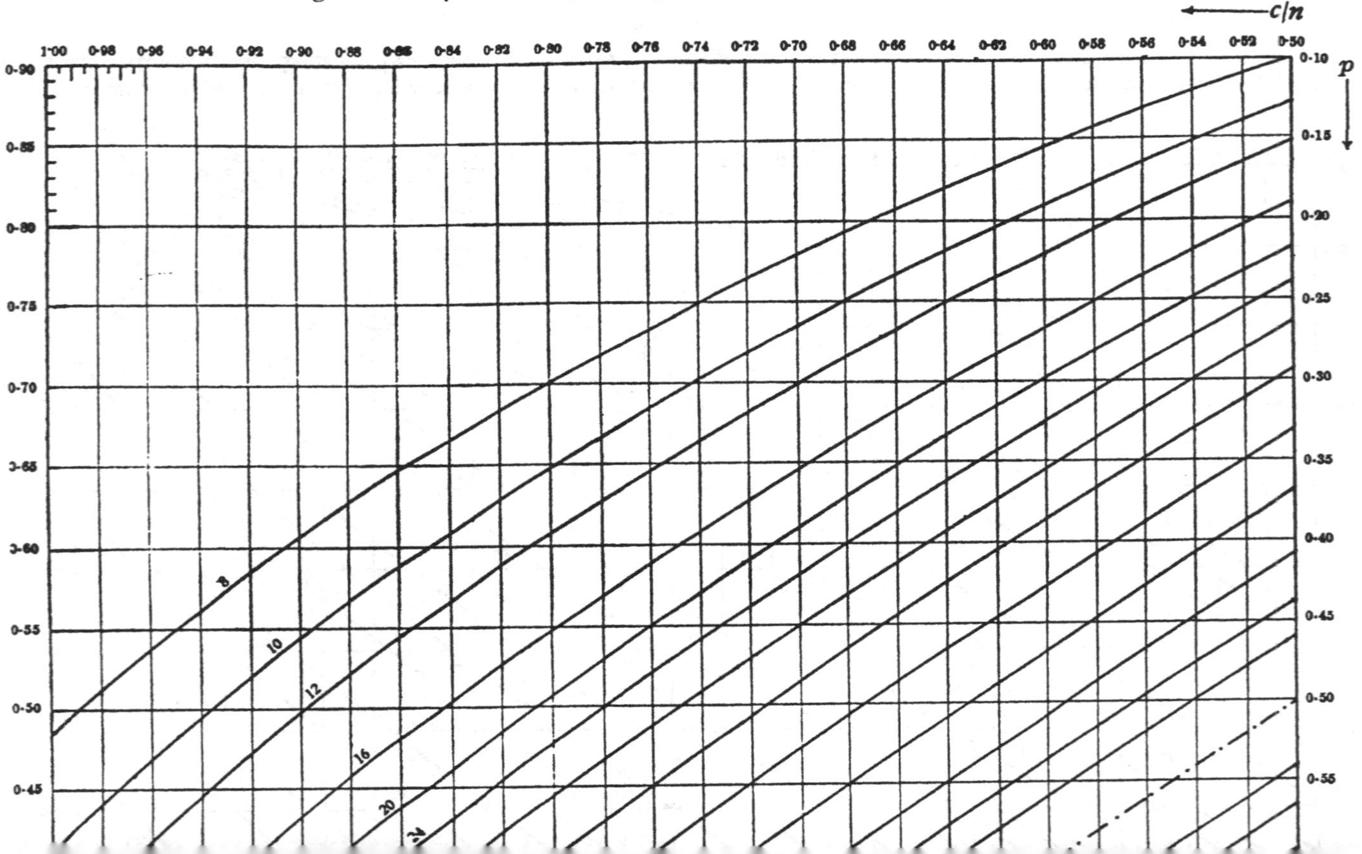

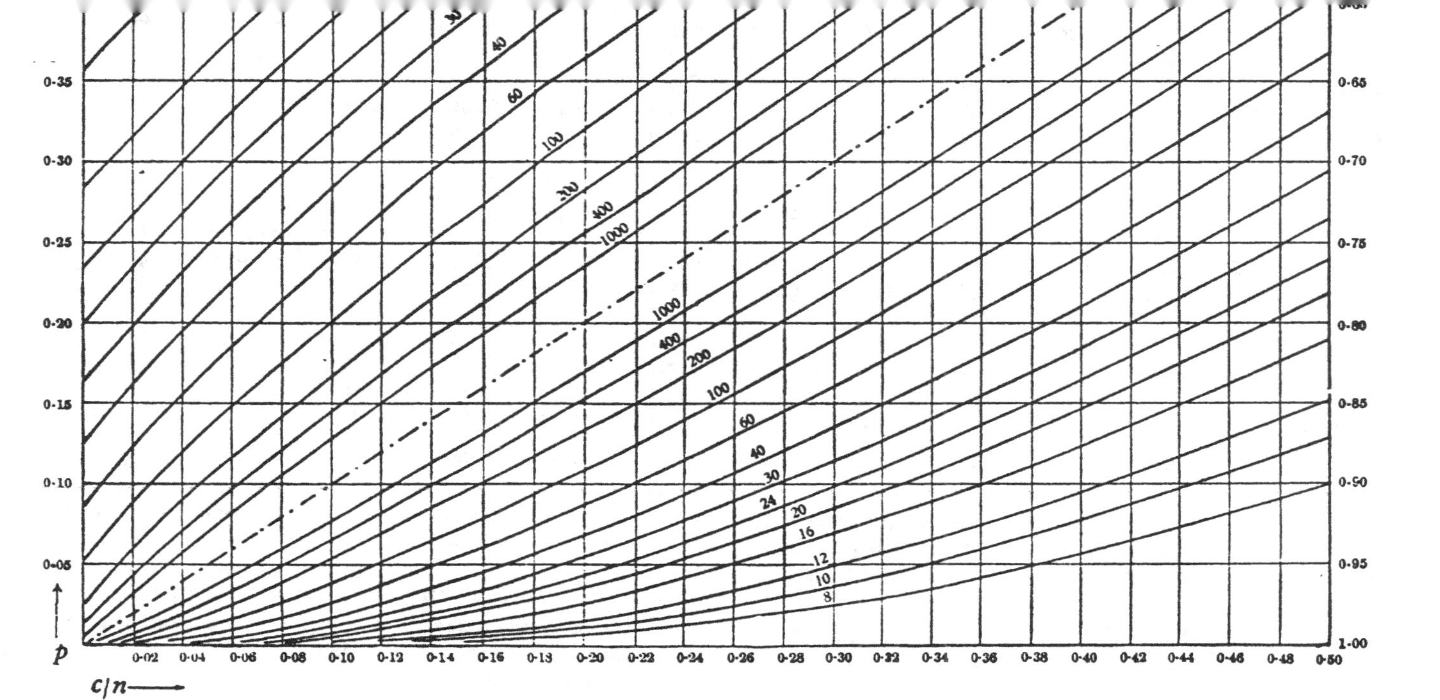

The numbers printed along the curves indicate the sample size n.

Note: the process of reading from the curves can be simplified with the help of the right-angled corner of a loose sheet of paper or thin card, along the edges of which are marked off the scales shown in the top left-hand corner of each Chart.

Table A2.1 *Percentage points of the* t-*distribution*

This table gives percentage points $t_\nu(P)$ defined by the equation

$$\frac{P}{100} = \frac{1}{\sqrt{\nu\pi}} \frac{\Gamma(\frac{1}{2}\nu+\frac{1}{2})}{\Gamma(\frac{1}{2}\nu)} \int_{t_\nu(P)}^{\infty} \frac{dt}{(1+t^2/\nu)^{\frac{1}{2}(\nu+1)}}.$$

Let X_1 and X_2 be independent random variables having a normal distribution with zero mean and unit variance and a χ^2-distribution with ν degrees of freedom respectively; then $t = X_1/\sqrt{X_2/\nu}$ has Student's t-distribution with ν degrees of freedom, and the probability that $t \geqslant t_\nu(P)$ is $P/100$. The lower percentage points are given by symmetry as $-t_\nu(P)$, and the probability that $|t| \geqslant t_\nu(P)$ is $2P/100$.

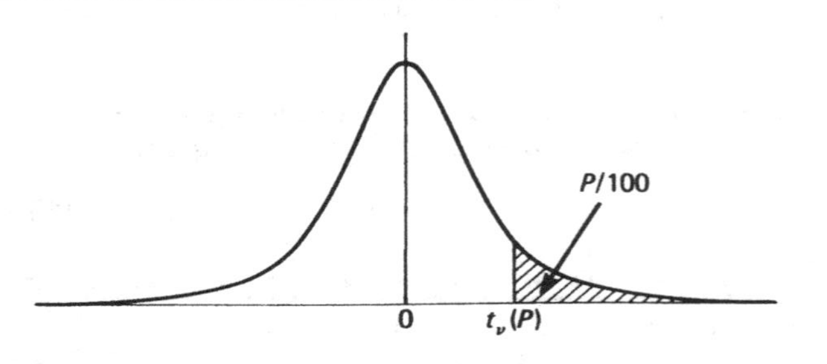

The limiting distribution of t as ν tends to infinity is the normal distribution with zero mean and unit variance. When ν is large interpolation in ν should be harmonic.

P	40	30	25	20	15	10	5	2·5	1	0·5	0·1	0·05
$\nu = 1$	0·3249	0·7265	1·0000	1·3764	1·963	3·078	6·314	12·71	31·82	63·66	318·3	636·6
2	0·2887	0·6172	0·8165	1·0607	1·386	1·886	2·920	4·303	6·965	9·925	22·33	31·60
3	0·2767	0·5844	0·7649	0·9785	1·250	1·638	2·353	3·182	4·541	5·841	10·21	12·92
4	0·2707	0·5686	0·7407	0·9410	1·190	1·533	2·132	2·776	3·747	4·604	7·173	8·610
5	0·2672	0·5594	0·7267	0·9195	1·156	1·476	2·015	2·571	3·365	4·032	5·893	6·869
6	0·2648	0·5534	0·7176	0·9057	1·134	1·440	1·943	2·447	3·143	3·707	5·208	5·959
7	0·2632	0·5491	0·7111	0·8960	1·119	1·415	1·895	2·365	2·998	3·499	4·785	5·408
8	0·2619	0·5459	0·7064	0·8889	1·108	1·397	1·860	2·306	2·896	3·355	4·501	5·041
9	0·2610	0·5435	0·7027	0·8834	1·100	1·383	1·833	2·262	2·821	3·250	4·297	4·781
10	0·2602	0·5415	0·6998	0·8791	1·093	1·372	1·812	2·228	2·764	3·169	4·144	4·587
11	0·2596	0·5399	0·6974	0·8755	1·088	1·363	1·796	2·201	2·718	3·106	4·025	4·437
12	0·2590	0·5386	0·6955	0·8726	1·083	1·356	1·782	2·179	2·681	3·055	3·930	4·318
13	0·2586	0·5375	0·6938	0·8702	1·079	1·350	1·771	2·160	2·650	3·012	3·852	4·221
14	0·2582	0·5366	0·6924	0·8681	1·076	1·345	1·761	2·145	2·624	2·977	3·787	4·140

15	0·2579	0·5357	0·6912	0·8662	1·074	1·341	1·753	2·131	2·602	2·947	3·733	4·073
16	0·2576	0·5350	0·6901	0·8647	1·071	1·337	1·746	2·120	2·583	2·921	3·686	4·015
17	0·2573	0·5344	0·6892	0·8633	1·069	1·333	1·740	2·110	2·567	2·898	3·646	3·965
18	0·2571	0·5338	0·6884	0·8620	1·067	1·330	1·734	2·101	2·552	2·878	3·610	3·922
19	0·2569	0·5333	0·6876	0·8610	1·066	1·328	1·729	2·093	2·539	2·861	3·579	3·883
20	0·2567	0·5329	0·6870	0·8600	1·064	1·325	1·725	2·086	2·528	2·845	3·552	3·850
21	0·2566	0·5325	0·6864	0·8591	1·063	1·323	1·721	2·080	2·518	2·831	3·527	3·819
22	0·2564	0·5321	0·6858	0·8583	1·061	1·321	1·717	2·074	2·508	2·819	3·505	3·792
23	0·2563	0·5317	0·6853	0·8575	1·060	1·319	1·714	2·069	2·500	2·807	3·485	3·768
24	0·2562	0·5314	0·6848	0·8569	1·059	1·318	1·711	2·064	2·492	2·797	3·467	3·745
25	0·2561	0·5312	0·6844	0·8562	1·058	1·316	1·708	2·060	2·485	2·787	3·450	3·725
26	0·2560	0·5309	0·6840	0·8557	1·058	1·315	1·706	2·056	2·479	2·779	3·435	3·707
27	0·2559	0·5306	0·6837	0·8551	1·057	1·314	1·703	2·052	2·473	2·771	3·421	3·690
28	0·2558	0·5304	0·6834	0·8546	1·056	1·313	1·701	2·048	2·467	2·763	3·408	3·674
29	0·2557	0·5302	0·6830	0·8542	1·055	1·311	1·699	2·045	2·462	2·756	3·396	3·659
30	0·2556	0·5300	0·6828	0·8538	1·055	1·310	1·697	2·042	2·457	2·750	3·385	3·646
32	0·2555	0·5297	0·6822	0·8530	1·054	1·309	1·694	2·037	2·449	2·738	3·365	3·622
34	0·2553	0·5294	0·6818	0·8523	1·052	1·307	1·691	2·032	2·441	2·728	3·348	3·601
36	0·2552	0·5291	0·6814	0·8517	1·052	1·306	1·688	2·028	2·434	2·719	3·333	3·582
38	0·2551	0·5288	0·6810	0·8512	1·051	1·304	1·686	2·024	2·429	2·712	3·319	3·566
40	0·2550	0·5286	0·6807	0·8507	1·050	1·303	1·684	2·021	2·423	2·704	3·307	3·551
50	0·2547	0·5278	0·6794	0·8489	1·047	1·299	1·676	2·009	2·403	2·678	3·261	3·496
60	0·2545	0·5272	0·6786	0·8477	1·045	1·296	1·671	2·000	2·390	2·660	3·232	3·460
120	0·2539	0·5258	0·6765	0·8446	1·041	1·289	1·658	1·980	2·358	2·617	3·160	3·373
∞	0·2533	0·5244	0·6745	0·8416	1·036	1·282	1·645	1·960	2·326	2·576	3·090	3·291

Taken from Table 10 of *New Cambridge statistical tables*, 2nd edn. (1996), edited by D.V. Lindley and W.F. Scott, with permission of Cambridge University Press.

Table A2.2. 5 *Per cent points of the* F *distribution*

If $F = \frac{X_1}{\nu_1} \Big/ \frac{X_2}{\nu_2}$, where X_1 and X_2 are independent random variables distributed as χ^2 with ν_1 and ν_2 degrees of freedom respectively, then the probabilities that $F \geqslant F(P)$ and that $F \leqslant F'(P)$ are both equal to $P/100$. Linear interpolation in ν_1 and ν_2 will generally be sufficiently accurate except when either $\nu_1 > 12$ or $\nu_2 > 40$, when harmonic interpolation should be used.

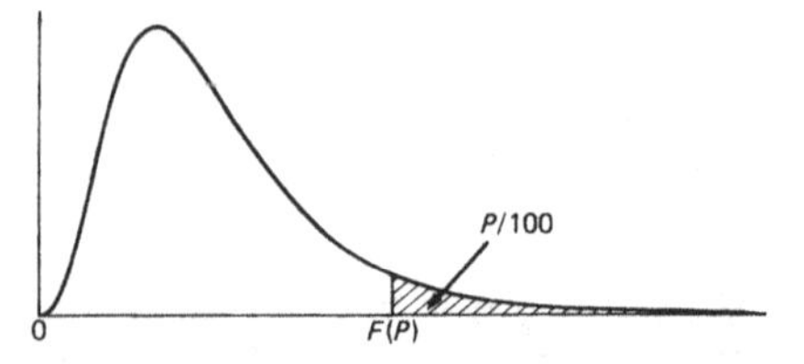

(This shape applies only when $\nu_1 \geqslant 3$. When $\nu_1 < 3$ the mode is at the origin.)

$\nu_1 =$	1	2	3	4	5	6	7	8	10	12	24	∞
$\nu_2 =$ 1	161·4	199·5	215·7	224·6	230·2	234·0	236·8	238·9	241·9	243·9	249·1	254·3
2	18·51	19·00	19·16	19·25	19·30	19·33	19·35	19·37	19·40	19·41	19·45	19·50
3	10·13	9·552	9·277	9·117	9·013	8·941	8·887	8·845	8·786	8·745	8·639	8·526
4	7·709	6·944	6·591	6·388	6·256	6·163	6·094	6·041	5·964	5·912	5·774	5·628
5	6·608	5·786	5·409	5·192	5·050	4·950	4·876	4·818	4·735	4·678	4·527	4·365
6	5·987	5·143	4·757	4·534	4·387	4·284	4·207	4·147	4·060	4·000	3·841	3·669
7	5·591	4·737	4·347	4·120	3·972	3·866	3·787	3·726	3·637	3·575	3·410	3·230
8	5·318	4·459	4·066	3·838	3·687	3·581	3·500	3·438	3·347	3·284	3·115	2·928
9	5·117	4·256	3·863	3·633	3·482	3·374	3·293	3·230	3·137	3·073	2·900	2·707
10	4·965	4·103	3·708	3·478	3·326	3·217	3·135	3·072	2·978	2·913	2·737	2·538
11	4·844	3·982	3·587	3·357	3·204	3·095	3·012	2·948	2·854	2·788	2·609	2·404
12	4·747	3·885	3·490	3·259	3·106	2·996	2·913	2·849	2·753	2·687	2·505	2·296
13	4·667	3·806	3·411	3·179	3·025	2·915	2·832	2·767	2·671	2·604	2·420	2·206
14	4·600	3·739	3·344	3·112	2·958	2·848	2·764	2·699	2·602	2·534	2·349	2·131
15	4·543	3·682	3·287	3·056	2·901	2·790	2·707	2·641	2·544	2·475	2·288	2·066
16	4·494	3·634	3·239	3·007	2·852	2·741	2·657	2·591	2·494	2·425	2·235	2·010
17	4·451	3·592	3·197	2·965	2·810	2·699	2·614	2·548	2·450	2·381	2·190	1·960
18	4·414	3·555	3·160	2·928	2·773	2·661	2·577	2·510	2·412	2·342	2·150	1·917
19	4·381	3·522	3·127	2·895	2·740	2·628	2·544	2·477	2·378	2·308	2·114	1·878
20	4·351	3·493	3·098	2·866	2·711	2·599	2·514	2·447	2·348	2·278	2·082	1·843
21	4·325	3·467	3·072	2·840	2·685	2·573	2·488	2·420	2·321	2·250	2·054	1·812
22	4·301	3·443	3·049	2·817	2·661	2·549	2·464	2·397	2·297	2·226	2·028	1·783
23	4·279	3·422	3·028	2·796	2·640	2·528	2·442	2·375	2·275	2·204	2·005	1·757
24	4·260	3·403	3·009	2·776	2·621	2·508	2·423	2·355	2·255	2·183	1·984	1·733
25	4·242	3·385	2·991	2·759	2·603	2·490	2·405	2·337	2·236	2·165	1·964	1·711
26	4·225	3·369	2·975	2·743	2·587	2·474	2·388	2·321	2·220	2·148	1·946	1·691
27	4·210	3·354	2·960	2·728	2·572	2·459	2·373	2·305	2·204	2·132	1·930	1·672
28	4·196	3·340	2·947	2·714	2·558	2·445	2·359	2·291	2·190	2·118	1·915	1·654
29	4·183	3·328	2·934	2·701	2·545	2·432	2·346	2·278	2·177	2·104	1·901	1·638
30	4·171	3·316	2·922	2·690	2·534	2·421	2·334	2·266	2·165	2·092	1·887	1·622
32	4·149	3·295	2·901	2·668	2·512	2·399	2·313	2·244	2·142	2·070	1·864	1·594
34	4·130	3·276	2·883	2·650	2·494	2·380	2·294	2·225	2·123	2·050	1·843	1·569
36	4·113	3·259	2·866	2·634	2·477	2·364	2·277	2·209	2·106	2·033	1·824	1·547
38	4·098	3·245	2·852	2·619	2·463	2·349	2·262	2·194	2·091	2·017	1·808	1·527
40	4·085	3·232	2·839	2·606	2·449	2·336	2·249	2·180	2·077	2·003	1·793	1·509
60	4·001	3·150	2·758	2·525	2·368	2·254	2·167	2·097	1·993	1·917	1·700	1·389
120	3·920	3·072	2·680	2·447	2·290	2·175	2·087	2·016	1·910	1·834	1·608	1·254
∞	3·841	2·996	2·605	2·372	2·214	2·099	2·010	1·938	1·831	1·752	1·517	1·000

Table A2.2. (*cont.*) 2.5 *Per cent points of the* F *distribution*

The function tabulated is $F(P) = F(P|\nu_1, \nu_2)$ defined by the equation

$$\frac{P}{100} = \frac{\Gamma(\frac{1}{2}\nu_1+\frac{1}{2}\nu_2)}{\Gamma(\frac{1}{2}\nu_1)\,\Gamma(\frac{1}{2}\nu_2)}\,\nu_1^{\frac{1}{2}\nu_1}\,\nu_2^{\frac{1}{2}\nu_2}\int_{F(P)}^{\infty}\frac{F^{\frac{1}{2}\nu_1-1}}{(\nu_2+\nu_1 F)^{\frac{1}{2}(\nu_1+\nu_2)}}\,dF,$$

for $P = 10$, 5, 2·5, 1, 0·5 and 0·1. The lower percentage points, that is the values $F'(P) = F'(P|\nu_1, \nu_2)$ such that the probability that $F \leqslant F'(P)$ is equal to $P/100$, may be found by the formula

$$F'(P|\nu_1, \nu_2) = 1/F(P|\nu_2, \nu_1).$$

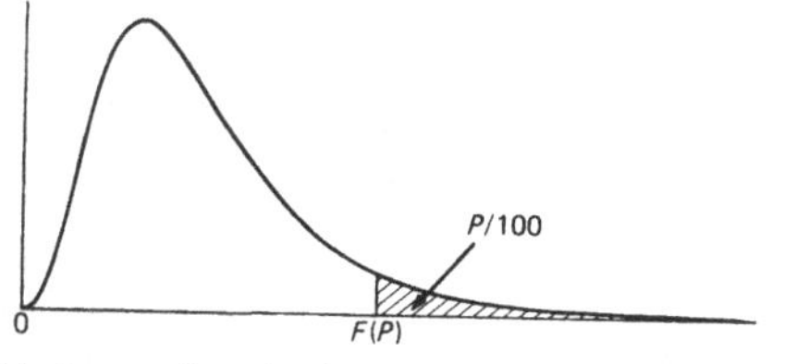

(This shape applies only when $\nu_1 \geqslant 3$. When $\nu_1 < 3$ the mode is at the origin.)

$\nu_1 =$	1	2	3	4	5	6	7	8	10	12	24	∞
$\nu_2 =$ 1	647·8	799·5	864·2	899·6	921·8	937·1	948·2	956·7	968·6	976·7	997·2	1018
2	38·51	39·00	39·17	39·25	39·30	39·33	39·36	39·37	39·40	39·41	39·46	39·50
3	17·44	16·04	15·44	15·10	14·88	14·73	14·62	14·54	14·42	14·34	14·12	13·90
4	12·22	10·65	9·979	9·605	9·364	9·197	9·074	8·980	8·844	8·751	8·511	8·257
5	10·01	8·434	7·764	7·388	7·146	6·978	6·853	6·757	6·619	6·525	6·278	6·015
6	8·813	7·260	6·599	6·227	5·988	5·820	5·695	5·600	5·461	5·366	5·117	4·849
7	8·073	6·542	5·890	5·523	5·285	5·119	4·995	4·899	4·761	4·666	4·415	4·142
8	7·571	6·059	5·416	5·053	4·817	4·652	4·529	4·433	4·295	4·200	3·947	3·670
9	7·209	5·715	5·078	4·718	4·484	4·320	4·197	4·102	3·964	3·868	3·614	3·333
10	6·937	5·456	4·826	4·468	4·236	4·072	3·950	3·855	3·717	3·621	3·365	3·080
11	6·724	5·256	4·630	4·275	4·044	3·881	3·759	3·664	3·526	3·430	3·173	2·883
12	6·554	5·096	4·474	4·121	3·891	3·728	3·607	3·512	3·374	3·277	3·019	2·725
13	6·414	4·965	4·347	3·996	3·767	3·604	3·483	3·388	3·250	3·153	2·893	2·595
14	6·298	4·857	4·242	3·892	3·663	3·501	3·380	3·285	3·147	3·050	2·789	2·487
15	6·200	4·765	4·153	3·804	3·576	3·415	3·293	3·199	3·060	2·963	2·701	2·395
16	6·115	4·687	4·077	3·729	3·502	3·341	3·219	3·125	2·986	2·889	2·625	2·316
17	6·042	4·619	4·011	3·665	3·438	3·277	3·156	3·061	2·922	2·825	2·560	2·247
18	5·978	4·560	3·954	3·608	3·382	3·221	3·100	3·005	2·866	2·769	2·503	2·187
19	5·922	4·508	3·903	3·559	3·333	3·172	3·051	2·956	2·817	2·720	2·452	2·133
20	5·871	4·461	3·859	3·515	3·289	3·128	3·007	2·913	2·774	2·676	2·408	2·085
21	5·827	4·420	3·819	3·475	3·250	3·090	2·969	2·874	2·735	2·637	2·368	2·042
22	5·786	4·383	3·783	3·440	3·215	3·055	2·934	2·839	2·700	2·602	2·331	2·003
23	5·750	4·349	3·750	3·408	3·183	3·023	2·902	2·808	2·668	2·570	2·299	1·968
24	5·717	4·319	3·721	3·379	3·155	2·995	2·874	2·779	2·640	2·541	2·269	1·935
25	5·686	4·291	3·694	3·353	3·129	2·969	2·848	2·753	2·613	2·515	2·242	1·906
26	5·659	4·265	3·670	3·329	3·105	2·945	2·824	2·729	2·590	2·491	2·217	1·878
27	5·633	4·242	3·647	3·307	3·083	2·923	2·802	2·707	2·568	2·469	2·195	1·853
28	5·610	4·221	3·626	3·286	3·063	2·903	2·782	2·687	2·547	2·448	2·174	1·829
29	5·588	4·201	3·607	3·267	3·044	2·884	2·763	2·669	2·529	2·430	2·154	1·807
30	5·568	4·182	3·589	3·250	3·026	2·867	2·746	2·651	2·511	2·412	2·136	1·787
32	5·531	4·149	3·557	3·218	2·995	2·836	2·715	2·620	2·480	2·381	2·103	1·750
34	5·499	4·120	3·529	3·191	2·968	2·808	2·688	2·593	2·453	2·353	2·075	1·717
36	5·471	4·094	3·505	3·167	2·944	2·785	2·664	2·569	2·429	2·329	2·049	1·687
38	5·446	4·071	3·483	3·145	2·923	2·763	2·643	2·548	2·407	2·307	2·027	1·661
40	5·424	4·051	3·463	3·126	2·904	2·744	2·624	2·529	2·388	2·288	2·007	1·637
60	5·286	3·925	3·343	3·008	2·786	2·627	2·507	2·412	2·270	2·169	1·882	1·482
120	5·152	3·805	3·227	2·894	2·674	2·515	2·395	2·299	2·157	2·055	1·760	1·310
∞	5·024	3·689	3·116	2·786	2·567	2·408	2·288	2·192	2·048	1·945	1·640	1·000

Table A2.2. (*cont.*) 1 *Per cent points of the* F *distribution*

If $F = \dfrac{X_1}{\nu_1}\Big/\dfrac{X_2}{\nu_2}$, where X_1 and X_2 are independent random variables distributed as χ^2 with ν_1 and ν_2 degrees of freedom respectively, then the probabilities that $F \geqslant F(P)$ and that $F \leqslant F'(P)$ are both equal to $P/100$. Linear interpolation in ν_1 or ν_2 will generally be sufficiently accurate except when either $\nu_1 > 12$ or $\nu_2 > 40$, when harmonic interpolation should be used.

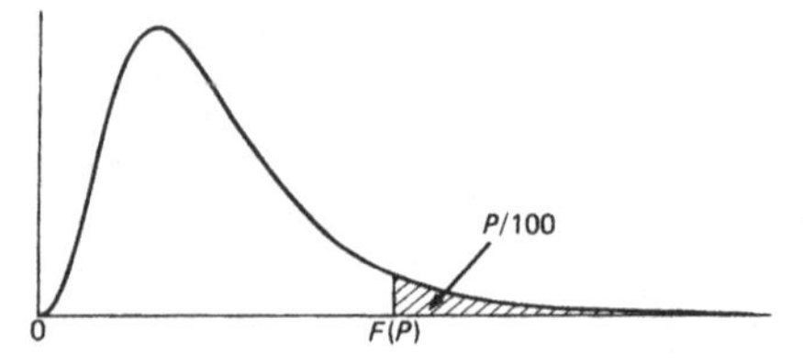

(This shape applies only when $\nu_1 \geqslant 3$. When $\nu_1 < 3$ the mode is at the origin.)

$\nu_1 =$	1	2	3	4	5	6	7	8	10	12	24	∞
$\nu_2 =$ 1	4052	4999	5403	5625	5764	5859	5928	5981	6056	6106	6235	6366
2	98·50	99·00	99·17	99·25	99·30	99·33	99·36	99·37	99·40	99·42	99·46	99·50
3	34·12	30·82	29·46	28·71	28·24	27·91	27·67	27·49	27·23	27·05	26·60	26·13
4	21·20	18·00	16·69	15·98	15·52	15·21	14·98	14·80	14·55	14·37	13·93	13·46
5	16·26	13·27	12·06	11·39	10·97	10·67	10·46	10·29	10·05	9·888	9·466	9·020
6	13·75	10·92	9·780	9·148	8·746	8·466	8·260	8·102	7·874	7·718	7·313	6·880
7	12·25	9·547	8·451	7·847	7·460	7·191	6·993	6·840	6·620	6·469	6·074	5·650
8	11·26	8·649	7·591	7·006	6·632	6·371	6·178	6·029	5·814	5·667	5·279	4·859
9	10·56	8·022	6·992	6·422	6·057	5·802	5·613	5·467	5·257	5·111	4·729	4·311
10	10·04	7·559	6·552	5·994	5·636	5·386	5·200	5·057	4·849	4·706	4·327	3·909
11	9·646	7·206	6·217	5·668	5·316	5·069	4·886	4·744	4·539	4·397	4·021	3·602
12	9·330	6·927	5·953	5·412	5·064	4·821	4·640	4·499	4·296	4·155	3·780	3·361
13	9·074	6·701	5·739	5·205	4·862	4·620	4·441	4·302	4·100	3·960	3·587	3·165
14	8·862	6·515	5·564	5·035	4·695	4·456	4·278	4·140	3·939	3·800	3·427	3·004
15	8·683	6·359	5·417	4·893	4·556	4·318	4·142	4·004	3·805	3·666	3·294	2·868
16	8·531	6·226	5·292	4·773	4·437	4·202	4·026	3·890	3·691	3·553	3·181	2·753
17	8·400	6·112	5·185	4·669	4·336	4·102	3·927	3·791	3·593	3·455	3·084	2·653
18	8·285	6·013	5·092	4·579	4·248	4·015	3·841	3·705	3·508	3·371	2·999	2·566
19	8·185	5·926	5·010	4·500	4·171	3·939	3·765	3·631	3·434	3·297	2·925	2·489
20	8·096	5·849	4·938	4·431	4·103	3·871	3·699	3·564	3·368	3·231	2·859	2·421
21	8·017	5·780	4·874	4·369	4·042	3·812	3·640	3·506	3·310	3·173	2·801	2·360
22	7·945	5·719	4·817	4·313	3·988	3·758	3·587	3·453	3·258	3·121	2·749	2·305
23	7·881	5·664	4·765	4·264	3·939	3·710	3·539	3·406	3·211	3·074	2·702	2·256
24	7·823	5·614	4·718	4·218	3·895	3·667	3·496	3·363	3·168	3·032	2·659	2·211
25	7·770	5·568	4·675	4·177	3·855	3·627	3·457	3·324	3·129	2·993	2·620	2·169
26	7·721	5·526	4·637	4·140	3·818	3·591	3·421	3·288	3·094	2·958	2·585	2·131
27	7·677	5·488	4·601	4·106	3·785	3·558	3·388	3·256	3·062	2·926	2·552	2·097
28	7·636	5·453	4·568	4·074	3·754	3·528	3·358	3·226	3·032	2·896	2·522	2·064
29	7·598	5·420	4·538	4·045	3·725	3·499	3·330	3·198	3·005	2·868	2·495	2·034
30	7·562	5·390	4·510	4·018	3·699	3·473	3·304	3·173	2·979	2·843	2·469	2·006
32	7·499	5·336	4·459	3·969	3·652	3·427	3·258	3·127	2·934	2·798	2·423	1·956
34	7·444	5·289	4·416	3·927	3·611	3·386	3·218	3·087	2·894	2·758	2·383	1·911
36	7·396	5·248	4·377	3·890	3·574	3·351	3·183	3·052	2·859	2·723	2·347	1·872
38	7·353	5·211	4·343	3·858	3·542	3·319	3·152	3·021	2·828	2·692	2·316	1·837
40	7·314	5·179	4·313	3·828	3·514	3·291	3·124	2·993	2·801	2·665	2·288	1·805
60	7·077	4·977	4·126	3·649	3·339	3·119	2·953	2·823	2·632	2·496	2·115	1·601
120	6·851	4·787	3·949	3·480	3·174	2·956	2·792	2·663	2·472	2·336	1·950	1·381
∞	6·635	4·605	3·782	3·319	3·017	2·802	2·639	2·511	2·321	2·185	1·791	1·000

Table A2.2. *(cont.)* 0.5 *Per cent points of the* F *distribution*

The function tabulated is $F(P) = F(P|\nu_1, \nu_2)$ defined by the equation

$$\frac{P}{100} = \frac{\Gamma(\frac{1}{2}\nu_1 + \frac{1}{2}\nu_2)}{\Gamma(\frac{1}{2}\nu_1)\,\Gamma(\frac{1}{2}\nu_2)}\,\nu_1^{\frac{1}{2}\nu_1}\,\nu_2^{\frac{1}{2}\nu_2}\int_{F(P)}^{\infty}\frac{F^{\frac{1}{2}\nu_1 - 1}}{(\nu_2 + \nu_1 F)^{\frac{1}{2}(\nu_1+\nu_2)}}\,dF,$$

for P = 10, 5, 2·5, 1, 0·5 and 0·1. The lower percentage points, that is the values $F'(P) = F'(P|\nu_1, \nu_2)$ such that the probability that $F \leqslant F'(P)$ is equal to $P/100$, may be found by the formula

$$F'(P|\nu_1, \nu_2) = 1/F(P|\nu_2, \nu_1).$$

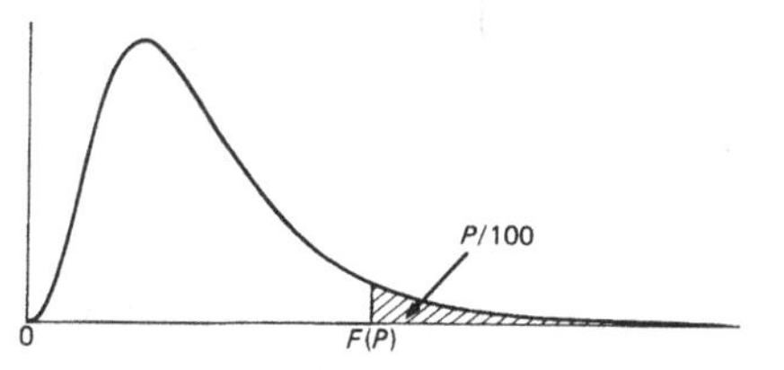

(This shape applies only when $\nu_1 \geqslant 3$. When $\nu_1 < 3$ the mode is at the origin.)

$\nu_1 =$	1	2	3	4	5	6	7	8	10	12	24	∞
$\nu_2 =$ 1	16211	20000	21615	22500	23056	23437	23715	23925	24224	24426	24940	25464
2	198·5	199·0	199·2	199·2	199·3	199·3	199·4	199·4	199·4	199·4	199·5	199·5
3	55·55	49·80	47·47	46·19	45·39	44·84	44·43	44·13	43·69	43·39	42·62	41·83
4	31·33	26·28	24·26	23·15	22·46	21·97	21·62	21·35	20·97	20·70	20·03	19·32
5	22·78	18·31	16·53	15·56	14·94	14·51	14·20	13·96	13·62	13·38	12·78	12·14
6	18·63	14·54	12·92	12·03	11·46	11·07	10·79	10·57	10·25	10·03	9·474	8·879
7	16·24	12·40	10·88	10·05	9·522	9·155	8·885	8·678	8·380	8·176	7·645	7·076
8	14·69	11·04	9·596	8·805	8·302	7·952	7·694	7·496	7·211	7·015	6·503	5·951
9	13·61	10·11	8·717	7·956	7·471	7·134	6·885	6·693	6·417	6·227	5·729	5·188
10	12·83	9·427	8·081	7·343	6·872	6·545	6·302	6·116	5·847	5·661	5·173	4·639
11	12·23	8·912	7·600	6·881	6·422	6·102	5·865	5·682	5·418	5·236	4·756	4·226
12	11·75	8·510	7·226	6·521	6·071	5·757	5·525	5·345	5·085	4·906	4·431	3·904
13	11·37	8·186	6·926	6·233	5·791	5·482	5·253	5·076	4·820	4·643	4·173	3·647
14	11·06	7·922	6·680	5·998	5·562	5·257	5·031	4·857	4·603	4·428	3·961	3·436
15	10·80	7·701	6·476	5·803	5·372	5·071	4·847	4·674	4·424	4·250	3·786	3·260
16	10·58	7·514	6·303	5·638	5·212	4·913	4·692	4·521	4·272	4·099	3·638	3·112
17	10·38	7·354	6·156	5·497	5·075	4·779	4·559	4·389	4·142	3·971	3·511	2·984
18	10·22	7·215	6·028	5·375	4·956	4·663	4·445	4·276	4·030	3·860	3·402	2·873
19	10·07	7·093	5·916	5·268	4·853	4·561	4·345	4·177	3·933	3·763	3·306	2·776
20	9·944	6·986	5·818	5·174	4·762	4·472	4·257	4·090	3·847	3·678	3·222	2·690
21	9·830	6·891	5·730	5·091	4·681	4·393	4·179	4·013	3·771	3·602	3·147	2·614
22	9·727	6·806	5·652	5·017	4·609	4·322	4·109	3·944	3·703	3·535	3·081	2·545
23	9·635	6·730	5·582	4·950	4·544	4·259	4·047	3·882	3·642	3·475	3·021	2·484
24	9·551	6·661	5·519	4·890	4·486	4·202	3·991	3·826	3·587	3·420	2·967	2·428
25	9·475	6·598	5·462	4·835	4·433	4·150	3·939	3·776	3·537	3·370	2·918	2·377
26	9·406	6·541	5·409	4·785	4·384	4·103	3·893	3·730	3·492	3·325	2·873	2·330
27	9·342	6·489	5·361	4·740	4·340	4·059	3·850	3·687	3·450	3·284	2·832	2·287
28	9·284	6·440	5·317	4·698	4·300	4·020	3·811	3·649	3·412	3·246	2·794	2·247
29	9·230	6·396	5·276	4·659	4·262	3·983	3·775	3·613	3·377	3·211	2·759	2·210
30	9·180	6·355	5·239	4·623	4·228	3·949	3·742	3·580	3·344	3·179	2·727	2·176
32	9·090	6·281	5·171	4·559	4·166	3·889	3·682	3·521	3·286	3·121	2·670	2·114
34	9·012	6·217	5·113	4·504	4·112	3·836	3·630	3·470	3·235	3·071	2·620	2·060
36	8·943	6·161	5·062	4·455	4·065	3·790	3·585	3·425	3·191	3·027	2·576	2·013
38	8·882	6·111	5·016	4·412	4·023	3·749	3·545	3·385	3·152	2·988	2·537	1·970
40	8·828	6·066	4·976	4·374	3·986	3·713	3·509	3·350	3·117	2·953	2·502	1·932
60	8·495	5·795	4·729	4·140	3·760	3·492	3·291	3·134	2·904	2·742	2·290	1·689
120	8·179	5·539	4·497	3·921	3·548	3·285	3·087	2·933	2·705	2·544	2·089	1·431
∞	7·879	5·298	4·279	3·715	3·350	3·091	2·897	2·744	2·519	2·358	1·898	1·000

Table A2.3 *Percentage points for Fisher's exact test*

	a	Probability			
		0·05	0·025	0·01	0·005
A=3 B=3	3	0 ·050	—	—	—
A=4 B=4	4	0 ·014	0 ·014	—	—
3	4	0 ·029	—	—	—
A=5 B=5	5	1 ·024	1 ·024	0 ·004	0 ·004
	4	0 ·024	0 ·024	—	—
4	5	1 ·048	0 ·008	0 ·008	—
	4	0 ·040	—	—	—
3	5	0 ·018	0 ·018	—	—
2	5	0 ·048	—	—	—
A=6 B=6	6	2 ·030	1 ·008	1 ·008	0 ·001
	5	1 ·040	0 ·008	0 ·008	—
	4	0 ·030	—	—	—
5	6	1 ·015+	0 ·015+	0 ·002	0 ·002
	5	0 ·013	0 ·013	—	—
	4	0 ·045+	—	—	—
4	6	1 ·033	0 ·005−	0 ·005−	0 ·005−

	a	Probability			
		0·05	0·025	0·01	0·005
A=8 B=8	8	4 ·038	3 ·013	2 ·003	2 ·003
	7	2 ·020	2 ·020	1 ·005+	0 ·001
	6	1 ·020	1 ·020	0 ·003	0 ·003
	5	0 ·013	0 ·013	—	—
	4	0 ·038	—	—	—
7	8	3 ·026	2 ·007	2 ·007	1 ·001
	7	2 ·035−	1 ·009	1 ·009	0 ·001
	6	1 ·032	0 ·006	0 ·006	—
	5	0 ·019	0 ·019	—	—
6	8	2 ·015−	2 ·015−	1 ·003	1 ·003
	7	1 ·016	1 ·016	0 ·002	0 ·002
	6	0 ·009	0 ·009	0 ·009	—
	5	0 ·028	—	—	—
5	8	2 ·035−	1 ·007	1 ·007	0 ·001
	7	1 ·032	0 ·005−	0 ·005−	0 ·005−
	6	0 ·016	0 ·016	—	—
	5	0 ·044	—	—	—
4	8	1 ·018	1 ·018	0 ·002	0 ·002
	7	0 ·010+	0 ·010+	—	—
	6	0 ·030	—	—	—
3	8	0 ·006	0 ·006	0 ·006	—
	7	0 ·024	0 ·024	—	—
2	8	0 ·022	0 ·022	—	—

	5	**0** ·048	—	—	—
2	6	**0** ·036	—	—	—
A=7 B=7	7	**3** ·035−	**2** ·010+	**1** ·002	**1** ·002
	6	**1** ·015−	**1** ·015−	**0** ·002	**0** ·002
	5	**0** ·010+	**0** ·010+	—	—
	4	**0** ·035−	—	—	—
6	7	**2** ·021	**2** ·021	**1** ·005−	**1** ·005−
	6	**1** ·025+	**0** ·004	**0** ·004	**0** ·004
	5	**0** ·016	**0** ·016	—	—
	4	**0** ·049	—	—	—
5	7	**2** ·045+	**1** ·010+	**0** ·001	**0** ·001
	6	**1** ·045+	**0** ·008	**0** ·008	—
	5	**0** ·027	—	—	—
4	7	**1** ·024	**1** ·024	**0** ·003	**0** ·003
	6	**0** ·015+	**0** ·015+	—	—
	5	**0** ·045+	—	—	—
3	7	**0** ·008	**0** ·008	**0** ·008	—
	6	**0** ·033	—	—	—
2	7	**0** ·028	—	—	—
A=9 B=9	9	**5** ·041	**4** ·015−	**3** ·005−	**3** ·005−
	8	**3** ·025−	**3** ·025−	**2** ·008	**1** ·002
	7	**2** ·028	**1** ·008	**1** ·008	**0** ·001
	6	**1** ·025−	**1** ·025−	**0** ·005−	**0** ·005−
	5	**0** ·015−	**0** ·015−	—	—
	4	**0** ·041	—	—	—
8	9	**4** ·029	**3** ·009	**3** ·009	**2** ·002
	8	**3** ·043	**2** ·013	**1** ·003	**1** ·003
	7	**2** ·044	**1** ·012	**0** ·002	**0** ·002
	6	**1** ·036	**0** ·007	**0** ·007	—
	5	**0** ·020	**0** ·020	—	—
7	9	**3** ·019	**3** ·019	**2** ·005−	**2** ·005−
	8	**2** ·024	**2** ·024	**1** ·006	**0** ·001
	7	**1** ·020	**1** ·020	**0** ·003	**0** ·003
	6	**0** ·010+	**0** ·010+	—	—
	5	**0** ·029	—	—	—
6	9	**3** ·044	**2** ·011	**1** ·002	**1** ·002
	8	**2** ·047	**1** ·011	**0** ·001	**0** ·001
	7	**1** ·035−	**0** ·006	**0** ·006	—
	6	**0** ·017	**0** ·017	—	—
	5	**0** ·042	—	—	—

The table shows: (1) In bold type, for given a, A and B, the value of b ($<a$) which is just significant at the probability level quoted (single-tail test).

(2) In small type, for given A, B and $r=a+b$, the exact probability (if there is independence) that b is equal to or less than the integer shown in bold type.

Table A2.3 (*continued*)

		Probability			
	a	0·05	0·025	0·01	0·005
A=9 B=5	9	2 ·027	1 ·005−	1 ·005−	1 ·005−
	8	1 ·023	1 ·023	0 ·003	0 ·003
	7	0 ·010+	0 ·010+	—	—
	6	0 ·028	—	—	—
4	9	1 ·014	1 ·014	0 ·001	0 ·001
	8	0 ·007	0 ·007	0 ·007	—
	7	0 ·021	0 ·021	—	—
	6	0 ·049	—	—	—
3	9	1 ·045+	0 ·005−	0 ·005−	0 ·005−
	8	0 ·018	0 ·018	—	—
	7	0 ·045+	—	—	—
2	9	0 ·018	0 ·018	—	—
A=10 B=10	10	6 ·043	5 ·016	4 ·005+	3 ·002
	9	4 ·029	3 ·010−	3 ·010−	2 ·003
	8	3 ·035−	2 ·012	1 ·003	1 ·003
	7	2 ·035−	1 ·010−	1 ·010−	0 ·002
	6	1 ·029	0 ·005+	0 ·005+	—
	5	0 ·016	0 ·016	—	—
	4	0 ·043	—	—	—
9	10	5 ·033	4 ·011	3 ·003	3 ·003
	9	4 ·050−	3 ·017	2 ·005−	2 ·005−

		Probability			
	a	0·05	0·025	0·01	0·005
A=10 B=4	10	1 ·011	1 ·011	0 ·001	0 ·001
	9	1 ·041	0 ·005−	0 ·005−	0 ·005−
	8	0 ·015−	0 ·015−	—	—
	7	0 ·035−	—	—	—
3	10	1 ·038	0 ·003	0 ·003	0 ·003
	9	0 ·014	0 ·014	—	—
	8	0 ·035−	—	—	—
2	10	0 ·015+	0 ·015+	—	—
	9	0 ·045+	—	—	—
A=11 B=11	11	7 ·045+	6 ·018	5 ·006	4 ·002
	10	5 ·032	4 ·012	3 ·004	3 ·004
	9	4 ·040	3 ·015−	2 ·004	2 ·004
	8	3 ·043	2 ·015−	1 ·004	1 ·004
	7	2 ·040	1 ·012	0 ·002	0 ·002
	6	1 ·032	0 ·006	0 ·006	—
	5	0 ·018	0 ·018	—	—
	4	0 ·045+	—	—	—
10	11	6 ·035+	5 ·012	4 ·004	4 ·004
	10	4 ·021	4 ·021	3 ·007	2 ·002
	9	3 ·024	3 ·024	2 ·007	1 ·002
	8	2 ·023	2 ·023	1 ·006	0 ·001
	7	1 ·017	1 ·017	0 ·003	0 ·003

	6	1 ·040	0 ·008	0 ·008	—
	5	0 ·022	0 ·022	—	—
8	10	4 ·023	4 ·023	3 ·007	2 ·002
	9	3 ·032	2 ·009	2 ·009	1 ·002
	8	2 ·031	1 ·008	1 ·008	0 ·001
	7	1 ·023	1 ·023	0 ·004	0 ·004
	6	0 ·011	0 ·011	—	—
	5	0 ·029	—	—	—
7	10	3 ·015−	3 ·015−	2 ·003	2 ·003
	9	2 ·018	2 ·018	1 ·004	1 ·004
	8	1 ·013	1 ·013	0 ·002	0 ·002
	7	1 ·036	0 ·006	0 ·006	—
	6	0 ·017	0 ·017	—	—
	5	0 ·041	—	—	—
6	10	3 ·036	2 ·008	2 ·008	1 ·001
	9	2 ·036	1 ·008	1 ·008	0 ·001
	8	1 ·024	1 ·024	0 ·003	0 ·003
	7	0 ·010+	0 ·010+	—	—
	6	0 ·026	—	—	—
5	10	2 ·022	2 ·022	1 ·004	1 ·004
	9	1 ·017	1 ·017	0 ·002	0 ·002
	8	1 ·047	0 ·007	0 ·007	—
	7	0 ·019	0 ·019	—	—
	6	0 ·042	—	—	—

	5	0 ·023	0 ·023	—	—
9	11	5 ·026	4 ·008	4 ·008	3 ·002
	10	4 ·038	3 ·012	2 ·003	2 ·003
	9	3 ·040	2 ·012	1 ·003	1 ·003
	8	2 ·035−	1 ·009	1 ·009	0 ·001
	7	1 ·025−	1 ·025−	0 ·004	0 ·004
	6	0 ·012	0 ·012	—	—
	5	0 ·030	—	—	—
8	11	4 ·018	4 ·018	3 ·005−	3 ·005−
	10	3 ·024	3 ·024	2 ·006	1 ·001
	9	2 ·022	2 ·022	1 ·005−	1 ·005−
	8	1 ·015−	1 ·015−	0 ·002	0 ·002
	7	1 ·037	0 ·007	0 ·007	—
	6	0 ·017	0 ·017	—	—
	5	0 ·040	—	—	—
7	11	4 ·043	3 ·011	2 ·002	2 ·002
	10	3 ·047	2 ·013	1 002	1 ·002
	9	2 ·039	1 ·009	1 ·009	0 ·001
	8	1 ·025−	1 ·025−	0 ·004	0 ·004
	7	0 ·010+	0 ·010+	—	—
	6	0 ·025−	0 ·025−	—	—
6	11	3 ·029	2 ·006	2 ·006	1 ·001
	10	2 ·028	1 ·005+	1 ·005+	0 ·001
	9	1 ·018	1 ·018	0 ·002	0 ·002

Table A2.3 (*continued*)

	a	Probability			
		0·05	0·025	0·01	0·005
A=11 B=6	8	1 ·043	0 ·007	0 ·007	—
	7	0 ·017	0 ·017	—	—
	6	0 ·037	—	—	—
5	11	2 ·018	2 ·018	1 ·003	1 ·003
	10	1 ·013	1 ·013	0 ·001	0 ·001
	9	1 ·036	0 ·005−	0 ·005−	0 ·005−
	8	0 ·013	0 ·013	—	—
	7	0 ·029	—	—	—
4	11	1 ·009	1 ·009	1 ·009	0 ·001
	10	1 ·033	0 ·004	0 ·004	0 ·004
	9	0 ·011	0 ·011	—	—
	8	0 ·026	—	—	—
3	11	1 ·033	0 ·003	0 ·003	0 ·003
	10	0 ·011	0 ·011	—	—
	9	0 ·027	—	—	—
2	11	0 ·013	0 ·013	—	—
	10	0 ·038	—	—	—
A=12 B=12	12	8 ·047	7 ·019	6 ·007	5 ·002
	11	6 ·034	5 ·014	4 ·005−	4 ·005−
	10	5 ·045−	4 ·018	3 ·006	2 ·002
	9	4 ·050−	3 ·020	2 ·006	1 ·001

	a	Probability			
		0·05	0·025	0·01	0·005
A=12 B=9	7	1 ·037	0 ·007	0 ·007	—
	6	0 ·017	0 ·017	—	—
	5	0 ·039	—	—	—
8	12	5 ·049	4 ·014	3 ·004	3 ·004
	11	3 ·018	3 ·018	2 ·004	2 ·004
	10	2 ·015+	2 ·015+	1 ·003	1 ·003
	9	2 ·040	1 ·010−	1 ·010−	0 ·001
	8	1 ·025−	1 ·025−	0 ·004	0 ·004
	7	0 ·010+	0 ·010+	—	—
	6	0 ·024	0 ·024	—	—
7	12	4 ·036	3 ·009	3 ·009	2 ·002
	11	3 ·038	2 ·010−	2 ·010−	1 ·002
	10	2 ·029	1 ·006	1 ·006	0 ·001
	9	1 ·017	1 ·017	0 ·002	0 ·002
	8	1 ·040	0 ·007	0 ·007	—
	7	0 ·016	0 ·016	—	—
	6	0 ·034	—	—	—
6	12	3 ·025−	3 ·025−	2 ·005−	2 ·005−
	11	2 ·022	2 ·022	1 ·004	1 ·004
	10	1 ·013	1 ·013	0 ·002	0 ·002
	9	1 ·032	0 ·005−	0 ·005−	0 ·005−
	8	0 ·011	0 ·011	—	—
	7	0 ·025−	0 ·025−	—	—
	6	0 ·050−	—	—	—

	6	**1** ·034	**0** ·007	**0** ·007	—
	5	**0** ·019	**0** ·019	—	—
	4	**0** ·047	—	—	—
11	12	**7** ·037	**6** ·014	**5** ·005−	**5** ·005−
	11	**5** ·024	**5** ·024	**4** ·008	**3** ·002
	10	**4** ·029	**3** ·010+	**2** ·003	**2** ·003
	9	**3** ·030	**2** ·009	**2** ·009	**1** ·002
	8	**2** ·026	**1** ·007	**1** ·007	**0** ·001
	7	**1** ·019	**1** ·019	**0** ·003	**0** ·003
	6	**1** ·045−	**0** ·009	**0** ·009	—
	5	**0** ·024	**0** ·024	—	—
10	12	**6** ·029	**5** ·010−	**5** ·010−	**4** ·003
	11	**5** ·043	**4** ·015+	**3** ·005−	**3** ·005−
	10	**4** ·048	**3** ·017	**2** ·005−	**2** ·005−
	9	**3** ·046	**2** ·015−	**1** ·004	**1** ·004
	8	**2** ·038	**1** ·010+	**0** ·002	**0** ·002
	7	**1** ·026	**0** ·005−	**0** ·005−	**0** ·005−
	6	**0** ·012	**0** ·012	—	—
	5	**0** ·030	—	—	—
9	12	**5** ·021	**5** ·021	**4** ·006	**3** ·002
	11	**4** ·029	**3** ·009	**3** ·009	**2** ·002
	10	**3** ·029	**2** ·008	**2** ·008	**1** ·002
	9	**2** ·024	**2** ·024	**1** ·006	**0** ·001
	8	**1** ·016	**1** ·016	**0** ·002	**0** ·002

	11	**1** ·010−	**1** ·010−	**1** ·010−	**0** ·001
	10	**1** ·028	**0** ·003	**0** ·003	**0** ·003
	9	**0** ·009	**0** ·009	**0** ·009	—
	8	**0** ·020	**0** ·020	—	—
	7	**0** ·041	—	—	—
4	12	**2** ·050	**1** ·007	**1** ·007	**0** ·001
	11	**1** ·027	**0** ·003	**0** ·003	**0** ·003
	10	**0** ·008	**0** ·008	**0** ·008	—
	9	**0** ·019	**0** ·019	—	—
	8	**0** ·038	—	—	—
3	12	**1** ·029	**0** ·002	**0** ·002	**0** ·002
	11	**0** ·009	**0** ·009	**0** ·009	—
	10	**0** ·022	**0** ·022	—	—
	9	**0** ·044	—	—	—
2	12	**0** ·011	**0** ·011	—	—
	11	**0** ·033	—	—	—
A=13 B=13	13	**9** ·048	**8** ·020	**7** ·007	**6** ·003
	12	**7** ·037	**6** ·015+	**5** ·006	**4** ·002
	11	**6** ·048	**5** ·021	**4** ·008	**3** ·002
	10	**4** ·024	**4** ·024	**3** ·008	**2** ·002
	9	**3** ·024	**3** ·024	**2** ·008	**1** ·002
	8	**2** ·021	**2** ·021	**1** ·006	**0** ·001

The table shows: (1) In bold type, for given a, A and B, the value of b ($<a$) which is just significant at the probability level quoted (single-tail test).

(2) In small type, for given A, B and $r=a+b$, the exact probability (if there is independence) that b is equal to or less than the integer shown in bold type.

Table A2.3 (*continued*)

	a	Probability 0·05	0·025	0·01	0·005
A=13 B=13	7	2 ·048	1 ·015+	0 ·003	0 ·003
	6	1 ·037	0 ·007	0 ·007	—
	5	0 ·020	0 ·020	—	—
	4	0 ·048	—	—	—
12	13	8 ·039	7 ·015−	6 ·005+	5 ·002
	12	6 ·027	5 ·010−	5 ·010−	4 ·003
	11	5 ·033	4 ·013	3 ·004	3 ·004
	10	4 ·036	3 ·013	2 ·004	2 ·004
	9	3 ·034	2 ·011	1 ·003	1 ·003
	8	2 ·029	1 ·008	1 ·008	0 ·001
	7	1 ·020	1 ·020	0 ·004	0 ·004
	6	1 ·046	0 ·010−	0 ·010−	—
	5	0 ·024	0 ·024	—	—
11	13	7 ·031	6 ·011	5 ·003	5 ·003
	12	6 ·048	5 ·018	4 ·006	3 ·002
	11	4 ·021	4 ·021	3 ·007	2 ·002
	10	3 ·021	3 ·021	2 ·006	1 ·001
	9	3 ·050−	2 ·017	1 ·004	1 ·004
	8	2 ·040	1 ·011	0 ·002	0 ·002
	7	1 ·027	0 ·005−	0 ·005−	0 ·005−
	6	0 ·013	0 ·013	—	—
	5	0 ·030	—	—	—
10	13	6 ·024	6 ·024	5 ·007	4 ·002

	a	Probability 0·05	0·025	0·01	0·005
A=13 B=7	11	2 ·022	2 ·022	1 ·004	1 ·004
	10	1 ·012	1 ·012	0 ·002	0 ·002
	9	1 ·029	0 ·004	0 ·004	0 ·004
	8	0 ·010+	0 ·010+	—	—
	7	0 ·022	0 ·022	—	—
	6	0 ·044	—	—	—
6	13	3 ·021	3 ·021	2 ·004	2 ·004
	12	2 ·017	2 ·017	1 ·003	1 ·003
	11	2 ·046	1 ·010−	1 ·010−	0 ·001
	10	1 ·024	1 ·024	0 ·003	0 ·003
	9	1 ·050−	0 ·008	0 ·008	—
	8	0 ·017	0 ·017	—	—
	7	0 ·034	—	—	—
5	13	2 ·012	2 ·012	1 ·002	1 ·002
	12	2 ·044	1 ·008	1 ·008	0 ·001
	11	1 ·022	1 ·022	0 ·002	0 ·002
	10	1 ·047	0 ·007	0 ·007	—
	9	0 ·015−	0 ·015−	—	—
	8	0 ·029	—	—	—
4	13	2 ·044	1 ·006	1 ·006	0 ·000
	12	1 ·022	1 ·022	0 ·002	0 ·002
	11	0 ·006	0 ·006	0 ·006	—
	10	0 ·015−	0 ·015−	—	—

		10	3 ·033	2 ·010+	1 ·002	1 ·002
		9	2 ·026	1 ·006	1 ·006	0 ·001
		8	1 ·017	1 ·017	0 ·003	0 ·003
		7	1 ·038	0 ·007	0 ·007	—
		6	0 ·017	0 ·017	—	—
		5	0 ·038	—	—	—
	9	13	5 ·017	5 ·017	4 ·005−	4 ·005−
		12	4 ·023	4 ·023	3 ·007	2 ·001
		11	3 ·022	3 ·022	2 ·006	1 ·001
		10	2 ·017	2 ·017	1 ·004	1 ·004
		9	2 ·040	1 ·010+	0 ·001	0 ·001
		8	1 ·025−	1 ·025−	0 ·004	0 ·004
		7	0 ·010+	0 ·010+	—	—
		6	0 ·023	0 ·023	—	—
		5	0 ·049	—	—	—
	8	13	5 ·042	4 ·012	3 ·003	3 ·003
		12	4 ·047	3 ·014	2 ·003	2 ·003
		11	3 ·041	2 ·011	1 ·002	1 ·002
		10	2 ·029	1 ·007	1 ·007	0 ·001
		9	1 ·017	1 ·017	0 ·002	0 ·002
		8	1 ·037	0 ·006	0 ·006	—
		7	0 ·015−	0 ·015−	—	—
		6	0 ·032	—	—	—
	7	13	4 ·031	3 ·007	3 ·007	2 ·001
		12	3 ·031	2 ·007	2 ·007	1 ·001
	3	13	1 ·025	1 ·025	0 ·002	0 ·002
		12	0 ·007	0 ·007	0 ·007	—
		11	0 ·018	0 ·018	—	—
		10	0 ·036	—	—	—
	2	13	0 ·010−	0 ·010−	0 ·010−	—
		12	0 ·029	—	—	—
A=14	B=14	14	10 ·049	9 ·020	8 ·008	7 ·003
		13	8 ·038	7 ·016	6 ·006	5 ·002
		12	6 ·023	6 ·023	5 ·009	4 ·003
		11	5 ·027	4 ·011	3 ·004	3 ·004
		10	4 ·028	3 ·011	2 ·003	2 ·003
		9	3 ·027	2 ·009	2 ·009	1 ·002
		8	2 ·023	2 ·023	1 ·006	0 ·001
		7	1 ·016	1 ·016	0 ·003	0 ·003
		6	1 ·038	0 ·008	0 ·008	—
		5	0 ·020	0 ·020	—	—
		4	0 ·049	—	—	—
	13	14	9 ·041	8 ·016	7 ·006	6 ·002
		13	7 ·029	6 ·011	5 ·004	5 ·004
		12	6 ·037	5 ·015+	4 ·005+	3 ·002
		11	5 ·041	4 ·017	3 ·006	2 ·001
		10	4 ·041	3 ·016	2 ·005−	2 ·005−
		9	3 ·038	2 ·013	1 ·003	1 ·003
		8	2 ·031	1 ·009	1 ·009	0 ·001

Table A2.3 (*continued*)

	a	Probability 0·05	0·025	0·01	0·005
A=14 B=13	7	1 ·021	1 ·021	0 ·004	0 ·004
	6	1 ·048	0 ·010+	—	—
	5	0 ·025−	0 ·025−	—	—
12	14	8 ·033	7 ·012	6 ·004	6 ·004
	13	6 ·021	6 ·021	5 ·007	4 ·002
	12	5 ·025+	4 ·009	4 ·009	3 ·003
	11	4 ·026	3 ·009	3 ·009	2 ·002
	10	3 ·024	3 ·024	2 ·007	1 ·002
	9	2 ·019	2 ·019	1 ·005−	1 ·005−
	8	2 ·042	1 ·012	0 ·002	0 ·002
	7	1 ·028	0 ·005+	0 ·005+	—
	6	0 ·013	0 ·013	—	—
	5	0 ·030	—	—	—
11	14	7 ·026	6 ·009	6 ·009	5 ·003
	13	6 ·039	5 ·014	4 ·004	4 ·004
	12	5 ·043	4 ·016	3 ·005−	3 ·005−
	11	4 ·042	3 ·015−	2 ·004	2 ·004
	10	3 ·036	2 ·011	1 ·003	1 ·003
	9	2 ·027	1 ·007	1 ·007	0 ·001
	8	1 ·017	1 ·017	0 ·003	0 ·003
	7	1 ·038	0 ·007	0 ·007	—
	6	0 ·017	0 ·017	—	—
	5	0 ·038	—	—	—

	a	Probability 0·05	0·025	0·01	0·005
A=14 B=7	14	4 ·026	3 ·006	3 ·006	2 ·001
	13	3 ·025	2 ·006	2 ·006	1 ·001
	12	2 ·017	2 ·017	1 ·003	1 ·003
	11	2 ·041	1 ·009	1 ·009	0 ·001
	10	1 ·021	1 ·021	0 ·003	0 ·003
	9	1 ·043	0 ·007	0 ·007	—
	8	0 ·015−	0 ·015−	—	—
	7	0 ·030	—	—	—
6	14	3 ·018	3 ·018	2 ·003	2 ·003
	13	2 ·014	2 ·014	1 ·002	1 ·002
	12	2 ·037	1 ·007	1 ·007	0 ·001
	11	1 ·018	1 ·018	0 ·002	0 ·002
	10	1 ·038	0 ·005+	0 ·005+	—
	9	0 ·012	0 ·012	—	—
	8	0 ·024	0 ·024	—	—
	7	0 ·044	—	—	—
5	14	2 ·010+	2 ·010+	1 ·001	1 ·001
	13	2 ·037	1 ·006	1 ·006	0 ·001
	12	1 ·017	1 ·017	0 ·002	0 ·002
	11	1 ·038	0 ·005−	0 ·005−	0 ·005−
	10	0 ·011	0 ·011	—	—
	9	0 ·022	0 ·022	—	—
	8	0 ·040	—	—	—

	12	4 ·028	3 ·009	3 ·009	2 ·002
	11	3 ·024	3 ·024	2 ·007	1 ·001
	10	2 ·018	2 ·018	1 ·004	1 ·004
	9	2 ·040	1 ·011	0 ·002	0 ·002
	8	1 ·024	1 ·024	0 ·004	0 ·004
	7	0 ·010−	0 ·010−	0 ·010−	—
	6	0 ·022	0 ·022	—	—
	5	0 ·047	—	—	—
9	14	6 ·047	5 ·014	4 ·004	4 ·004
	13	4 ·018	4 ·018	3 ·005−	3 ·005−
	12	3 ·017	3 ·017	2 ·004	2 ·004
	11	3 ·042	2 ·012	1 ·002	1 ·002
	10	2 ·029	1 ·007	1 ·007	0 ·001
	9	1 ·017	1 ·017	0 ·002	0 ·002
	8	1 ·036	0 ·006	0 ·006	—
	7	0 ·014	0 ·014	—	—
	6	0 ·030	—	—	—
8	14	5 ·036	4 ·010−	4 ·010−	3 ·002
	13	4 ·039	3 ·011	2 ·002	2 ·002
	12	3 ·032	2 ·008	2 ·008	1 ·001
	11	2 ·022	2 ·022	1 ·005−	1 ·005−
	10	2 ·048	1 ·012	0 ·002	0 ·002
	9	1 ·026	0 ·004	0 ·004	0 ·004
	8	0 ·009	0 ·009	0 ·009	—
	7	0 ·020	0 ·020	—	—
	6	0 ·040	—	—	—

	13	1 ·019	1 ·019	0 ·002	0 ·002
	12	1 ·044	0 ·005−	0 ·005−	0 ·005−
	11	0 ·011	0 ·011	—	—
	10	0 ·023	0 ·023	—	—
	9	0 ·041	—	—	—
3	14	1 ·022	1 ·022	0 ·001	0 ·001
	13	0 ·006	0 ·006	0 ·006	—
	12	0 ·015−	0 ·015−	—	—
	11	0 ·029	—	—	—
2	14	0 ·008	0 ·008	0 ·008	—
	13	0 ·025	0 ·025	—	—
	12	0 ·050	—	—	—
A=15 B=15	15	11 ·050−	10 ·021	9 ·008	8 ·003
	14	9 ·040	8 ·018	7 ·007	6 ·003
	13	7 ·025+	6 ·010+	5 ·004	5 ·004
	12	6 ·030	5 ·013	4 ·005−	4 ·005−
	11	5 ·033	4 ·013	3 ·005−	3 ·005−
	10	4 ·033	3 ·013	2 ·004	2 ·004
	9	3 ·030	2 ·010+	1 ·003	1 ·003
	8	2 ·025+	1 ·007	1 ·007	0 ·001
	7	1 ·018	1 ·018	0 ·003	0 ·003
	6	1 ·040	0 ·008	0 ·008	—
	5	0 ·021	0 ·021	—	—
	4	0 ·050−	—	—	—

The table shows: (1) In bold type, for given a, A and B, the value of b ($<a$) which is just significant at the probability level quoted (single-tail test).

(2) In small type, for given A, B and $r=a+b$, the exact probability (if there is independence) that b is equal to or less than the integer shown in bold type.

Table A2.3 (*continued*)

	a	Probability			
		0·05	0·025	0·01	0·005
A=15 B=14	15	10 ·042	9 ·017	8 ·006	7 ·002
	14	8 ·031	7 ·013	6 ·005−	6 ·005−
	13	7 ·041	6 ·017	5 ·007	4 ·002
	12	6 ·046	5 ·020	4 ·007	3 ·002
	11	5 ·048	4 ·020	3 ·007	2 ·002
	10	4 ·046	3 ·018	2 ·006	1 ·001
	9	3 ·041	2 ·014	1 ·004	1 ·004
	8	2 ·033	1 ·009	1 ·009	0 ·001
	7	1 ·022	1 ·022	0 ·004	0 ·004
	6	1 ·049	0 ·011	—	—
	5	0 ·025+	—	—	—
13	15	9 ·035−	8 ·013	7 ·005−	7 ·005−
	14	7 ·023	7 ·023	6 ·009	5 ·003
	13	6 ·029	5 ·011	4 ·004	4 ·004
	12	5 ·031	4 ·012	3 ·004	3 ·004
	11	4 ·030	3 ·011	2 ·003	2 ·003
	10	3 ·026	2 ·008	2 ·008	1 ·002
	9	2 ·020	2 ·020	1 ·005+	0 ·001
	8	2 ·043	1 ·013	0 ·002	0 ·002
	7	1 ·029	0 ·005+	0 ·005−	—
	6	0 ·013	0 ·013	—	—
	5	0 ·031	—	—	—
12	15	8 ·028	7 ·010−	7 ·010−	6 ·003
	14	7 ·043	6 ·016	5 ·006	4 ·002

	a	Probability			
		0·05	0·025	0·01	0·005
A=15 B=9	13	4 ·042	3 ·013	2 ·003	2 ·003
	12	3 ·032	2 ·009	2 ·009	1 ·002
	11	2 ·021	2 ·021	1 ·005−	1 ·005−
	10	2 ·045−	1 ·011	0 ·002	0 ·002
	9	1 ·024	1 ·024	0 ·004	0 ·004
	8	1 ·048	0 ·009	0 ·009	—
	7	0 ·019	0 ·019	—	—
	6	0 ·037	—	—	—
8	15	5 ·032	4 ·008	4 ·008	3 ·002
	14	4 ·033	3 ·009	3 ·009	2 ·002
	13	3 ·026	2 ·006	2 ·006	1 ·001
	12	2 ·017	2 ·017	1 ·003	1 ·003
	11	2 ·037	1 ·008	1 ·008	0 ·001
	10	1 ·019	1 ·019	0 ·003	0 ·003
	9	1 ·038	0 ·006	0 ·006	—
	8	0 ·013	0 ·013	—	—
	7	0 ·026	—	—	—
	6	0 ·050−	—	—	—
7	15	4 ·023	4 ·023	3 ·005−	3 ·005−
	14	3 ·021	3 ·021	2 ·004	2 ·004
	13	2 ·014	2 ·014	1 ·002	1 ·002
	12	2 ·032	1 ·007	1 ·007	0 ·001
	11	1 ·015+	1 ·015+	0 ·002	0 ·002
	10	1 ·032	0 ·005−	0 ·005−	0 ·005−
	9	0 ·010+	0 ·010+	—	—

	12	[illegible]	[illegible]	[illegible]	[illegible]
	11	4 ·045+	3 ·017	2 ·005−	2 ·005−
	10	3 ·038	2 ·012	1 ·003	1 ·003
	9	2 ·028	1 ·007	1 ·007	0 ·001
	8	1 ·018	1 ·018	0 ·003	0 ·003
	7	1 ·038	0 ·007	0 ·007	—
	6	0 ·017	0 ·017	—	—
	5	0 ·037	—	—	—
11	15	7 ·022	7 ·022	6 ·007	5 ·002
	14	6 ·032	5 ·011	4 ·003	4 ·003
	13	5 ·034	4 ·012	3 ·003	3 ·003
	12	4 ·032	3 ·010+	2 ·003	2 ·003
	11	3 ·026	2 ·008	2 ·008	1 ·002
	10	2 ·019	2 ·019	1 ·004	1 ·004
	9	2 ·040	1 ·011	0 ·002	0 ·002
	8	1 ·024	1 ·024	0 ·004	0 ·004
	7	1 ·049	0 ·010−	0 ·010−	—
	6	0 ·022	0 ·022	—	—
	5	0 ·046	—	—	—
10	15	6 ·017	6 ·017	5 ·005−	5 ·005−
	14	5 ·023	5 ·023	4 ·007	3 ·002
	13	4 ·022	4 ·022	3 ·007	2 ·001
	12	3 ·018	3 ·018	2 ·005−	2 ·005−
	11	3 ·042	2 ·013	1 ·003	1 ·003
	10	2 ·029	1 ·007	1 ·007	0 ·001
	9	1 ·016	1 ·016	0 ·002	0 ·002
	8	1 ·034	0 ·006	0 ·006	—
	7	0 ·013	0 ·013	—	—
	6	0 ·028	—	—	—
9	15	6 ·042	5 ·012	4 ·003	4 ·003
	14	5 ·047	4 ·015−	3 ·004	3 ·004

	7	0 ·038	—	—	—
6	15	3 ·015+	3 ·015+	2 ·003	2 ·003
	14	2 ·011	2 ·011	1 ·002	1 ·002
	13	2 ·031	1 ·006	1 ·006	0 ·001
	12	1 ·014	1 ·014	0 ·002	0 ·002
	11	1 ·029	0 ·004	0 ·004	0 ·004
	10	0 ·009	0 ·009	0 ·009	—
	9	0 ·017	0 ·017	—	—
	8	0 ·032	—	—	—
5	15	2 ·009	2 ·009	2 ·009	1 ·001
	14	2 ·032	1 ·005−	1 ·005−	1 ·005−
	13	1 ·014	1 ·014	0 ·001	0 ·001
	12	1 ·031	0 ·004	0 ·004	0 ·004
	11	0 ·008	0 ·008	0 ·008	—
	10	0 ·016	0 ·016	—	—
	9	0 ·030	—	—	—
4	15	2 ·035+	1 ·004	1 ·004	1 ·004
	14	1 ·016	1 ·016	0 ·001	0 ·001
	13	1 ·037	0 ·004	0 ·004	0 ·004
	12	0 ·009	0 ·009	0 ·009	—
	11	0 ·018	0 ·018	—	—
	10	0 ·033	—	—	—
3	15	1 ·020	1 ·020	0 ·001	0 ·001
	14	0 ·005−	0 ·005−	0 ·005−	0 ·005−
	13	0 ·012	0 ·012	—	—
	12	0 ·025−	0 ·025−	—	—
	11	0 ·043	—	—	—
2	15	0 ·007	0 ·007	0 ·007	—
	14	0 ·022	0 ·022	—	—
	13	0 ·044	—	—	—

Reproduction of Table 38 from *Biometrika tables for statisticians*, volume 1, third edition (1966), edited by E.S. Pearson and H.O. Hartley, with permission of the Biometrika Trustees.

Table A2.4 *Percentage points for the signed rank statistic T*

The percentiles listed cover the range α = .005 to .125 for every sample size up to $n = 20$. Values $T_{(+)}$ are such that the probability is α that the signed rank statistic is less than or equal to $T_{(+)}$. The values $T_{(-)}$ are such that the probability is α that T is greater than or equal to $T_{(-)}$.

$T_{(+)}$	$T_{(-)}$	α
	$n = 1$	
0	1	.500
	$n = 2$	
0	3	.250
	$n = 3$	
0	6	.125
	$n = 4$	
0	10	.062
1	9	.125
	$n = 5$	
0	15	.031
1	14	.062
2	13	.094
3	12	.156
	$n = 6$	
0	21	.016
1	20	.031
2	19	.047
3	18	.078
4	17	.109
5	16	.156
	$n = 7$	
0	28	.008
1	27	.016
2	26	.023
3	25	.039
4	24	.055
5	23	.078
6	22	.109
7	21	.148
	$n = 8$	
0	36	.004
1	35	.008
2	34	.012
3	33	.020
4	32	.027
5	31	.039
6	30	.055
7	29	.074
8	28	.098
9	27	.125
	$n = 9$	
1	44	.004
2	43	.006
3	42	.010
	$n = 9$ (*Cont.*)	
4	41	.014
5	40	.020
6	39	.027
7	38	.037
8	37	.049
9	36	.064
10	35	.082
11	34	.102
12	33	.125
	$n = 10$	
3	52	.005
4	51	.007
5	50	.010
6	49	.014
7	48	.019
8	47	.024
9	46	.032
10	45	.042
11	44	.053
12	43	.065
13	42	.080
14	41	.097
15	40	.116
16	39	.138
	$n = 11$	
5	61	.005
6	60	.007
7	59	.009
8	58	.012
9	57	.016
10	56	.021
11	55	.027
12	54	.034
13	53	.042
14	52	.051
15	51	.062
16	50	.074
17	49	.087
18	48	.103
19	47	.120
20	46	.139
	$n = 12$	
7	71	.005
8	70	.006
	$n = 12$ (*Cont.*)	
9	69	.008
10	68	.010
11	67	.013
12	66	.017
13	65	.021
14	64	.026
15	63	.032
16	62	.039
17	61	.046
18	60	.055
19	59	.065
20	58	.076
21	57	.088
22	56	.102
23	55	.117
24	54	.133
	$n = 13$	
9	82	.004
10	81	.005
11	80	.007
12	79	.009
13	78	.011
14	77	.013
15	76	.016
16	75	.020
17	74	.024
18	73	.029
19	72	.034
20	71	.040
21	70	.047
22	69	.055
23	68	.064
24	67	.073
25	66	.084
26	65	.095
27	64	.108
28	63	.122
29	62	.137
	$n = 14$	
12	93	.004
13	92	.005
14	91	.007
15	90	.008
16	89	.010
	$n = 14$ (*Cont.*)	
17	88	.012
18	87	.015
19	86	.018
20	85	.021
21	84	.025
22	83	.029
23	82	.034
24	81	.039
25	80	.045
26	79	.052
27	78	.059
28	77	.068
29	76	.077
30	75	.086
31	74	.097
32	73	.108
33	72	.121
34	71	.134
	$n = 15$	
15	105	.004
16	104	.005
17	103	.006
18	102	.008
19	101	.009
20	100	.011
21	99	.013
22	98	.015
23	97	.018
24	96	.021
25	95	.024
26	94	.028
27	93	.032
28	92	.036
29	91	.042
30	90	.047
31	89	.053
32	88	.060
33	87	.068
34	86	.076
35	85	.084
36	84	.094
37	83	.104
38	82	.115
39	81	.126

Table A2.4 (*continued*)

$T_{(+)}$	$T_{(-)}$	α
$n = 16$		
19	117	.005
20	116	.005
21	115	.007
22	114	.008
23	113	.009
24	112	.011
25	111	.012
26	110	.014
27	109	.017
28	108	.019
29	107	.022
30	106	.025
31	105	.029
32	104	.033
33	103	.037
34	102	.042
35	101	.047
36	100	.052
37	99	.058
38	98	.065
39	97	.072
40	96	.080
41	95	.088
42	94	.096
43	93	.106
44	92	.116
45	91	.126
46	90	.137
$n = 17$		
23	130	.005
24	129	.005
25	128	.006
26	127	.007
27	126	.009
28	125	.010
29	124	.012
30	123	.013
31	122	.015
32	121	.017
33	120	.020
34	119	.022
35	118	.025
36	117	.028
37	116	.032
38	115	.036
39	114	.040
40	113	.044
41	112	.049
42	111	.054
43	110	.060
44	109	.066
45	108	.073
46	107	.080
47	106	.087
48	105	.095
49	104	.103
50	103	.112
51	102	.122
52	101	.132
$n = 18$		
27	144	.004
28	143	.005
29	142	.006
30	141	.007
31	140	.008
32	139	.009
33	138	.010
34	137	.012
35	136	.013
36	135	.015
37	134	.017
38	133	.019
39	132	.022
40	131	.024
41	130	.027
42	129	.030
43	128	.033
44	127	.037
45	126	.041
46	125	.045
47	124	.049
48	123	.054
49	122	.059
50	121	.065
51	120	.071
52	119	.077
53	118	.084
54	117	.091
55	116	.098
56	115	.106
57	114	.114
58	113	.123
59	112	.132
$n = 19$		
32	158	.005
33	157	.005
34	156	.006
35	155	.007
36	154	.008
37	153	.009
38	152	.010
39	151	.011
40	150	.013
41	149	.014
42	148	.016
43	147	.018
44	146	.020
45	145	.022
46	144	.025
47	143	.027
48	142	.030
49	141	.033
50	140	.036
51	139	.040
52	138	.044
53	137	.048
54	136	.052
55	135	.057
56	134	.062
57	133	.067
58	132	.072
59	131	.078
60	130	.084
61	129	.091
62	128	.098
63	127	.105
64	126	.113
65	125	.121
66	124	.129
$n = 20$		
37	173	.005
38	172	.005
39	171	.006
40	170	.007
41	169	.008
42	168	.009
43	167	.010
44	166	.011
45	165	.012
46	164	.013
47	163	.015
48	162	.016
49	161	.018
50	160	.020
51	159	.022
52	158	.024
53	157	.027
54	156	.029
55	155	.032
56	154	.035
57	153	.038
58	152	.041
59	151	.045
60	150	.049
61	149	.053
62	148	.057
63	147	.062
64	146	.066
65	145	.071
66	144	.077
67	143	.082
68	142	.088
69	141	.095
70	140	.101
71	139	.108
72	138	.115
73	137	.123
74	136	.131

Reproduction of Table A-21 from *Introduction to statistical analysis*, third edition, by Dixon and Massey, with permission of McGraw-Hill.

Table A2.5 *Percentage points for the rank sum statistic T*

The values of T_1, T_2, and α are such that if the n_1 and n_2 observations are chosen at random from the same population the chance that the rank sum T of the n_1 observations in the smaller sample is equal to or less than T_1 is α and the chance that T is equal to or greater than T_2 is α. The sample sizes are shown in parentheses (n_1, n_2).

T_1	T_2	α
	(1,1)	
1	2	.500
	(1,2)	
1	3	.333
2	2	.667
	(1,3)	
1	4	.250
2	3	.500
	(1,4)	
1	5	.200
2	4	.400
3	3	.600
	(1,5)	
1	6	.167
2	5	.333
3	4	.500
	(1,6)	
1	7	.143
2	6	.286
3	5	.428
4	4	.571
	(1,7)	
1	8	.125
2	7	.250
3	6	.375
4	5	.500
	(1,8)	
1	9	.111
2	8	.222
3	7	.333
4	6	.444
5	5	.556
	(1,9)	
1	10	.100
2	9	.200
3	8	.300
4	7	.400
5	6	.500
	(1,10)	
1	11	.091
2	10	.182
3	9	.273
4	8	.364
5	7	.455
6	6	.545

T_1	T_2	α
	(2,2)	
3	7	.167
4	6	.333
5	5	.667
	(2,3)	
3	9	.100
4	8	.200
5	7	.400
6	6	.600
	(2,4)	
3	11	.067
4	10	.133
5	9	.267
6	8	.400
7	7	.600
	(2,5)	
3	13	.047
4	12	.095
5	11	.190
6	10	.286
7	9	.429
8	8	.571
	(2,6)	
3	15	.036
4	14	.071
5	13	.143
6	12	.214
7	11	.321
8	10	.429
9	9	.571
	(2,7)	
3	17	.028
4	16	.056
5	15	.111
6	14	.167
7	13	.250
8	12	.333
9	11	.444
10	10	.556
	(2,8)	
3	19	.022
4	18	.044
5	17	.089
6	16	.133
7	15	.200

T_1	T_2	α
	(2,8) (*Cont.*)	
8	14	.267
9	13	.356
10	12	.444
11	11	.556
	(2,9)	
3	21	.018
4	20	.036
5	19	.073
6	18	.109
7	17	.164
8	16	.218
9	15	.291
10	14	.364
11	13	.455
12	12	.545
	(2,10)	
3	23	.015
4	22	.030
5	21	.061
6	20	.091
7	19	.136
8	18	.182
9	17	.242
10	16	.303
11	15	.379
12	14	.455
13	13	.545
	(3,3)	
6	15	.050
7	14	.100
8	13	.200
9	12	.350
10	11	.500
	(3,4)	
6	18	.028
7	17	.057
8	16	.114
9	15	.200
10	14	.314
11	13	.429
12	12	.571
	(3,5)	
6	21	.018
7	20	.036

T_1	T_2	α
	(3,5) (*Cont.*)	
8	19	.071
9	18	.125
10	17	.196
11	16	.286
12	15	.393
13	14	.500
	(3,6)	
6	24	.012
7	23	.024
8	22	.048
9	21	.083
10	20	.131
11	19	.190
12	18	.274
13	17	.357
14	16	.452
15	15	.548
	(3,7)	
6	27	.008
7	26	.017
8	25	.033
9	24	.058
10	23	.092
11	22	.133
12	21	.192
13	20	.258
14	19	.333
15	18	.417
16	17	.500
	(3,8)	
6	30	.006
7	29	.012
8	28	.024
9	27	.042
10	26	.067
11	25	.097
12	24	.139
13	23	.188
14	22	.248
15	21	.315
16	20	.387
17	19	.461
18	18	.539

Table A2.5 (*continued*)

T_1	T_2	α
	(3,9)	
6	33	.005
7	32	.009
8	31	.018
9	30	.032
10	29	.050
11	28	.073
12	27	.105
13	26	.141
14	25	.186
15	24	.241
16	23	.300
17	22	.363
18	21	.432
19	20	.500
	(3,10)	
6	36	.003
7	35	.007
8	34	.014
9	33	.024
10	32	.038
11	31	.056
12	30	.080
13	29	.108
14	28	.143
15	27	.185
16	26	.234
17	25	.287
18	24	.346
19	23	.406
20	22	.469
21	21	.531
	(4,4)	
10	26	.014
11	25	.029
12	24	.057
13	23	.100
14	22	.171
15	21	.243
16	20	.343
17	19	.443
18	18	.557
	(4,5)	
10	30	.008
11	29	.016
12	28	.032
13	27	.056
14	26	.095
15	25	.143
16	24	.206
	(4,5) (*Cont.*)	
17	23	.278
18	22	.365
19	21	.452
20	20	.548
	(4,6)	
10	34	.005
11	33	.010
12	32	.019
13	31	.033
14	30	.057
15	29	.086
16	28	.129
17	27	.176
18	26	.238
19	25	.305
20	24	.381
21	23	.457
22	22	.545
	(4,7)	
10	38	.003
11	37	.006
12	36	.012
13	35	.021
14	34	.036
15	33	.055
16	32	.082
17	31	.115
18	30	.158
19	29	.206
20	28	.264
21	27	.324
22	26	.394
23	25	.464
24	24	.538
	(4,8)	
10	42	.002
11	41	.004
12	40	.008
13	39	.014
14	38	.024
15	37	.036
16	36	.055
17	35	.077
18	34	.107
19	33	.141
20	32	.184
21	31	.230
22	30	.285
23	29	.341
	(4,8) (*Cont.*)	
24	28	.404
25	27	.467
26	26	.533
	(4,9)	
10	46	.001
11	45	.003
12	44	.006
13	43	.010
14	42	.017
15	41	.025
16	40	.038
17	39	.053
18	38	.074
19	37	.099
20	36	.130
21	35	.165
22	34	.207
23	33	.252
24	32	.302
25	31	.355
26	30	.413
27	29	.470
28	28	.530
	(4,10)	
10	50	.001
11	49	.002
12	48	.004
13	47	.007
14	46	.012
15	45	.018
16	44	.026
17	43	.038
18	42	.053
19	41	.071
20	40	.094
21	39	.120
22	38	.152
23	37	.187
24	36	.227
25	35	.270
26	34	.318
27	33	.367
28	32	.420
29	31	.473
30	30	.527
	(5,5)	
15	40	.004
16	39	.008
17	38	.016
	(5,5) (*Cont.*)	
18	37	.028
19	36	.048
20	35	.075
21	34	.111
22	33	.155
23	32	.210
24	31	.274
25	30	.345
26	29	.421
27	28	.500
	(5,6)	
15	45	.002
16	44	.004
17	43	.009
18	42	.015
19	41	.026
20	40	.041
21	39	.063
22	38	.089
23	37	.123
24	36	.165
25	35	.214
26	34	.268
27	33	.331
28	32	.396
29	31	.465
30	30	.535
	(5,7)	
15	50	.001
16	49	.003
17	48	.005
18	47	.009
19	46	.015
20	45	.024
21	44	.037
22	43	.053
23	42	.074
24	41	.101
25	40	.134
26	39	.172
27	38	.216
28	37	.265
29	36	.319
30	35	.378
31	34	.438
32	33	.500
	(5,8)	
15	55	.001
16	54	.002

Table A2.5 (*continued*)

T_1	T_2	α
(5,8) (*Cont.*)		
17	53	.003
18	52	.005
19	51	.009
20	50	.015
21	49	.023
22	48	.033
23	47	.047
24	46	.064
25	45	.085
26	44	.111
27	43	.142
28	42	.177
29	41	.217
30	40	.262
31	39	.311
32	38	.362
33	37	.416
34	36	.472
35	35	.528
(5,9)		
15	60	.000
16	59	.001
17	58	.002
18	57	.003
19	56	.006
20	55	.009
21	54	.014
22	53	.021
23	52	.030
24	51	.041
25	50	.056
26	49	.073
27	48	.095
28	47	.120
29	46	.149
30	45	.182
31	44	.219
32	43	.259
33	42	.303
34	41	.350
35	40	.399
36	39	.449
37	38	.500
(5,10)		
15	65	.000
16	64	.001
17	63	.001
18	62	.002
19	61	.004

T_1	T_2	α
(5,10) (*Cont.*)		
20	60	.006
21	59	.010
22	58	.014
23	57	.020
24	56	.028
25	55	.038
26	54	.050
27	53	.065
28	52	.082
29	51	.103
30	50	.127
31	49	.155
32	48	.185
33	47	.220
34	46	.257
35	45	.297
36	44	.339
37	43	.384
38	42	.430
39	41	.477
40	40	.523
(6,6)		
21	57	.001
22	56	.002
23	55	.004
24	54	.008
25	53	.013
26	52	.021
27	51	.032
28	50	.047
29	49	.066
30	48	.090
31	47	.120
32	46	.155
33	45	.197
34	44	.242
35	43	.294
36	42	.350
37	41	.409
38	40	.469
39	39	.531
(6,7)		
21	63	.001
22	62	.001
23	61	.002
24	60	.004
25	59	.007
26	58	.011
27	57	.017

T_1	T_2	α
(6,7) (*Cont.*)		
28	56	.026
29	55	.037
30	54	.051
31	53	.069
32	52	.090
33	51	.117
34	50	.147
35	49	.183
36	48	.223
37	47	.267
38	46	.314
39	45	.365
40	44	.418
41	43	.473
42	42	.527
(6,8)		
21	69	.000
22	68	.001
23	67	.001
24	66	.002
25	65	.004
26	64	.006
27	63	.010
28	62	.015
29	61	.021
30	60	.030
31	59	.041
32	58	.054
33	57	.071
34	56	.091
35	55	.114
36	54	.141
37	53	.172
38	52	.207
39	51	.245
40	50	.286
41	49	.331
42	48	.377
43	47	.426
44	46	.475
45	45	.525
(6,9)		
21	75	.000
22	74	.000
23	73	.001
24	72	.001
25	71	.002
26	70	.004
27	69	.006

T_1	T_2	α
(6,9) (*Cont.*)		
28	68	.009
29	67	.013
30	66	.018
31	65	.025
32	64	.033
33	63	.044
34	62	.057
35	61	.072
36	60	.091
37	59	.112
38	58	.136
39	57	.164
40	56	.194
41	55	.228
42	54	.264
43	53	.303
44	52	.344
45	51	.388
46	50	.432
47	49	.477
48	48	.523
(6,10)		
21	81	.000
22	80	.000
23	79	.000
24	78	.001
25	77	.001
26	76	.002
27	75	.004
28	74	.005
29	73	.008
30	72	.011
31	71	.016
32	70	.021
33	69	.028
34	68	.036
35	67	.047
36	66	.059
37	65	.074
38	64	.090
39	63	.110
40	62	.132
41	61	.157
42	60	.184
43	59	.214
44	58	.246
45	57	.281
46	56	.318
47	55	.356

Table A2.5 (*continued*)

T_1	T_2	α
(6,10) (*Cont.*)		
48	54	.396
49	53	.437
50	52	.479
51	51	.521
(7,7)		
28	77	.000
29	76	.001
30	75	.001
31	74	.002
32	73	.003
33	72	.006
34	71	.009
35	70	.013
36	69	.019
37	68	.027
38	67	.036
39	66	.049
40	65	.064
41	64	.082
42	63	.104
43	62	.130
44	61	.159
45	60	.191
46	59	.228
47	58	.267
48	57	.310
49	56	.355
50	55	.402
51	54	.451
52	53	.500
(7,8)		
28	84	.000
29	83	.000
30	82	.001
31	81	.001
32	80	.002
33	79	.003
34	78	.005
35	77	.007
36	76	.010
37	75	.014
38	74	.020
39	73	.027
40	72	.036
41	71	.047
42	70	.060
43	69	.076
44	68	.095
45	67	.116

T_1	T_2	α
(7,8) (*Cont.*)		
46	66	.140
47	65	.168
48	64	.198
49	63	.232
50	62	.268
51	61	.306
52	60	.347
53	59	.389
54	58	.433
55	57	.478
56	56	.522
(7,9)		
28	91	.000
29	90	.000
30	89	.000
31	88	.001
32	87	.001
33	86	.002
34	85	.003
35	84	.004
36	83	.006
37	82	.008
38	81	.011
39	80	.016
40	79	.021
41	78	.027
42	77	.036
43	76	.045
44	75	.057
45	74	.071
46	73	.087
47	72	.105
48	71	.126
49	70	.150
50	69	.175
51	68	.204
52	67	.235
53	66	.268
54	65	.303
55	64	.340
56	63	.379
57	62	.419
58	61	.459
59	60	.500
(7,10)		
28	98	.000
29	97	.000
30	96	.000
31	95	.000

T_1	T_2	α
(7,10) (*Cont.*)		
32	94	.001
33	93	.001
34	92	.001
35	91	.002
36	90	.003
37	89	.005
38	88	.007
39	87	.009
40	86	.012
41	85	.017
42	84	.022
43	83	.028
44	82	.035
45	81	.044
46	80	.054
47	79	.067
48	78	.081
49	77	.097
50	76	.115
51	75	.135
52	74	.157
53	73	.182
54	72	.209
55	71	.237
56	70	.268
57	69	.300
58	68	.335
59	67	.370
60	66	.406
61	65	.443
62	64	.481
63	63	.519
(8,8)		
36	100	.000
37	99	.000
38	98	.000
39	97	.001
40	96	.001
41	95	.001
42	94	.002
43	93	.003
44	92	.005
45	91	.007
46	90	.010
47	89	.014
48	88	.019
49	87	.025
50	86	.032
51	85	.041

T_1	T_2	α
(8,8) (*Cont.*)		
52	84	.052
53	83	.065
54	82	.080
55	81	.097
56	80	.117
57	79	.139
58	78	.164
59	77	.191
60	76	.221
61	75	.253
62	74	.287
63	73	.323
64	72	.360
65	71	.399
66	70	.439
67	69	.480
68	68	.520
(8,9)		
36	108	.000
40	104	.000
41	103	.001
42	102	.001
43	101	.002
44	100	.003
45	99	.004
46	98	.006
47	97	.008
48	96	.010
49	95	.014
50	94	.018
51	93	.023
52	92	.030
53	91	.037
54	90	.046
55	89	.057
56	88	.069
57	87	.084
58	86	.100
59	85	.118
60	84	.138
61	83	.161
62	82	.185
63	81	.212
64	80	.240
65	79	.271
66	78	.303
67	77	.336
68	76	.371
69	75	.407

Table A2.5 (*continued*)

T_1	T_2	α
(8,9) (*Cont.*)		
70	74	.444
71	73	.481
72	72	.519
(8,10)		
36	116	.000
41	111	.000
42	110	.001
43	109	.001
44	108	.002
45	107	.002
46	106	.003
47	105	.004
48	104	.006
49	103	.008
50	102	.010
51	101	.013
52	100	.017
53	99	.022
54	98	.027
55	97	.034
56	96	.042
57	95	.051
58	94	.061
59	93	.073
60	92	.086
61	91	.102
62	90	.118
63	89	.137
64	88	.158
65	87	.180
66	86	.204
67	85	.230
68	84	.257
69	83	.286
70	82	.317
71	81	.348
72	80	.381
73	79	.414
74	78	.448
75	77	.483
76	76	.517

T_1	T_2	α
(9,9)		
45	126	.000
50	121	.000
51	120	.001
52	119	.001
53	118	.001
54	117	.002
55	116	.003
56	115	.004
57	114	.005
58	113	.007
59	112	.009
60	111	.012
61	110	.016
62	109	.020
63	108	.025
64	107	.031
65	106	.039
66	105	.047
67	104	.057
68	103	.068
69	102	.081
70	101	.095
71	100	.111
72	99	.129
73	98	.149
74	97	.170
75	96	.193
76	95	.218
77	94	.245
78	93	.273
79	92	.302
80	91	.333
81	90	.365
82	89	.398
83	88	.432
84	87	.466
85	86	.500
(9,10)		
45	135	.000
52	128	.000
53	127	.001

T_1	T_2	α
(9,10) (*Cont.*)		
54	126	.001
55	125	.001
56	124	.002
57	123	.003
58	122	.004
59	121	.005
60	120	.007
61	119	.009
62	118	.011
63	117	.014
64	116	.017
65	115	.022
66	114	.027
67	113	.033
68	112	.039
69	111	.047
70	110	.056
71	109	.067
72	108	.078
73	107	.091
74	106	.106
75	105	.121
76	104	.139
77	103	.158
78	102	.178
79	101	.200
80	100	.223
81	99	.248
82	98	.274
83	97	.302
84	96	.330
85	95	.360
86	94	.390
87	93	.421
88	92	.452
89	91	.484
90	90	.516
(10,10)		
55	155	.000
63	147	.000
64	146	.001

T_1	T_2	α
(10,10) (*Cont.*)		
65	145	.001
66	144	.001
67	143	.001
68	142	.002
69	141	.003
70	140	.003
71	139	.004
72	138	.006
73	137	.007
74	136	.009
75	135	.012
76	134	.014
77	133	.018
78	132	.022
79	131	.026
80	130	.032
81	129	.038
82	128	.045
83	127	.053
84	126	.062
85	125	.072
86	124	.083
87	123	.095
88	122	.109
89	121	.124
90	120	.140
91	119	.157
92	118	.176
93	117	.197
94	116	.218
95	115	.241
96	114	.264
97	113	.289
98	112	.315
99	111	.342
100	110	.370
101	109	.398
102	108	.427
103	107	.456
104	106	.485
105	105	.515

For sample sizes greater than 10 the chance that the statistic T will be less than or equal to an integer k is given approximately by the area under the standard normal curve to the left of

$$z = \frac{k + \frac{1}{2} - n_1(n_1 + n_2 + 1)/2}{\sqrt{n_1 n_2 (n_1 + n_2 + 1)/12}}$$

Reproduction of Table A-19 from *Introduction to statistical analysis*, third edition, by Dixon and Massey, with permission of McGraw-Hill.

Table A2.6. *The binomial distribution function*

$n = 2$	$r = 0$	1
p = 0·01	0·9801	0·9999
·02	·9604	·9996
·03	·9409	·9991
·04	·9216	·9984
0·05	0·9025	0·9975
·06	·8836	·9964
·07	·8649	·9951
·08	·8464	·9936
·09	·8281	·9919
0·10	0·8100	0·9900
·11	·7921	·9879
·12	·7744	·9856
·13	·7569	·9831
·14	·7396	·9804
0·15	0·7225	0·9775
·16	·7056	·9744
·17	·6889	·9711
·18	·6724	·9676
·19	·6561	·9639
0·20	0·6400	0·9600
·21	·6241	·9559
·22	·6084	·9516
·23	·5929	·9471
·24	·5776	·9424
0·25	·5625	0·9375
·26	·5476	·9324
·27	·5329	·9271
·28	·5184	·9216
·29	·5041	·9159
0·30	0·4900	0·9100
·31	·4761	·9039
·32	·4624	·8976
·33	·4489	·8911
·34	·4356	·8844
0·35	0·4225	0·8775
·36	·4096	·8704
·37	·3969	·8631
·38	·3844	·8556
·39	·3721	·8479
0·40	0·3600	0·8400
·41	·3481	·8319
·42	·3364	·8236
·43	·3249	·8151
·44	·3136	·8064
0·45	0·3025	0·7975
·46	·2916	·7884
·47	·2809	·7791
·48	·2704	·7696
·49	·2601	·7599
0·50	0·2500	0·7500

$n = 3$	$r = 0$	1	2
p = 0·01	0·9703	0·9997	
·02	·9412	·9988	
·03	·9127	·9974	
·04	·8847	·9953	0·9999
0·05	0·8574	0·9928	0·9999
·06	·8306	·9896	·9998
·07	·8044	·9860	·9997
·08	·7787	·9818	·9995
·09	·7536	·9772	·9993
0·10	0·7290	0·9720	0·9990
·11	·7050	·9664	·9987
·12	·6815	·9603	·9983
·13	·6585	·9537	·9978
·14	·6361	·9467	·9973
0·15	0·6141	0·9393	0·9966
·16	·5927	·9314	·9959
·17	·5718	·9231	·9951
·18	·5514	·9145	·9942
·19	·5314	·9054	·9931
0·20	0·5120	0·8960	0·9920
·21	·4930	·8862	·9907
·22	·4746	·8761	·9894
·23	·4565	·8656	·9878
·24	·4390	·8548	·9862
0·25	0·4219	0·8438	0·9844
·26	·4052	·8324	·9824
·27	·3890	·8207	·9803
·28	·3732	·8087	·9780
·29	·3579	·7965	·9756
0·30	0·3430	0·7840	0·9730
·31	·3285	·7713	·9702
·32	·3144	·7583	·9672
·33	·3008	·7452	·9641
·34	·2875	·7318	·9607
0·35	0·2746	0·7182	0·9571
·36	·2621	·7045	·9533
·37	·2500	·6906	·9493
·38	·2383	·6765	·9451
·39	·2270	·6623	·9407
0·40	0·2160	0·6480	0·9360
·41	·2054	·6335	·9311
·42	·1951	·6190	·9259
·43	·1852	·6043	·9205
·44	·1756	·5896	·9148
0·45	0·1664	0·5748	0·9089
·46	·1575	·5599	·9027
·47	·1489	·5449	·8962
·48	·1406	·5300	·8894
·49	·1327	·5150	·8824
0·50	0·1250	0·5000	0·8750

The function tabulated is

$$F(r|n, p) = \sum_{t=0}^{r} \binom{n}{t} p^t (1-p)^{n-t}$$

for $r = 0, 1, \ldots, n-1$, $n \leqslant 20$ and $p \leqslant 0{\cdot}5$; n is sometimes referred to as the index and p as the parameter of the distribution. $F(r|n, p)$ is the probability that X, the number of occurrences in n independent trials of an event with probability p of occurrence in each trial, is less than or equal to r; that is,

$$\Pr\{X \leqslant r\} = F(r|n, p).$$

Note that

$$\Pr\{X \geqslant r\} = 1 - \Pr\{X \leqslant r-1\}$$
$$= 1 - F(r-1|n, p).$$

$F(n|n, p) = 1$, and the values for $p > 0{\cdot}5$ may be found using the result

$$F(r|n, p) = 1 - F(n-r-1|n, 1-p).$$

The probability of *exactly* r occurrences, $\Pr\{X = r\}$, is equal to

$$F(r|n, p) - F(r-1|n, p) = \binom{n}{r} p^r (1-p)^{n-r}.$$

Linear interpolation in p is satisfactory over much of the table but there are places where quadratic interpolation is necessary for high accuracy. When $r = 0$, 1 or $n-1$ a direct calculation is to be preferred:

$$F(0|n, p) = (1-p)^n,$$
$$F(1|n, p) = (1-p)^{n-1}[1 + (n-1)p]$$

and $\quad F(n-1|n, p) = 1 - p^n.$

For $n > 20$ the number of occurrences X is approximately normally distributed with mean np and variance $np(1-p)$; hence, including $\frac{1}{2}$ for continuity, we have

$$F(r|n, p) \doteqdot \Phi(s)$$

where $s = \dfrac{r + \frac{1}{2} - np}{\sqrt{np(1-p)}}$ and $\Phi(s)$ is the normal distribution function. The approximation can usually be improved by using the formula

$$F(r|n, p) \doteqdot \Phi(s) - \frac{\gamma}{6\sqrt{2\pi}} e^{-\frac{1}{2}s^2}(s^2 - 1)$$

where $\gamma = \dfrac{1-2p}{\sqrt{np(1-p)}}$.

An alternative approximation for $n > 20$ when p is small and np is of moderate size is to use the Poisson distribution:

$$F(r|n, p) \doteqdot F(r|\mu)$$

where $\mu = np$ and $F(r|\mu)$ is the Poisson distribution function. If $1-p$ is small and $n(1-p)$ is of moderate size a similar approximation gives

$$F(r|n, p) \doteqdot 1 - F(n-r-1|\mu)$$

where $\mu = n(1-p)$.

Omitted entries to the left and right of tabulated values are 0 and 1 respectively, to four decimal places.

Table A2.6. (*cont.*)

$n = 4$	$r = 0$	1	2	3
p = 0·01	0·9606	0·9994		
·02	·9224	·9977		
·03	·8853	·9948	0·9999	
·04	·8493	·9909	·9998	
0·05	0·8145	0·9860	0·9995	
·06	·7807	·9801	·9992	
·07	·7481	·9733	·9987	
·08	·7164	·9656	·9981	
·09	·6857	·9570	·9973	0·9999
0·10	0·6561	0·9477	0·9963	0·9999
·11	·6274	·9376	·9951	·9999
·12	·5997	·9268	·9937	·9998
·13	·5729	·9153	·9921	·9997
·14	·5470	·9032	·9902	·9996
0·15	0·5220	0·8905	0·9880	0·9995
·16	·4979	·8772	·9856	·9993
·17	·4746	·8634	·9829	·9992
·18	·4521	·8491	·9798	·9990
·19	·4305	·8344	·9765	·9987
0·20	0·4096	0·8192	0·9728	0·9984
·21	·3895	·8037	·9688	·9981
·22	·3702	·7878	·9644	·9977
·23	·3515	·7715	·9597	·9972
·24	·3336	·7550	·9547	·9967
0·25	0·3164	0·7383	0·9492	0·9961
·26	·2999	·7213	·9434	·9954
·27	·2840	·7041	·9372	·9947
·28	·2687	·6868	·9306	·9939
·29	·2541	·6693	·9237	·9929
0·30	0·2401	0·6517	0·9163	0·9919
·31	·2267	·6340	·9085	·9908
·32	·2138	·6163	·9004	·9895
·33	·2015	·5985	·8918	·9881
·34	·1897	·5807	·8829	·9866
0·35	0·1785	0·5630	0·8735	0·9850
·36	·1678	·5453	·8638	·9832
·37	·1575	·5276	·8536	·9813
·38	·1478	·5100	·8431	·9791
·39	·1385	·4925	·8321	·9769
0·40	0·1296	0·4752	0·8208	0·9744
·41	·1212	·4580	·8091	·9717
·42	·1132	·4410	·7970	·9689
·43	·1056	·4241	·7845	·9658
·44	·0983	·4074	·7717	·9625
0·45	0·0915	0·3910	0·7585	0·9590
·46	·0850	·3748	·7450	·9552
·47	·0789	·3588	·7311	·9512
·48	·0731	·3431	·7169	·9469
·49	·0677	·3276	·7023	·9424
0·50	0·0625	0·3125	0·6875	0·9375

$n = 5$	$r = 0$	1	2	3	4
p = 0·01	0·9510	0·9990			
·02	·9039	·9962	0·9999		
·03	·8587	·9915	·9997		
·04	·8154	·9852	·9994		
0·05	0·7738	0·9774	0·9988		
·06	·7339	·9681	·9980	0·9999	
·07	·6957	·9575	·9969	·9999	
·08	·6591	·9456	·9955	·9998	
·09	·6240	·9326	·9937	·9997	
0·10	0·5905	0·9185	0·9914	0·9995	
·11	·5584	·9035	·9888	·9993	
·12	·5277	·8875	·9857	·9991	
·13	·4984	·8708	·9821	·9987	
·14	·4704	·8533	·9780	·9983	0·9999
0·15	0·4437	0·8352	0·9734	0·9978	0·9999
·16	·4182	·8165	·9682	·9971	·9999
·17	·3939	·7973	·9625	·9964	·9999
·18	·3707	·7776	·9563	·9955	·9998
·19	·3487	·7576	·9495	·9945	·9998
0·20	0·3277	0·7373	0·9421	0·9933	0·9997
·21	·3077	·7167	·9341	·9919	·9996
·22	·2887	·6959	·9256	·9903	·9995
·23	·2707	·6749	·9164	·9886	·9994
·24	·2536	·6539	·9067	·9866	·9992
0·25	0·2373	0·6328	0·8965	0·9844	0·9990
·26	·2219	·6117	·8857	·9819	·9988
·27	·2073	·5907	·8743	·9792	·9986
·28	·1935	·5697	·8624	·9762	·9983
·29	·1804	·5489	·8499	·9728	·9979
0·30	0·1681	0·5282	0·8369	0·9692	0·9976
·31	·1564	·5077	·8234	·9653	·9971
·32	·1454	·4875	·8095	·9610	·9966
·33	·1350	·4675	·7950	·9564	·9961
·34	·1252	·4478	·7801	·9514	·9955
0·35	0·1160	0·4284	0·7648	0·9460	0·9947
·36	·1074	·4094	·7491	·9402	·9940
·37	·0992	·3907	·7330	·9340	·9931
·38	·0916	·3724	·7165	·9274	·9921
·39	·0845	·3545	·6997	·9204	·9910
0·40	0·0778	0·3370	0·6826	0·9130	0·9898
·41	·0715	·3199	·6651	·9051	·9884
·42	·0656	·3033	·6475	·8967	·9869
·43	·0602	·2871	·6295	·8879	·9853
·44	·0551	·2714	·6114	·8786	·9835
0·45	0·0503	0·2562	0·5931	0·8688	0·9815
·46	·0459	·2415	·5747	·8585	·9794
·47	·0418	·2272	·5561	·8478	·9771
·48	·0380	·2135	·5375	·8365	·9745
·49	·0345	·2002	·5187	·8248	·9718
0·50	0·0313	0·1875	0·5000	0·8125	0·9688

Taken from Table 1 of *New Cambridge statistical tables*, 2nd edn. (1996), edited by D.V. Lindley and W.F. Scott, with permission of Cambridge University Press.

Appendix Three

Notes for Appendix One

Box 3

A1.10 This is the standard equation for stratified sampling. To show that it also applies to multistage sampling with unequal-sized primary units, consider Cochran (1977, Eq. 11.25), which gives his estimator for the population total. M_0 = the number of units in the population so dividing his Eq. 11.25 by M_0 gives the estimated mean per population unit. Also, the estimated total for the ith primary unit equals $M_i\bar{y}_i$. Changing to 'w notation' yields our Eq. A1.10.

We also use Eq. A1.10 as the estimator for multistage sampling with unequal-size primary units. Cochran (1977) presents two estimators of the population *total* for this case, his Eqs. 11.21 and 11.25. To convert them to estimates of the mean / population unit, we divide by the number of units in the population, which may be written $N\bar{M}$. Dividing the middle expression in Eq. 11.21 by $N\bar{M}$ yields

$$\bar{y} = \frac{1}{n\bar{M}} \Sigma M_i \bar{y}_i. \tag{A3.1}$$

Dividing Eq. 11.25 by $N\bar{M} = M_0$ yields (from the middle expression of Eq.11.25)

$$\bar{y} = \frac{\Sigma M_i \bar{y}_i}{\Sigma M_i}$$

$$= \frac{1}{n\bar{M}} \Sigma M_i \bar{y}_i. \tag{A3.2}$$

Thus, Cochran's Eqs. 11.21 and 11.25 have the same *form* and this form is the same as our Eq. A3.1.

There is a possible difference, however. We may use either the sample mean for the average size of primary units, or, if we know it, the population mean. If we use the sample mean, then we have a ratio estimator since both the numerator and the denominator would vary from sample to sample. Cochran (1977) considers this case in Section 11.8, so we should derive our variance from this Section if we use the sample mean for the average size of primary units. If instead we use the population mean, then we should derive our variance from Section 11.7 of Cochran (1977).

We expect the ratio estimator, with the sample mean for the average size, to be a better estimator in general. If we happen to get large primary units, then the numerator will probably be larger than average. By using the sample average – which will also be larger than average – in the denominator, we get a good estimate of the mean per population unit. This advantage would be lost if we used the population mean for the average size. Furthermore, we usually do not know the population mean. Thus, for both reasons, we stress the ratio estimator (i.e., we *define* $\overline{w}$ as the sample mean, except in C2a, Box 3).

A1.11 The standard formula for stratified sampling except that we replaced $W_h = N_h/N$ with $N_h/n\overline{N}_h$ and, substituted w for N. This allows us to *define* w as either the number, or proportion, of units in a stratum.

A1.12 Equation 5.16 (Cochran 1977) for *df* is

$$n_c = \frac{\left[\sum \frac{N_h(N_h - n_h)}{n_h} s_h^2\right]^2}{\sum \left[\frac{N_h(N_h - n_h)}{n_h} s_h^2\right]^2 / (n_h - 1)} \qquad \text{(A3.3)}$$

Replace $N_h - n_h$ with $N_h(1 - n_h/N_h)$ and then drop the finite population correction (fpc). Divide top and bottom by N^2. This yields

$$n_c = \frac{\left[\sum w_h^2 v(\overline{y}_h)\right]^2}{\sum \frac{[w_h^2 v(\overline{y}_h)]^2}{n_h - 1}} \qquad \text{(A3.4)}$$

which is our expression with $g_i = w_h^{\,2}\, se(\overline{y}_h)^2 = w_h^{\,2}\, v(\overline{y}_h)$.

A1.13 This formula for double sampling comes from Cochran (1977) Eq. 12.24′ which has three terms. In the first, we replace s_h^2/n_h with $[se(\overline{y})]^2$ to make the equation more closely resemble our version with stratified sampling. We have omitted the second term because N is assumed to be large. In the third term, we have replaced g'/n' with $(1/n^* - 1/N^*)$ which follows easily if we let $g' = (N - n')/N$ rather than $(N - n')/(N - 1)$. The reason for our version is to make it clearer that the difference between $se(\bar{y})$ with double sampling and $se(\bar{y})$ with stratified sampling depends on how well we know the stratum sizes (e.g., if $n' = N$, the third term drops out). We use n^* rather than n'.

A1.14 and A1.15 Expression A1.14 is a special case of Eq. A1.18 so we derive Eq. A1.14 first. It comes from Cochran (1977) Eq. 11.30 which is his recommended estimate for the variance of the ratio estimator of the population total (Eq. 11.25). To obtain the estimated mean/population unit, we divided by M_0 (see note about Eq. A1.10). We therefore divide his variance (Eq. 11.30) by $M_0^2 = (N\bar{M})^2$ getting

$$v(\bar{y}) = \frac{1}{n\bar{M}^2}\left[\frac{N^2(1-f_1)\sum M_i^2(\bar{y}_i - \bar{y})^2}{n(n-1)}\right] + \frac{1}{n\bar{M}^2}\left[\frac{N}{n}\sum\frac{M_i^2(1-f_{2i})s_{2i}^2}{m_i}\right], \tag{A3.5}$$

where we use $\bar{y}$ instead of his $\hat{\bar{\bar{y}}}_R$. Moving $(N\bar{M})^2$ inside

$$v(\bar{y}) = \left(1 - \frac{n}{N}\right)\frac{1}{n\bar{M}^2}\sum\frac{M_i^2(\bar{y}_i - \bar{y})^2}{n-1} + \frac{1}{nN\bar{M}^2}\sum M_i^2(1-f_{2i})\frac{s_{2i}^2}{m_i} \tag{A3.6}$$

Multiplying the right-most term by n/n and changing to 'w' notation yields

$$v(\bar{y}) = \left(1 - \frac{n}{N}\right)\frac{1}{n\bar{w}^2}\sum\frac{w_i^2(\bar{y}_i - \bar{y})^2}{n-1} + \left(\frac{n}{N}\right)\frac{1}{(n\bar{w})^2}\sum w_i^2(1-f_{2i})\frac{s_{2i}^2}{m_i}, \tag{A3.7}$$

which is the square of Eq. A1.18 with the first part equal to our Eq. A1.14 and the second part equal to Eq. A1.11 (with $se(y_i)$ calculated as in Box 2, including the fpc).

There remains one problem however. Cochran proved Eq. 11.30 for the case of two-stage sampling and simple random sampling within primary units whereas we want expressions for any kind of sampling plan within primary units. We thus still have to show that Eq. 11.30 would hold for any plan. Cochran invokes Theorem 11.2 to prove Eq. 11.30. The theorem obtains the left term in Eq. 11.30 by writing out an unbiased estimator from *one-stage* sampling (i.e., primary units censused – see text following Eq. 11.26). This part of the proof would be the same for any plan as long as it yielded unbiased estimates within primary units. To this first term we add $\Sigma w_{is}\hat{\sigma}^2_{2i}$ where w_{is} depends only on the size of the ith primary unit and $\hat{\sigma}^2_{2i}$ is an unbiased estimate of the variance of $\hat{Y}_i$ (text following Eq. 11.16). Thus, the theorem proved by Cochran is general, requiring only that the estimated means ($\bar{y}_i$) and variances $\{[se(\bar{y}_i)]^2\}$ within primary units be unbiased.

A1.16 and 'Note' following A1.15 The 'note' states that if the mean size of primary units in the entire population is known, then it may be substituted in our Eq. A1.10 for the estimated mean/population unit. The statements above show that when this is done, the estimator is equivalent to Cochran's estimator, Eq. 11.21, used to estimate the population mean, rather than total, and that therefore the appropriate formula to estimate the variance is Cochran's equation from that section (Eq. 11.24) modified to estimate the mean, rather than the total. We show below that with this modification, Cochran's Eq. 11.24 is the same as Eq. A1.16 ignoring the term in $1/N$.

First divide his Eq. 11.24 by $(N\bar{M})^2$ (since we divided his estimate of the population total by $N\bar{M}$ to get the population mean).

The comments above show that Cochran's Section 11.7 pertains to cases in which the known mean size per primary unit for the population is used. This gives

$$v(\bar{y}) = \left(1 - \frac{n}{N}\right)\frac{1}{n\bar{M}^2}\frac{\Sigma(Y_i - \bar{Y}_u)^2}{n-1} + \frac{1}{nN\bar{M}^2}\Sigma\frac{M_i^2(1-f_{2i})s_{2i}^2}{m_i} \quad \text{(A3.8)}$$

where Y_i and $\bar{Y}_u$ are *estimates*. The right term drops out since we consider N to be large.

Now Y_i, the estimated total for the ith primary unit, equals $M_i\bar{y}_i$ and $\bar{Y}_u$, the average of the Y_i, equals $(1/n)\Sigma M_i\bar{y}_i = \bar{M}\bar{y}$ since

$$\bar{y} = \frac{1}{n\bar{M}}\Sigma M_i\bar{y}_i. \quad \text{(A3.9)}$$

where $\bar{y}$ is the estimated population mean. Continuing

$$\Sigma(Y_i - \bar{Y}_u)^2 = \Sigma(M_i\bar{y}_i - \bar{M}\bar{y})^2$$
$$= \bar{M}^2\Sigma\left(\frac{M_i\bar{y}_i}{\bar{M}} - \bar{y}\right)^2, \quad \text{(A3.10)}$$

and thus

$$v(\bar{y}) = \frac{\sum\left(\frac{M_i\bar{y}_i}{\bar{M}} - \bar{y}\right)^2}{n(n-1)}$$

$$= \frac{\sum(y_i^* - \bar{y})^2}{n(n-1)}. \qquad \text{(A3.11)}$$

Box 5

A1.26 For one-tailed tests, Snedecor and Cochran (1980 p. 119) use

$$z_c = \frac{|r - nP| - 0.5}{\sqrt{nPQ}} \qquad \text{(A3.12)}$$

where P is their value under the null hypothesis and r = number of 'successes'. They use lower-case p's and q's for the parameters but we use upper case to distinguish parameters from estimates. Write

$$z_c = \frac{|np - nP| - 0.5}{\sqrt{nPQ}}$$

$$= \frac{|p - P|}{\sqrt{PQ/n}} - \frac{0.5}{\sqrt{nPQ}}, \qquad \text{(A3.13)}$$

which is our Eq. A1.26 except that we use c in place of 0.5. For the two-tailed test (Snedecor and Cochran 1980 p. 119), they replace 0.5 in Eq. A3.12 above with another quantity, which we call c. If the fractional part, f, is ≤ 0.5, then all of it is subtracted ($c = f$); if $f > 0.5$, then the fractional part of $|r - nP|$ is reduced to 0.5 (i.e., $c = f - 0.5$).

A1.27 Our equation is directly from Cochran, Eq. 3.19.

A1.28 and A1.29 The upper limit is given by Hollander and Wolfe (1973 p. 24) as

$$P_L = \frac{B}{B + (n - B + 1)f(\alpha/2|\nu_1,\nu_2)}, \qquad \text{(A3.14)}$$

where $B = np$. Thus

$$P_L = \frac{1}{1 + \frac{(n - np + 1)f(\alpha/2|\nu_1,\nu_2)}{np}} \qquad \text{(A3.15)}$$

$$= \frac{1}{1 + \frac{(q + 1/n)f(\alpha/2|\nu_1,\nu_2)}{p}},$$

where

$$\nu_1 = 2(n - B + 1) = 2(nq + 1)$$

$$\nu_2 = 2B = 2np. \qquad \text{(A3.16)}$$

For the upper limit, we have

$$P_U(n,B) = 1 - P_L(n, n - B), \qquad \text{(A3.17)}$$

where the notation for P_U means 'use the formula for P_L but substitute '*n-B*' everywhere *B* appears' (D. Wolfe, personal communication). Thus,

$$P_U = 1 - \frac{n - B}{n - B + (n - n + B + 1)F(\alpha/2|2(n - n + B + 1, 2(n - B))} \qquad \text{(A3.18)}$$

$$= 1 - \frac{1}{1 + \left(\frac{B+1}{n-B}\right)F(\alpha/2|2(B + 1, 2(n - B))}$$

$$= 1 - \frac{1}{1 + \left(\frac{np+1}{n(1-p)}\right)F(\alpha/2|2(np + 1), 2(n - np))},$$

which may be written

$$P_U = 1 - \frac{1}{1 + \left(\frac{np+1}{nq}\right)F(\alpha/2|\nu_1, \nu_2)}, \qquad \text{(A3.19)}$$

where

$$\nu_1 = 2(np + 1) \qquad \text{(A3.20)}$$

$$\nu_2 = 2nq.$$

Now let $(np+1)/nq = a$, so that we have

$$P_U = 1 - \frac{1}{1 + af_{\alpha/2, \nu_1, \nu_2}} \qquad \text{(A3.21)}$$

Combining terms leads to $P_U = 1/(1 + 1/af)$ from which we obtain

$$P_U = \frac{1}{1 + \frac{nq}{(np+1)F(\alpha/2|\nu_1, \nu_2)}} \qquad \text{(A3.22)}$$

$$= \frac{1}{1 + \frac{q}{(p + 1/n)F(\alpha/2|\nu_1, \nu_2)}},$$

where

$$\nu_1 = 2(np+1)$$
$$\nu_2 = 2nq. \quad \text{(A3.23)}$$

Box 6

Our formula for hypothesis tests is the same as that of Snedecor and Cochran (1980 p. 89), except that we use Y_α in place of $(\mu_1 - \mu_2)$ and we say the test statistic must be $\geq$ the tabled values of t rather than defining the test statistic as t and then saying it has to exceed the tabled values.

Hypothesis tests

For $n_1 \neq n_2$, Snedecor and Cochran (1980 Section 6.7) give the 'pooled' variance as

$$s^2 = \frac{\sum x_1^2 + \sum x_2^2}{2(n-1)}$$
$$= \frac{1}{2}\left[s_1^2 + s_2^2\right], \quad \text{(A3.24)}$$

and the test statistic as

$$t = \frac{\sqrt{n}[(\bar{X}_1 - \bar{X}_2) - \mu]}{\sqrt{2}s} \quad \text{(A3.25)}$$
$$= \frac{(\bar{X}_1 - \bar{X}_2) - \mu}{\frac{\sqrt{2}}{\sqrt{n}}\sqrt{\frac{1}{2}(s_1^2 + s_2^2)}}$$
$$= \frac{(\bar{X}_1 - \bar{X}_2) - \mu}{\sqrt{\frac{s_1^2}{n} + \frac{s_2^2}{n}}}$$
$$= \frac{(\bar{X}_1 - \bar{X}_2) - \mu}{\sqrt{se_1^2 + se_2^2}}$$

Thus, even though they call the variance pooled, it is the same as the unpooled variance. Also, they give df as $2(n-1)$ which equals $df_1 + df_2$. This shows why Eq. A1.35 applies to hypothesis testing with equal sample sizes.

For $n_1 \neq n_2$, Snedecor and Cochran (1980) give (Sec. 6.9)

$$v(diff.) = \left(\frac{n_1 + n_2}{n_1 n_2}\right)\sigma^2, \quad \text{(A3.26)}$$

where σ^2 is estimated by s^2 as

$$s^2 = \frac{\sum x_1^2 + \sum x_2^2}{n_1 + n_2 - 2}, \tag{A3.27}$$

and $\sum x_1^2 = n(n-1)se_1^2$. This leads immediately to our formula (which admittedly is cumbersome, but uses the standard errors, which we assume the user will already have calculated). The df are $n_1 + n_2 - 2 = df_1 + df_2$.

Confidence intervals (A1.32, A1.33) They *do not* say always to use the non pooled variance estimate. They only say we might be hesitant to use it if the means are 'very different' and we are computing confidence limits (Section 6.11, Snedecor and Cochran 1980, item no. 2). Their example in Section 6.9 uses the pooled s^2 (with $n_1 \neq n_2$) for the confidence interval. The variances in this case were similar (457 and 425).

For unequal variances (Snedecor and Cochran 1980 Section 6.11) they give the test statistic as

$$t' = \frac{\bar{X}_1 - \bar{X}_2}{\sqrt{\frac{s_1^2}{n_1} + \frac{s_2^2}{n_2}}} \tag{A3.28}$$

$$= \frac{\bar{X}_1 - \bar{X}_2}{\sqrt{se_1^2 + se_2^2}}.$$

This shows that Eq. A1.35 applies to confidence intervals, if variances are assumed to be unequal. For df they give

$$df = \frac{[s_1^2/n_1 + s_2^2/n_2]^2}{\frac{(s_1^2/n_1)^2}{n_1 - 1} + \frac{(s_2^2/n_2)^2}{n_2 - 1}}$$

$$= \frac{(se_1^2 + se_2^2)^2}{\frac{se_1^4}{n_1 - 1} + \frac{se_2^4}{n_2 - 1}}, \tag{A3.29}$$

which equals our expression.

Box 7

Our equations come mainly from Fleiss (1981) as shown below.

Our equation	Fleiss's equation
A1.37	2.5
A1.38	Formula in Table 2.6
A1.39	8.2 and 8.3 (his '$b+c$' = our $n_{sf}+n_{fs}$)
A1.40	2.14 (independent) and 8.15 (paired)
A1.41	2.13 (except we use $n-1$ in the denominator.
A1.42	8.14

Most of the equations can also be found in Snedecor and Cochran (1980), though they use $1/2n$ in one place as the correction for continuity, and they use a notation we find harder to follow.

The recommendation about when the large-sample procedure may be used for confidence intervals is from Fleiss (1981 p. 29). We do not find any corresponding guideline for paired samples, so we have used the same guideline for independent and paired data.

Box 8

We used Snedecor and Cochran (1980 pp. 141–2) for the Wilcoxin test except that we suggest omitting pairs with equal value and thus differences of 0 (in accordance with Castellan and Siegel, 1988 p. 88) whereas they do not include this recommendation. Our T for large samples is the same as theirs (8.6.1 and 8.6.2) except that we substitute their formula for μ into σ. Also, we use the Table for values up to 25. We used Snedecor and Cochran (1980 pp. 144–5) for the Mann-Whitney test.

Box 10

Part D Our $K_{\alpha,P'}$ is Snedecor and Cochran's (1980)$(Z_\alpha+Z_\beta)^2$. We find their description a little hard to follow, but the examples they give below 6.14.1 may be used to verify the figures we give in the Table.

A1.56 is their Eq. 6.14.4 '+2'. Our '+2' comes from their guidelines below 6.14.4 which recommend increasing the initial value by from 1 to 3 depending on the level of significance and on whether the data are paired. We used 2 because it is simpler and the sample size guidelines are approximate anyway.

References

Aitkin, M.A., Anderson, D., Francis, B. and Hinde, J. (1989). *Statistical modelling in GLIM*. Oxford University Press, Oxford, pp. 1–374.

Alexander, H.M., Slade, N.A. and Kettle, W.D. (1997) Application of mark-recapture models to estimation of the population size of plants. *Ecology* **78**:1230–7.

Anderson, D.R. and Southwell, C. (1995). Estimates of macropod density from line transect surveys relative to analyst expertise. *J. Wildl. Manage.* **59**:852–7.

Arthur, S.M., Manly, B.F.J., MacDonald, L.L. and Garner, G.W. (1996). Assessing habitat selection when availability changes. *Ecology* **77**:215–27.

Bakeman, R. and Gottman, J. (1986). *Observing interactions: an introduction to sequential analysis*. Cambridge University Press, Cambridge, pp. 1–221.

Bantock, C.R., Bayley, J.A. and Harvey, P.H. (1976). Simultaneous selective predation on two features of a mixed sibling species population. *Evolution* **29**:636–49.

Bart, J. and Robson, D.S. (1982). Estimating survivorship when the subjects are visited periodically. *Ecology* **63**:1078–90.

Bart J. and Schoultz, J.D. (1984). Reliability of singing bird surveys: changes in observer efficiency with avian density. *Auk* **101**:307–18.

Batschelet, E. (1981). *Circular statistics in biology*. Academic Press, New York, pp. 1–371.

Bergerud, A.T. and Ballard, W.B. (1988). Wolf predation on caribou: the Nelchina herd case history, a different interpretation. *J. Wildl. Manage.* **52**:344–57.

Brownie, C., Anderson, D.R., Burnham, K.P. and Robson, D.S. (1985). *Statistical inference from band recovery data – a handbook*. US Fish and Wildlife Service Resource Publication 156.

Brownie, C., Hines, J.E., Nichols, J.D., Pollock, K.H. and Hestbeck, J.B. (1993). Capture–recapture studies for multiple strata including non-Markovian transitions. *Biometrics* **49**:1173–87.

Buckland, S.T., Anderson, D.R., Burnham, K.P. and Laake, J.L. (1993). *Distance sampling: estimating abundance of biological populations*. Chapman and Hall, New York, pp. 1–446.

Bunck, C.M., Chen, C.-L. and Pollock, K.H. (1995). Robustness of survival estimates from radio-telemetry studies with uncertain relocation of individuals. *J. Wildl. Manage.* **59**:790–4.

Burnham, K.P. and Rexstad, E.A. (1993). Modeling heterogeneity in survival rates of banded waterfowl. *Biometrics* **49**:1194–208.

Burnham, K.P, Anderson, D.R., White, G.C., Brownie, C. and Pollock, K.H. (1987). *Design and analysis methods for fish survival experiments based on release-recapture.* American Fisheries Society Monograph 5, pp. 1–437.

Castellen, N.J. and Siegel, S. (1988). *Nonparametric statistics for the behavioral sciences.* McGraw-Hill, New York, pp. 1–399.

Catchpole, C.K. (1989). Pseudoreplication and external validity: playback experiments in avian bioacoustics. *Trends Ecol. Evol.* **4**:286–7.

Chesson, J. (1978). Measuring preference in selective predation. *Ecology* **59**:211–15.

Clarke, M.F. and Kramer, D.L. (1994). The placement, recovery, and loss of scatter hoards by eastern chipmunks, *Tamias striatus*. *Behav. Ecol.* **5**:353–61.

Clobert, J., Lebreton, J.-D., Allaine, D. and Gaillard, J.M. (1994). The estimation of age-specific breeding probabilities from recaptures or resightings in vertebrate populations: II. longitudinal models. *Biometrics* **50**:375–87.

Cochran, W.G. (1977). *Sampling techniques*, third edn. John Wiley and Sons, New York, pp. 1–428.

Cohen, J. (1988). *Statistical power analysis for the behavioral sciences*, second. edn. Lawrence-Erlbaum, Hillsdale, New Jersey.

Cohen, J. and Cohen, P. (1983). *Applied multiple regression / correlation analysis for the behavioral sciences*, second edn. Lawrence Erlbaum Associates, Hillsdale, New Jersey, pp. 1–545.

Compton, B.B., Mackie, R.J. and Dusek, G.L. (1988). Factors influencing distribution of white-tailed deer in riparian habitats. *J. Wildl. Manage.* **52**:544–8.

Cooch, E.G., Pradel, R. and Nur, N. (1996). *A practical guide to mark-recapture analysis using SURGE.* Centre d'Ecologie Fonctionelle et Evolutive, CNRS, Montpellier, France, pp. 1–125.

Cormack, R.M. (1964). Estimates of survival from the sighting of marked animals. *Biometrika* **51**:429–38.

Cox, D.R. and Oakes, D. (1984). *Analysis of survival data.* Chapman and Hall, New York, pp. 1–201.

Cox, D.R. and Snell, E.J. (1989). *The analysis of binary data*, second edn. Chapman and Hall, New York, pp. 1–236.

Crowder, M.J. and Hand, D.J. (1990). *Analysis of repeated measures.* Chapman and Hall, New York, pp. 1–256.

Diernfeld, E.S., Sandfort, C.E. and Satterfield, W.C. (1989). Influence of diet on plasma vitamin E in captive peregrine falcons. *J. Wildl. Manage.* **53**: 160–4.

Dixon, F.J. and Massey W.J. Jr. (1969). *Introduction to statistical analysis*, third edn. McGraw-Hill, New York

Dugatkin, L.A. and Wilson, D.S. (1992). The prerequisites for strategic behaviour in bluegill sunfish, *Lepomis macrochirus*. *Anim. Behav.* **44**:223–30.

Efron, B. and Tibshirani, R.J. (1993). *An introduction to the bootstrap*. Chapman and Hall, New York, pp. 1–436.

Elston, D.A., Illius, A.W. and Gordon, I.J. (1996). Assessment of preference among a range of options using log ratio analysis. *Ecology* 77:2538–48.

Fancy, S.G. (1997). A new approach for analyzing bird densities from variable circular-plot counts. *Pac. Sci.* **51**:107–14.

Fleiss, J.L. (1981). *Statistical methods for rates and proportions*. John Wiley and Sons, New York, pp. 1–223

Flint, P.L., Pollock, K.H., Thomas, D. and Sedinger, J.S. (1995). Estimating pre-fledging survival: allowing for brood mixing and dependence among brood mates. *J. Wildl. Manage.* **59**:448–55.

Fryxell, J.M., Mercer, W.E. and Gellately, R.B. (1988). Population dynamics of Newfoundland moose using cohort analysis. *J. Wildl. Manage.* **52**:14–21.

Gauch, H.G. Jr. (1982). *Multivariate analysis in community ecology*. Cambridge University Press, Cambridge, pp. 1–298.

Geissler, P.H. and Sauer, J.R. (1990). Topics in route-regression analysis. In: J.R. Sauer and S. Droege (eds.) *Survey designs and statistical methods for the estimation of avian population trends*. US Fish and Wildlife Service Biology Report 90(1).

Gese, E.M., Rongstad, O.J. and Mytton, W.R. (1988). Home range and habitat use of coyotes in southeastern Colorado. *J. Wildl. Manage.* **52**:640–6.

Gottman, J.M. and Roy, A.K. (1990). *Sequential analysis: a guide for behavioral researchers*. Cambridge University Press, Cambridge, pp. 1–275.

Green, R.H. (1979). *Sampling design and statistical methods for environmental biologists*. John Wiley and Sons, New York, pp. 1–257.

Greenwood, J.J.D. (1993). Statistical power. *Anim. Behav.* **46**:1011–12.

Haccou, P. and Meelis, E. (1992). *Statistical analysis of behavioral data: an approach based on time-structured models*. Oxford University Press, Oxford, pp. 1–396.

Heisey, D.M. and Fuller, T.K. (1985). Evaluation of survival and cause-specific mortality rates using telemetry data. *J. Wildl. Manage.* **49**:668–74.

Hollander, M. and Wolfe, D.A. (1973). *Nonparametric statistical methods*. John Wiley and Sons, New York, pp. 1–503.

Holm, B.A., Johnson, R.J., Jensen, D.D. and Stroup, W.W. (1988). Responses of deer mice to methiocarb and thiram seed treatments. *J. Wildl. Manage.* **52**:497–502.

Hurlbert, S.H. (1984). Pseudoreplication and the design of ecological field experiments. *Ecol. Monogr.* **54**:187–211.

James, F.C., McCulloch, C.E. and Wiedenfeld, D.A. (1996). New approaches to the analysis of population trends in land birds. *Ecology* **77**:13–27.

Jolly, G.M. (1965). Explicit estimates from capture-recapture data with both death and immigration-stochastic model. *Biometrika* **52**:225–47.

Johnson, D.B., Guthery, F.S. and Koerth, N.E. (1989). Grackle damage to grapefruit in the lower Rio Grande Valley. *Wildl. Soc. Bull.* **17**:46–50.

Johnson, D.H. (1979). Estimating nest success: the Mayfield method and an alternative. *Auk* **96**:651–61.

Johnson, D.H. (1995). Statistical sirens: the alure of nonparametrics. *Ecology* **76**: 1998–2000.

Johnson, D.H. and Shaffer, T.L. (1990). Estimating nest success: when Mayfield wins. *Auk* **107**:595–600.

Johnson, R.A. and Wichern, D.W. (1982). *Applied multivariate statistical analysis.* Prentice-Hall, New Jersey, pp. 1–594.

Kaplan, E.L. and Meier, P. (1958). Nonparametric estimation from incomplete observations. *J. Am. Stat. Assoc.* **53**:457–81.

Kendall, W.L. and Nichols, J.D. (1995). On the use of secondary capture-recapture samples to estimate temporary emigration and breeding proportions. *J. Appl. Stat.* **22**:751–62.

Kendall, W.L., Nichols, J.D. and Hines, J.E. (1997). Estimating temporary emigration using capture–recapture data with Pollock's robust design. *Ecology* **78**:563–78.

Knick, S.T. and Dyer, D.L. (1997). Distribution of black-tailed jackrabbit habitat determined by GIS in southwestern Idaho. *J. Wildl. Manage.* **61**: 75–85.

Kotz, S. and Johnson, N.L. (1988). *Encyclopedia of statistical sciences*. John Wiley and Sons, New York, pp. 1–762.

Kroodsma, D.E. (1986). Design of song playback experiments. *Auk* **103**: 640–2.

Kroodsma, D.E. (1989a). Suggested experimental designs for song playbacks. *Anim. Behav.* **37**:600–9.

Kroodsma, D.E. (1989b). Inappropriate experimental designs impede progress in bioaccoustic research: a reply. *Anim. Behav.* **38**:717–19.

Kroodsma, D.E. (1990). Using appropriate experimental designs for intended hypotheses in 'song' playbacks, with examples for testing effects of song repertoire size. *Anim. Behav.* **40**:1138–50.

Lamprecht, J. and Hofer, H. (1994). Cooperation among sunfish: do they have the cognitive abilities? *Anim. Behav.* **47**:1457–8.

Lancia, R.A., Nichols, J.D. and Pollock, K.H. (1994). Estimating the number of animals in wildlife populations. In: T.A. Bookhaut (ed.) *Research and management techniques for wildlife and habitats*, fifth edn., pp. 215–53. The Wildlife Society, Bethesda, Maryland, pp. 1–740.

Lebreton, J.-D., Burnham, K.P., Clobert, J. and Anderson, D.R. (1992). Modeling survival and testing biological hypotheses using marked animals: a unified approach with case studies. *Ecol. Mono.* **62**:67–88.

Lebreton, J.-D., Pradel, R. and Clobert, J. (1993). Estimating and comparing survival rates in animal populations. *Trends Ecol. Evol.* **8**:91–5.

Lehmann, E.L. (1975). *Nonparametrics: statistical methods based on ranks*. McGraw-Hill, New York, pp. 1–467.

Lehner, P.N. (1996). *Handbook of ethological methods*. Cambridge University Press, Cambridge, pp. 1–672.

Leuschner, W.A., Ritchie, V.P. and Stauffer, D.F. (1989). Options on wildlife: responses of resource managers and wildlife users in southeastern United States. *Wildl. Soc. Bull.* **17**:24–9.

Li, C.C. (1975). *Path analysis: a primer*. Boxwood Press, Pacific Grove, California, pp. 1–346.

Lindley, D.V. and Scott, W.F. (1996). *New Cambridge statistical tables*, second edn. Cambridge University Press, Cambridge.

Lombardi, C.M. and Hurlbert, S.H. (1996). Sunfish cognition and pseudoreplication. *Anim. Behav.* **52**:419–22.

Machlis, L., Dood, P.W.D. and Fentress, J.C. (1985). The pooling fallacy: problems arising when individuals contribute more than one observation to the data set. *Z. Tierpsychol.* **68**:201–14.

Manly, B.F.J., Miller, P. and Cook, L.M. (1972). Analysis of a selective predation experiment. *Am. Nat.* **106**:719–36.

Manly, B.F.J., McDonald, L.L. and Thomas, D.L. (1993). *Resource selection by animals: statistical design and analysis for field studies.* Chapman and Hall, New York, pp. 177.

Mardia, K.V. (1972). *Statistics of directional data*. Academic Press, New York, pp. 1–357.

Martin, P. and Bateson, P. (1993). *Measuring behaviour: an introductory guide*, second edn. Cambridge University Press, Cambridge, pp. 1–222.

Mayfield, H. (1961). Nesting success calculated from exposure. *Wilson Bull.* **73**:255–61.

Mayfield, H. (1975). Suggestions for calculating nest success. *Wilson Bull.* **87**:456–66.

McCullagh, P. and Nelder, J.A. (1989). *Generalized linear models.* Chapman and Hall, New York, pp. 1–511.

McGregor, P.K., Catchpole, C.K., Dabelsteen, T., Falls, J.B., Fusani, L., Gerhardt, H.C., Gilbert, F., Horn, A.G., Klump, G.M., Kroodsma, D.E., Lambrechts, M.M., McComb, K.E., Nelson, D.A., Pepperberg, I.M., Ratcliffe, L., Searcy, W.A. and Weary, D.M. (1992). Design of playback experiments: the Thornbridge Hall Nato ARW consensus. In: P.K. McGregor (ed.) *Playback and studies of animal communication*, pp. 1–9. NATO Advanced Science Institute Series A, Vol 228. Plenum Press, New York, pp. 1–231.

Miller, H.W. and Johnson, D.H. (1978). Interpreting the results of nesting studies. *J. Wildl. Manage.* **42**:471–6.

Mood, A.M., Graybill, F.A., and Boes, D.C. (1974). *Introduction to the theory of statistics.* McGraw-Hill, New York, pp. 1–564.

Moore, D.S. and McCabe, G.P. (1993). *Introduction to the practice of statistics*, second edn. W.H. Freeman, New York, pp. 1–854.

Neter, J. and Wasserman, W. (1974). *Applied linear statistical models.* Richard D. Irwin, Homewood, Illinois, pp. 1–842.

Neter, J., Wasserman, W. and Kutner, M.H. (1983). *Applied linear regression models.* Richard D. Irwin, Homewood, Illinois pp. 1–547.

Nichols, J.D. (1992). Capture-recapture models: Using marked animals to study population dynamics. *BioScience* **42**:94–102.

Nichols, J.D. and Pollock, K.H. (1990). Estimation of recruitment from immigration versus *in situ* reproduction using Pollock's robust design. *Ecology* **71**:21–6.

Nichols, J.D., Brownie, C., Hines, J.E., Pollock, K.H. and Hestbeck, J.B. (1993). The estimation of exchanges among populations or subpopulations. In: J.-D. Lebreton and P.M. North (eds.) *Marked individuals in the study of bird populations*, pp. 265–280. Advances in Life Sciences, Birkhauser Verlag, Berlin, pp. 1–397.

Nichols, J.D., Hines, J.E. and Blums, P. (1997). Tests for senescent decline in annual survival probabilities in Common Pochards. *Ecology* **78**: 1009–18.

Nixon, C.M., Hansen, L.P. and Brewer, P.A. (1988). Characteristics of winter habitats used by deer in Illinois. *J. Wildl. Manage.* **52**:552–5.

Otis, D.L., Burnham, K.P., White, G.C. and Anderson, D.R. (1978). Statistical inference from capture data on closed animal populations. *Wildl. Monogr.* **62**:1–135.

Pollock, K.H. (1981). Capture-recapture models allowing for age-dependent survival and capture rates. *Biometrics* **37**:521–9.

Pollock, K.H. (1982). A capture-recapture design robust to unequal probability of capture. *J. Wildl. Manage.* **46**:752–7.

Pollock, K.H., Nichols, J.D., Brownie, C. and Hines, J.E. (1990). Statistical inference for capture-recapture experiments. *Wildl. Monogr.* **107**:1–97.

Pollock, K.H., Winterstein, S.R., Bunck, C.M. and Curtis, P.D. (1989a). Survival analysis in telemetry studies: the staggered entry design. *J. Wildl. Manage.* **53**:7–15.

Pollock, K.H., Winterstein, S.R. and Conroy, M.J. (1989b). Estimation and analysis of survival distributions for radio-tagged animals. *Biometrics* **45**:99–109.

Pradel, R., Johnson, A.R., Viallefont, A., Nager, R.G. and Cezilly, F. (1997). Local recruitment in the greater flamingo: a new approach using capture–mark–recapture data. *Ecology* **78**:1431–45.

Quang, P.X. and Becker, E.F. (1996). Line transect sampling under varying conditions with application and aerial surveys. *Ecology* **77**:1297–302.

Ralph, C.J. and Scott, J.M. (eds.). (1981) Estimating numbers of terrestrial birds. *Stud. Avian Biol.* **6**:1– 630.

Randles, R.H. and Wolfe, D.A. (1979). *Introduction to the theory of nonparametric statistics.* John Wiley and Sons, New York, pp. 1–450.

Renecker, L.A. and Hudson, R.J. (1989). Seasonal activity budgets of moose in aspen-dominated boreal forests. *J. Wildl. Manage.* **53**:296–302.

Rexstad, E. and Burnham, K. (1991). *Users' guide for interactive program CAPTURE.* Colo. Coop. Fish and Wildl. Res. Unit, Colorado State University, Fort Collins, pp. 1–29.

Rice, J.A. (1995). *Mathematical statistics and data analysis*, second edn. W.H. Freeman, New York, pp. 1–651.

Samuel, M.D. and Fuller, M.R. (1994). Wildlife radiotelemetry. In: T.A. Bookout (ed.) *Research and management techniques for wildlife and habitats*, fifth edn., The Wildlife Society, Bethesda, Maryland, pp. 370–418.

Sauer, J.R., Peterjohn, B.G. and Link, W.A. (1994). Observer differences in the North American Breeding Bird Survey. *Auk* **111**:50–62.

Schaeffer, R.L., Mendenhall, W. and Ott, L. (1979). *Elementary survey sampling*, second edn. Duxbury Press, North Scituate, Massachusetts, pp. 1–278.

Searcy, W.A. (1989). Pseudoreplication, external validity and the design of playback experiments. *Anim. Behav.* **38**:715–17.

Seber, G.A.F. (1965). A note on the multiple recapture census. *Biometrika* **52**:249–59.

Seber, G.A.F. (1977). *Linear regression analysis.* John Wiley and Sons, New York, pp. 1–465.

Skalski, J.R. and Robson, D.S. (1992). *Techniques for wildlife investigations: design and analysis of capture data.* Academic Press, San Diego, California, pp. 1–237.

Skalski, J.R., Hoffman, A. and Smith, S.G. (1993). Testing the significance of individual and cohort-level covariates in animal survival studies. In: J.-D. Lebreton and P.M. North (eds.) *Marked individuals in the study of bird populations*, pp. 265–280. Advances in Life Sciences, Birkhauser Verlag, Berlin, pp. 1–397.

Smith, S.M. (1995). Distribution-free and robust statistical methods: viable alternatives to parametric statistics. *Ecology* **76**:1997–8.

Snedecor, G.W. and Cochran, W.G. (1980). *Statistical methods*, seventh edn. Iowa State University Press, Ames, Iowa, pp. 1–507.

Sokal, R.R. and Rohlf, F.J. (1981). *Biometry: the principles and practices of statistics in biological research,* second edn. W.H. Freeman, New York, pp. 1–859.

Steel, R.G.D. and Torrie, J.H. (1980). *Principles and procedures of statistics*, second edn. McGraw-Hill, New York, pp. 1–633.

Steidl, R.J., Hayes, J.P. and Schauber, E. (1997). Statistical power analysis in wildlife research. *J. Wildl. Manage.* **61**:270–9.

Stewart-Oaten, A. (1995). Rules and judgements in statistics: three examples. *Ecology* **76**:2001–9.

Sutherland, W.J. (1996). *Ecological census techniques: a handbook.* Cambridge University Press, Cambridge, pp. 1–336.

Taylor, B.L. and Gerrodette, T. (1993). The uses of statistical power in convservation biology: the vaquita and northern spotted owl. *Conserv. Biol.* **7**: 489–500.

ter Braak, C.J.F. (1995a). Calibration. In: R.H.G. Jongman, C.J.F. ter Braak and O.F.R. van Tongeren (eds.), *Data analysis in community and landscape ecology.* Cambridge University Press, Cambridge, pp. 78–90.

ter Braak, C.J.F. (1995b). Ordination. In: R.H.G. Jongman, C.J.F. ter Braak and O.F.R. van Tongeren (eds.), *Data analysis in community and landscape ecology.* Cambridge University Press, Cambridge, pp. 91–173.

Thery, M. and Vehrencamp S.L. (1995). Light patterns as cues for mate choice in the lekking white-throated manakin (*Corapipo gutturalis*). *Auk* **112**:133–45.

Thomas, J.W., Forsman, E.D., Lint, J.B., Meslow, E.C., Noon, B.R. and Verner, J. (1990). *A conservation strategy for the northern spotted owl.* US Government Printing Office Document No. 1990–791–171 / 20026, pp. 1–427.

Thomas, L. (1996). Monitoring long-term population change: why are there so many analysis methods? *Ecology* **77**:49–58.

Thompson, S.K. (1992). *Sampling.* John Wiley and Sons, New York, pp. 1–343.

Thomson, J.D., Weiblen, G., Thomson, B.A., Alfaro, S. and Legendre, P. (1996). Untangling multiple factors in spatial distributions: lilies, gophers, and rocks. *Ecology* **77**:1698–715.

Trent, T.T. and Rongstad, O.J. (1974). Home range and survival of cottontail rabbits in southwestern Wisconsin. *J. Wildl. Manage.* **38**:459–72.

Van Ballenberghe, V. (1989). Wolf predation on the Nelchina caribou herd: a comment. *J. Wildl. Manage.* **53**:243–50.

White, G.C. (1983). Numerical estimation of survival rates from band-recovery and biotelemetry data. *J. Wildl. Manage.* **47**:716–28.

White, G.C. (1996). NOREMARK: population estimation from mark-resighting surveys. *Wildl. Soc. Bull.* **24**:50–2.

White, G.C. and Garrott, R.A. (1990). *Analysis of wildlife radio-tracking data.* Academic Press, New York, pp. 1–383.

White, G.C., Burnham K.P., Otis, D.L. and Anderson, D.R. (1978). *User's manual for program CAPTURE.* Utah State University Press, Logan, pp. 1–40.

Wilson, D.S. and Dugatkin, L.A. (1996). A reply to Lombardi and Hurlbert. *Anim. Behav.* **52**:422–5.

Wright, S. (1921). Correlation and causation. *J. Agric. Res.* **20**:557–85.

Index

abundance, monitoring 200–18
  estimators 209–10
  index methods for 211–16
  outlier and missing data, effect of 210
  pseudoreplication in 216–17
  trend, definition of 201–9
  using polynomial regression 202–9
  using route regression 201, 206–9
assignment process 4
assumptions
  failure of 75–8

behavior, sampling 190–9
  allocation of effort 192–6
  data acquisition 196–7
  data analysis 197–9
  defining bouts 190–2
bias
  definition of 8–11
  in estimated standard deviation 28
  in relation to standard error 40
  Taylor Series approximation 38–9
bouts 190–2

capture–recapture methods 219–27
  capture histories 220–2
  models 220–2
  rationale 219–20
  SURGE 223–6
coefficient of variation 31–2
community analysis 253–4
computer programs, examples of
  bias of the estimated standard deviations 29
  for defining bouts 192
  multistage sampling 122, 131
  sampling hypothetical populations 107, 132
  unbiased variance 107
confidence intervals 58–65, 257–8, 265–9
  for a difference 60, 267
  for a ratio 63–5, 272
  for a single parameter 59, 264
  for proportions 265–6, 268–9
  meaning of 58
  relation to hypothesis 61–3
  utility of 60–1
constant, definition of 6
correction for continuity 49
correlation
  partial 166–7
  simple 148–54
covariance 25–6
  definition 25–6
covariance calculating formula 28

degrees of freedom 55–7
distributions 17–21
  binomial 18–20
  bivariate normal 155
  F 56–7, 59
  multinomial 20–1
  normal 31–2, 49–50, 59
  t 56

error 8–11
estimation, methods of 14–46
expected values 21–36
  definition 22
  rules for 23–4, 32–6

finite population correction 97–9
focal animal sampling 193

index methods 211–16
influential points 152
internal and external validity 2

maximum likelihood estimation (MLE) 42–5
  rationale 42

standard errors for 43–5
Mayfield method 41, 231–6
mean square error 156

notation 15–16
for complex designs 137–9

outliers 152

P-values 49–50
paired data 72–5, 264
parameter, definition of 5
path analysis 250–1
population
definition of 1–5, 16, 122–4
extrapolation from sampled 11–12
inferences about 5–11
variation within 30–2
population units, definition of 6–8
power 65–8, 273–7, 319–20
primary sampling units 90
proportions
as means 7
formulas for 257–69
pseudoreplication 177–89
in monitoring abundance 216–17
recommendations for avoiding 188

random variable, definition of 6
ratios 63–5, 272
regression 148–76
multiple 159–76
assumptions 167–8
categorical variables 168–72
definitions 163–4
inferences from 161–6
models for 159–61
multistage sampling and 174–6
partial correlation 166–7
stepwise 172–4
simple linear 154–9
assumptions 157–9
correlation coefficient 150–2
inferences from 155–6
models for 154–5
with categorical variables 168–72
resource selection 238–47
definitions 239
multivariate analyses 244–6
use-availability analyses 239–44

sample
analogs 15
distinct 22
sample size estimates 65–8 (*also see* power)
sampling error definition of 8–9
sampling methods 85–147
adaptive 248–9
cluster 94
comparison of methods 131–2
double sampling 131, 135
unequal probability 262
line transect 249
multistage 89–95, 109–24, 311–15
alternate definitions of the population 122–4
and independent selection 110–12
constructing primary units 117–18
definition of the population in doubling sampling 131
formulas for 257–63, 311–15
problems with few primary units 120–2
weighting of estimates 112–20
nonrandom 87, 101–2, 139–46
ratio and regression methods 131–5
simple random 87, 99–101, 259–60
stages in 86–7, 89–90
stratified 94–7, 125–31, 261–3
construction of strata 128
doubling sampling with 131, 261
formulas for 125–8, 257–8, 261–3
poststratification 129–31
use of 125, 128–9
systematic 87, 102–7
unequal probability 107–9
with visual estimates 135–6
scatterplots 148–54
sequential analysis 251–3
significance
biological 54
statistical 49–55
standard deviation 26–30
standard error 26–30 (*also see* sampling methods)
for sampling plans 8–11
of a difference 34, 54
of a linear combination 35–6
of a random variable times a constant 32
of a sum 32–4
of maximum likelihood estimations 43–4
superpopulation 3–5
survivorship, estimation of 228–37
Kaplan–Meier estimator 229–31
nesting success 231–6
staggered entry design 230

tables, statistical 279–308
Taylor Series approximation 36–41
tests, statistical 47–58, 61–3, 68–84
and correction for continuity 70
Bonferroni 82–3
chi-square 72, 82
Duncan's 72, 82
exact 59
Fisher's exact 72
formulas for 257–8, 264–72, 315–19

tests, statistical (*cont.*)
  G 72
  interpretation of results 50–2, 54–5
  Kruskal–Wallis 82
  multiple comparisons 81–4, 257–8, 270–1
  nonparametric 78–81
  one sample 48–52
  parametric 68–78
  Student–Neuman–Keuls 82
  t 68–78
  Tukey's 82
  two sample 52–3
  when assumptions are not met 75–8
  with paired data 72–5
transformations
  affine 34–6
  linear 34–6

Variables
  Bernoulli random 17–18
  types of 16–17
variance 24–5, 32–4
  calculating formula 28
  definition 24–5
  homogeneity of 154
  of a linear combination 35–6
  of constant times a random variable 32
  of mean with simple random sampling 27
  of product 33
  of sum 32
  Taylor series approximation for 38–9

Visual estimates 135–7

Printed in the United States
By Bookmasters